Parallel Processing Algorithms for GIS

Parallel Processing Algorithms for GIS

Edited by

Richard Healey
Department of Geography
University of Portsmouth
Portsmouth

Steve Dowers,
Bruce Gittings,
Mike Mineter
Department of Geography
The University of Edinburgh
Edinburgh

CRC Press
Taylor & Francis Group
Boca Raton   London   New York

CRC Press is an imprint of the
Taylor & Francis Group, an **informa** business

UK   Taylor & Francis Ltd., 1 Gunpowder Square, London EC4A 3DE
USA   Taylor & Francis Inc., 1900 Frost Road, Suite 101, Bristol, PA 19007

British Library Cataloguing in Publication Data

A catalogue record for this book is available from the British Library
ISBN 0-7484-0508-9 (cased)
ISBN 0-7484-0509-7 (paperback)

Library of Congress Cataloging in Publication Data, are available

Cover design by Jim Wilkie based on Parallel Landscape: Processors from a Cray T3E, draped over an Edinburgh landscape. Original photograph by P. Tuffy, Edinburgh Parallel Computing Centre; overlay by R.A. Fillinger, B.M. Gittings and S. Dowers, Department of Geography, University of Edinburgh.

Printed in Great Britain by T.J. International, Padstow, UK.

This book is dedicated to Anne, Karen, Bettina, Catherine, Christopher, Helen, John and Veronica.

Contents

PART THREE: PARALLELISING FUNDAMENTAL GIS OPERATIONS

PART FOUR: APPLICATION OF PARALLEL PROCESSING

Editors' Biographical Details

Richard Healey is Professor of Geography at the University of Portsmouth. Previously he taught at the University of Edinburgh, where he was Course Director of the Masters Programme in GIS for ten years. In addition to parallel processing and GIS, his research interests include database management, simulation modelling and both socio-economic and environmental applications of GIS. He has recently edited a special issue of IJGIS on parallel processing.

Steve Dowers is Principal Computing Officer in the Department of Geography, University of Edinburgh, and Director of the Planning and Data Management Service. He maintains research interests in recreation facility planning, supply demand modelling, benchmarking GIS performance, parallelising core GIS operations and the development of SQL to support spatio-temporal data. He was a member of the ISO Database Languages Rapporteur Group.

Bruce Gittings is Director of the Parallel Architectures Laboratory for Geographical Information Systems and jointly responsible for the Masters Programme in GIS at the University of Edinburgh. He serves on the Council of the UK Association for Geographic Information and maintains research interests in database management, performance, parallel computing, internetworking and user interfacing issues relating to GIS and geographical databases.

Mike Mineter completed his doctorate in nuclear structure physics at Oxford University, and then spent nine years developing commercial real-time software, in particular for major financial trading organisations. In 1992 he moved to the Parallel Architectures Laboratory in the Department of Geography, University of Edinburgh, to facilitate the exploitation of high performance computers in geographical information processing.

Acknowledgements

The Department of Trade and Industry (DTI) and the Science and Engineering Research Council (SERC) provided core funding for the Parallel Applications Programme, which supported the work described in Chapters 4 to 13 and 15. We are greatly appreciative of this support.

Further, we wish to record our deep appreciation to the following companies, who were partners in the Parallel Applications Programme and provided matching industrial funding:

British Gas Plc, Edinburgh
Digital Equipment Company Ltd, Reading
ESRI-UK Ltd, Watford
GIMMS (GIS) Ltd, Edinburgh
Laser-Scan Ltd, Cambridge
Meiko Ltd, Bristol
Oracle Corporation UK Ltd, Bracknell
Smallworld Systems Ltd, Cambridge

We are most grateful for their assistance.

The editors gratefully acknowledge funding provided under the ESRC Regional Research Laboratory Initiative, the ESRC/NERC Joint Programme in Geographical Information Handling, together with generous funding provided by the Digital Equipment Corp. (under its European External Research Programme Grant UK-038) and from the University of Edinburgh, which provided the seed-corn for much of this work.

Particular thanks are due to Dr Nick Radcliffe of Quadstone Systems. While a member of Edinburgh Parallel Computing Centre, Nick made a substantial contribution to the Parallel Applications Programme, in terms of project management and technical expertise.

We also wish to record our thanks to Tom Waugh, who helped considerably with the Parallel Applications Programme work.

We are grateful to many other colleagues from the Edinburgh Parallel Computing Centre who gave technical advice and assistance directly and indirectly to the project. Particular mention must be made of Simon Chapple, Lyndon Clarke, Andy Sanwell, Evan Welsh and Andrew Wilson.

The chapter by Paul Densham and Marc Armstrong is a contribution to Initiative 17 (Collaborative Spatial Decision-Making) of the National Center for Geographic Information and Analysis which is supported by the National Science Foundation (SBR-88-10917). The authors wish to acknowledge the substantial contributions made by their respective collaborators in previous work: Yuemin Ding, Claire E. Pavlik, Demetrius Rokos and Richard Marciano. Armstrong also wishes to acknowledge the receipt of a faculty developmental award from The University of Iowa.

Michael J. Mineter acknowledges support from a NERC grant during the final editing of the book, and from the University of Edinburgh during the preparation of the text. The latter comprised a research grant from the Faculty of Social Sciences and a grant from the Principal's Bridging Fund.

Mette J. Tranter (and her supervisors, Bruce M. Gittings and Richard G. Healey) are grateful to the Ordnance Survey for their contribution to this work through the funding of a doctoral studentship.

Jane Boygle's help in correcting the word-processed text, proofs and general editorial assistance was enormously valued.

Comments from several colleagues in the Department of Geography, University of Edinburgh, led to improvements in draft chapters; especially valued were comments from Claire Jarvis, Nick Hulton and Neil Stuart.

All trademarks are hereby acknowledged.

Contributors

Marc P. Armstrong
Departments of Geography and Computer Science and Programme in Applied Mathematical and Computational Sciences, 316 Jessup Hall, The University of Iowa, Iowa City, IA 52242, USA.
Email: marc-armstrong@uiowa.edu

Paul J. Densham
Centre for Advanced Spatial Analysis, Department of Geography, University College London, 26 Bedford Way, London WC1H 0AP, United Kingdom.
Email: pdensham@geog.ucl.ac.uk

Steve Dowers
Department of Geography, The University of Edinburgh, Drummond Street, Edinburgh EH8 9XP, United Kingdom.
Email: sd@geo.ed.ac.uk

Bruce M. Gittings
Department of Geography, The University of Edinburgh, Drummond Street, Edinburgh EH8 9XP, United Kingdom.
Email: bruce@geo.ed.ac.uk

Timothy J. Harding
Quadstone Ltd, 16 Chester Street, Edinburgh EH3 7RA, United Kingdom.
Email: tjh@quadstone.co.uk

Richard G. Healey
Department of Geography, University of Portsmouth, Buckingham Building, Lion Terrace, Portsmouth PO1 3HE, United Kingdom.
Email: healeyrg@geog.port.ac.uk

Sara Hopkins
3L Ltd, 86/92 Causewayside, Edinburgh EH9 1PY, United Kingdom.
Email: sh@threel.co.uk

Paola Magillo
Dipartimento di Informatica e Science dell'Informazione, Universita di Genova, Via Dodecaneso 35 - 16146 Genova, Italy.
Email: magillo@disi.unige.it

Michael J. Mineter
Department of Geography, The University of Edinburgh, Drummond Street,
Edinburgh EH8 9XP, United Kingdom.
Email: mjm@geo.ed.ac.uk

Enrico Puppo
Instituto per la Mathematica Applicata, Consiglio Nazionale delle Ricerche, Via
De Marini, 6, Torre di Francia, 16149 Genova, Italy.
Email: puppo@ima.ge.cnr.it

Mark Sawyer
Edinburgh Parallel Computing Centre, The University of Edinburgh,
James Clerk Maxwell Building, Edinburgh EH9 3JZ, United Kingdom.
Email: cms@epcc.ed.ac.uk

Terence M. Sloan
Edinburgh Parallel Computing Centre, The University of Edinburgh, James Clerk
Maxwell Building, Edinburgh EH9 3JZ, United Kingdom.
Email: tms@epcc.ed.ac.uk

Mette J. Tranter
Department of Geography, The University of Edinburgh, Drummond Street,
Edinburgh EH8 9XP, United Kingdom.
Email: mjt@geo.ed.ac.uk

Shari M. Trewin
Edinburgh Parallel Computing Centre, The University of Edinburgh, James Clerk
Maxwell Building, Edinburgh EH9 3JZ, United Kingdom.
Email: shari@epcc.ed.ac.uk

Graeme G. Wilkinson
School of Computer Science and Electronic Systems, Kingston University,
Penrhyn Rd, Kingston upon Thames KT1 2EE, United Kingdom.
Email: g.wilkinson@kingston.ac.uk

1
Introduction

R.G. Healey, B.M. Gittings, S. Dowers and M.J. Mineter

1.1 GIS Challenges

Over the last fifteen years the field of GIS has been transformed. Once a Cinderella subject, it was untouched by all but the most quantitatively avant-garde of departments of Geography, and a small group of far-sighted public sector bodies. From these small beginnings it has now become a multi-billion dollar industry and a major player within the broader field of information technology.

In the course of its evolution it has negotiated the restrictions of limited processing power and mass storage, the tedium of manual digitising and the dearth of reliable software, to become a fully fledged technology, widely deployed across a range of application fields. Indeed, its success and that of the sister technology of satellite remote sensing now threaten to engulf us in digital data, if not information, which we can scarcely store, let alone analyse, in any potentially useful time frame. Although the makers of computer chips continue to surprise us with advances in processor speed, cache performance and graphics capability, they are no match for the growth of available GIS and remote sensing data. Further to this, the spectre of real-time GIS applications (Xiong & Marble, 1996) which would add sub-second response times to analytical operations on large and highly dynamic datasets, is also starting to haunt us.

Faced by these immense data resources, many of them freely available over the Internet, and yesterday's workstation that capital budget restrictions prevent us from replacing, it is difficult to stretch beyond moderate sized datasets and limited analytical operations if our processing capabilities are not to be overwhelmed. The more exploratory, combinatorial or interactive kinds of analysis are therefore not accessible to the extent we would wish.

It is very important that the field should not lose impetus, now that rich data resources are available. The expectations of potential users are high, yet the throughput of useful results from GIS analysis is often limited. Processor, memory, disk and network constraints still temper the enthusiasm of even the most energetic and insomniac of postgraduate students. Similarly, contract deadlines tick past before image datasets can be classified, generalised and vectorised, while dozens of workstations on the network stand idle overnight. Power on the desktop has brought excellent interactivity to the user interface, but it has yet to be harnessed enterprise-wide for cost-effective GIS processing.

One further emerging area of application is likely to hold new and significant challenges. The increasing use of real-time mapping and analysis in conjunction with the World-Wide Web (WWW), with millions of potential users, will result in enormous demands for multi-streamed performance. Already many web-servers are seen to be receiving tens of thousands of accesses every day, and as the sophistication of these accesses develop, with demands for complex database queries and the mapping of results, so the performance of the system must service these requests rapidly to satisfy a user community at present frustrated by lack of network bandwidth.

Already Web sites such as Digital's AltaVista search engine use powerful multi-processor servers to satisfy current levels of use, yet with only simple and unsophisticated queries. It cannot be long until the demands made of GIS software, customised into vertical markets such as tourist information systems, reach and rapidly exceed these levels.

Given these demands for both high performance computation and data input/output (I/O), it is now a matter of concern that the GIS community is still largely restricted to algorithmic approaches originally developed when only serial processors were available, with performance orders of magnitude lower than contemporary hardware.

1.2 Parallel Opportunities and Problems

The literature has for some years been drawing attention to the 'promise of general-purpose parallel computing' (Hack, 1989). This has been based on a number of arguments. Firstly, and most obviously, the ability to distribute components of a large computational task across a number of processors ought to result in more rapid throughput than if only a single processor is utilised (Nicol & Willard, 1988). Secondly, irrespective of how long it takes before fundamental limits to the speed-up of serial processors are approached, any performance gains available can be multiplied dramatically by running the processors in parallel. Thirdly, the trend towards the use of commodity processors, even if their interconnects are proprietary, has made parallel machines more maintainable, lower cost and more able to run versions of standard systems software such as UNIX.

However, algorithms developed for serial machines may not run effectively in a parallel environment and vice versa, so utilisation of parallel technology may be less straightforward than at first thought. More specifically, it requires consideration of the trade-offs between:

- the architecture of individual processing nodes

- the granularity of the machine (no. of nodes)

- communication bandwidth and the latency (the time between the sending and receipt of messages between processors)

- the topology of the interconnection between processors

- the extent of overlap (or not) of communication and computation. (Fox, 1991)

Unfortunately, the issues involved in moving to a parallel computing environment are not limited to the purely algorithmic. In a critical assessment of the challenges of scalable parallel computing, Bell (1994) has referred to the four 'flat tyres' of the parallel bandwagon, caused by lack of: systems software, skilled programmers, good heuristics for the design and use of parallel algorithms, and good parallelisable applications. To this, regrettably, we might add the spare tyre of commercial failures among parallel hardware vendors, which have not enhanced confidence in the potential user community. However, it should be noted that failure has been more common among vendors of massively parallel machines, where each component processor has local memory, than among the suppliers of shared memory multi-processors, where common memory is accessed by all processors across a bus (see Chapter 2).

It could be argued that the field of parallel processing, in both hardware and software components, is simply relearning at an accelerated pace the earlier lessons of computing as a whole: namely, that viability is a function of generality (Bell, *op. cit*). More specifically, without general purpose hardware components, and standardised, re-usable software component libraries, soaring development and maintenance costs will rapidly render even the best-funded of projects uneconomic.

Yet, parallel computing technology and GIS need to be brought together. The increasing complexity of the applications, the volumes of data and the escalating costs of designing higher and higher performance serial processors mean that the adoption of parallelism is inevitable. Until recently, the costs of redesigning software to take advantage of this technology (due to the lack of tools, standards and generic operating system support) has deterred the GIS community from embracing parallelism, but this is now changing.

1.3 GIS and Parallel Processing

With GIS applications clamouring for enhanced throughput, and the potential offered by parallel processing, bringing the technologies together would seem to offer substantial benefits, if successful. However, several earlier points counsel caution in relation to such an ambition. At a very general level, while GIS operations are both compute and I/O intensive, until very recently parallel processing has concentrated much more on performance enhancements of the former rather than the latter type. This situation is now changing with machines such as the CRAY T3E. Secondly, GIS data structures are complex, with varying numbers of variable length data records, spread across linked files which may be very large (see Chapter 6). Extensive pre-processing may therefore be necessary within the parallel environment, if subsequent computation is to make optimal use of available processors. Thirdly, GIS operations may be multi-stage, requiring the application of several algorithms in turn. These algorithms may contain sequential components. Fourthly, computation and I/O may be interleaved during the same operation. This may be a function of the operation itself, or of the very large size of the datasets, such that it cannot be assumed they will fit entirely into the memory (either shared or distributed among processors) that may be available at any given time. Fifthly, it may be appropriate to link the code for individual operations to a proprietary database manager, for storage or manipulation of co-ordinate or attribute data.

The range of issues involved, from database management to applied computational geometry, indicates that the task of making GIS operations, as opposed to individual algorithms, fully functional in a parallel environment is a daunting one. A large number of design decisions have to be made, where the trade-offs may be difficult to calculate with any precision, before any approach to implementation can even be considered. However, what is essential at this stage in the development of the field is that these issues and problems are clearly explained and documented, so future work can build on it in a structured manner. This is the approach adopted in this book. The main focus is on documentation of algorithm and software design issues, to a much greater level than is common is GIS texts, with implementation examples to illustrate how the research could be taken forward in future.

A final point under this heading is that, if there is a shortage of skilled staff in parallel processing, there are even fewer individuals with expertise in the complex and sometimes arcane world of GIS algorithms. To find those with both types of skill is therefore even more difficult, and makes interdisciplinary collaboration essential. Even with this, the time requirements and costs of work of this kind are considerable and relatively few groups will be able to entertain it. This is all the more reason for proper documentation of work that is done, so no available effort is wasted.

1.4 The Context of this Book

The subsequent sections of this introduction outline the background to the project that funded much of the work reported in this volume, followed by an outline structure of the ensuing chapters.

1.4.1 Project Background

This book has its origins in a large research project on parallel GIS funded at the University of Edinburgh by the Department of Trade and Industry (DTI) and SERC, as part of their Parallel Applications Programme, with substantial contributions from a number of industrial partners, who are acknowledged individually at the beginning of this volume. Owing to the size of the interdisciplinary research team, comprising both parallel processing and GIS specialists, the appropriate format for wider dissemination of the results of the project was an edited collection of papers, rather than a book written by just one or two individuals. However, unlike most edited collections, a sustained argument is developed through the first fourteen chapters, before broadening out the coverage in the later chapters to consider a range of related topics.

1.4.2 Project Aims

The DTI/SERC Parallel Applications Programme started in 1991 from the premise that parallel processing was now a sufficiently mature field for the funding emphasis to move from fundamental research towards industrial applications. While this was perhaps true in certain areas, it was difficult to argue this view in relation to GIS, where, at the time, the published output in relation to the application of parallel processing methods was extremely limited. Indeed, with hindsight, a view much closer to that put forward by Kuck (1994) would have been more appropriate. He identified the need to distinguish in parallel processing between computer architectures (i.e. hardware), compilers, applications and problem solving environments. In each of these areas there are three stages, namely research and development, commercialisation and commodity availability. While architecture and compiler developments were well down the road to commercialisation by the early 1990s, the same could not be said of applications and problem solving environments. These were barely at the beginning of a process of commercialisation that could be expected to extend until 2020 and beyond. The results of the GIS project reported here fit more comfortably into Kuck's realistic assessments of timescales than an over-enthusiastic push for 'near-market' activity that is not based on sound research foundations.

Given the very limited amount of previous work on parallel processing in GIS, the main focus of the project was necessarily on algorithm design rather than implementation and performance testing. Restriction in the range of algorithms considered was also essential and the target GIS operations were chosen in consultation with the industrial partners. Initially vector polygon overlay and raster-vector conversionwere identified as the most important algorithms for parallelisation, because of the well-known processing bottlenecks associated with them in a serial environment. Vector-raster conversion was added for completeness although its processing demands were not as great.

Additional aspects of the project brief were a mixture of the fundamental and the practical. It was apparent, firstly, that detailed consideration would need to be given to the problem of parallel I/O, since GIS operations are both compute and I/O intensive. Secondly, the formats of available GIS datasets and their representation of map topology needed to be addressed, with particular reference to how parallel operations might be interfaced to existing vendor GIS products.

Thirdly, while the need for new algorithms was most closely identified with distributed memory massively parallel machines (see Chapter 2), the commercial impact of such machines was limited compared to the less powerful, but more widely available, shared memory multi-processors. The latter should therefore not be ignored during the course of the research programme. Finally, although parallel processing was closely identified with highly compute intensive problems, a second area of application was also coming to prominence. This involved the use of parallel machines as database servers. Since database management plays a major part in GIS, additional investigation of the potential role of parallel database servers would enhance understanding of this aspect of parallel technology in relation to GIS in general.

The requirement for industrially relevant yet essentially fundamental research has therefore resulted in a distinctive flavour to the work reported here. While algorithms themselves require detailed understanding of computational geometry and parallel processing techniques, the design process is also informed by an appreciation of the software engineering imperative of developing modular, re-usable code, and the many practical aspects of GIS data management and manipulation. In particular, when designing algorithms for three related operations, the opportunity to demonstrate the interlinkages between module components has been taken wherever it is appropriate to do so. This topic is explored in detail in Chapter 7. Further to this, design assumptions, algorithm stages and possible implementation issues are set out in much greater detail than is common in the literature. This is largely to enable the implications of parallelisation to be grasped by the reader, but it is hoped that it will also be a useful contribution to the relatively sparse literature on GIS algorithms more generally.

1.5 Intended Audience

From the foregoing it is apparent that the book is aimed at a postgraduate or professional software development audience, rather than being an introductory text, although significant portions of it should be accessible to final year undergraduates with the relevant background. It is also aimed at the GIS and the Computing Science/Software Engineering communities. At the risk of failing to meet the needs of either group, it endeavours to give sufficient introduction to both GIS and parallel processing issues for the subsequent argument to be meaningful, regardless of disciplinary starting point. Within the project team this process of interdisciplinary education was itself an extended process and a single volume cannot hope to cover all the issues, other than by giving pointers to the relevant literature. Nevertheless, it is intended that the sustained argument, developed through the main body of the text, will give colleagues in parallel processing a clear sense of the scale of problems posed by large GIS application. Likewise it should give to colleagues in GIS a grasp of the implications and challenges of moving well tried, if not entirely trusted, algorithms from the serial to the parallel domain.

1.6 Outline Structure

The book is divided into five parts, the first three of which are based on the work of the DTI/SERC project, while the fourth contains contributions from other research groups working in cognate areas, and the fifth presents conclusions.

Part 1 provides an introduction to the concepts, terminology and techniques of parallel processing, both in general and with particular reference to GIS. Chapters 2 and 3 cover hardware and software developments, including the recently established message passing interface (MPI) standard. Space precludes them from being either exhaustive or comprehensive, and further elaboration of the points raised can be found in the standard texts. They are designed instead to provide the necessary basis for the GIS reader to grasp the fundamental issues, before proceeding to the detailed investigation of later chapters. Chapter 4 examines the high level programming paradigms and software engineering issues which underlie parallel software development. Particular emphasis is given to the problem of designing modular re-usable software libraries.

The second part situates the general problem of parallel software design in a GIS applications context (Chapters 5 and 7). Additional chapters address the characteristics of GIS data structures for both vector (Chapters 6, 8 and 9) and raster data (Chapter 10). In the case of vector data, alternative representations are examined and an example generic data format (NTF) is chosen for use in later chapters. This format also allows data transfer to and from the formats of most commercial GIS vendors. Chapters 8 and 9 develop the design for vector data management in a parallel environment beginning with the sort/merge requirements for data pre-processing (Chapter 8) and continuing with the creation of vector topological data structures in the following chapter. The latter are an essential precursor to more advanced processing such as polygon overlay.

In part 3, the focus shifts to the design of algorithm sequences for the major GIS operations of vector-raster conversion (Chapter 11), raster-vector conversion (Chapter 12) and vector polygon overlay (Chapter 13). The interrelationship between components of these algorithms and the scope for software re-use are explored. These chapters build directly on the material contained in the earlier chapters. Chapter 14 is an implementation case study, based around a raster generalisation problem, which illustrates some of the principles described in the first two sections and utilises one of the parallel software libraries described in Chapter 4.

Part 4 contains several chapters on related parallel processing developments in the GIS/Remote Sensing field. The first (Chapter 15) is concerned with the role of parallel database technology in a GIS environment.

Subsequent chapters provide a review of research progress to date in their respective application areas and deal with both algorithmic and implementation issues. These application areas include terrain modelling and visualisation (Chapter 16), aspects of spatial analysis (Chapter 17) and Remote Sensing (Chapter 18).

Part 5 presents conclusions. It is proposed that the development of parallel processing for geographical applications is significant in a range of application areas, and that the creation of software libraries and frameworks is necessary to exploit this potential. Overall, it is hoped that the reader will gain from the different contributions a much clearer picture of both the technical problems and the benefits/costs of developing GIS algorithms in a parallel environment.

1.7 References

Bell, G., 1994, Scalable, parallel computers: alternatives, issues, and challenges. *International Journal of Parallel Programming*, 22, **1**, 3-46.

Fox, G.C., 1991, Achievements and prospects for parallel computing. *Concurrency, Practice and Experience*, 3 (6), 725-739.

Hack, J.J., 1989, On the promise of general purpose parallel computing. *Parallel Computing*, **10**, 261-275.

Kuck, D.J., 1994, What do users of parallel computer systems really need? *International Journal of Parallel Programming*, 22 (1), 99-127.

Nicol, D.M. & Willard, F.H., 1988, Problem size, parallel architecture and optimal speed-up. *Journal of Parallel and Distributed Computing*, 5, 404-420.

Xiong, D. & Marble, D.F., 1996, Strategies for real-time spatial analysis using massively parallel SIMD computers: an application to urban traffic flow analysis. *International Journal of GIS*, **10** (6), 769-789.

Part One

Parallel Processing Technology in Context

2

The Development of Hardware for Parallel Processing

M. Sawyer

The basic programming model for the computer, the von Neumann model, describes the functionality of a computer in terms of a single processing unit. The simplicity of the model has considerable attraction to the designers of both hardware and software; machines are simpler to build if they contain one of everything, and are undoubtedly easier to program.

There has been a continuous rise in the speed at which processors can operate, with a doubling of performance being achieved approximately every month. This trend has gone some way to meeting the increasing demand for processing. However, this rise in performance will ultimately be limited by the speed at which the millions of electronic switches contained within a processor can operate. Predictions of when this limit will be reached vary, but most estimates agree that by the year 2010 it will not be possible to increase further the performance of a single processor. Parallel computing - the use of more than one processor to solve a problem - is one way in which this limit could be overcome. Even today, the use of multiple processors offers a convenient means of achieving more processing power than is currently available from a single processor.

This chapter gives an overview of the development of parallel processing hardware. The technological advances which led to parallel processing becoming viable are explained, and important architectural concepts are illustrated with examples of actual machines. The importance of I/O speed as well as processing power is discussed.

2.1 The Evolution of Parallel Processing

The idea of using more than one processing element to work on a program is not a recent concept; this was clearly considered by Babbage during the design of his calculating engines (Morrison & Morrison, 1961). More recently, the first electronic computer, the Electronic Numerical Integrator and Computer (ENIAC) (Goldstine, 1946) was built from a number of separate units which could operate independently.

To examine how parallel computing has evolved from being an interesting theoretical idea to a practical and desirable proposition, it is helpful to examine the changes in technology which have taken place since the first computers.

The first community of computer users were scientists, chiefly interested in getting the machine to perform correctly, with reliability of operation being at least as important as speed. Computers were highly specialist pieces of equipment and predictions about the future of computing were not optimistic about their widespread use.

To explain why these predictions were inaccurate, we must allow for the fact that early computers were built using expensive and unreliable components, such as valves (vacuum tubes). The prospect of building a computer with multiple processors using such technology was simply too daunting for any serious attempts to be made. Contrast this with the technology of today: the development of solid-state devices in the 1960s, through integrated circuits and then VLSI production techniques in the 1970s have improved the reliability and reduced the cost of computer components to a remarkable extent. The result has been that within fifteen years of the first stored-program computer, a machine capable of many times the performance could be purchased for a fraction of the cost. Moreover, the generation of computers available in the 1960s were genuinely useful general-purpose machines, capable of being used to solve scientific problems and to serve commercial applications. In addition to the improvements in computer hardware, significant developments were also taking place in terms of programming techniques and a body of software becoming available to undertake data-processing operations, in addition to scientific applications.

With an established market and reliable machines, computers quickly became an indispensable tool in science and business. Some users, primarily scientists, began to demand better processing performance in order to solve larger and more complex problems. This demand was satisfied by increasingly sophisticated processing hardware; advances in technology and manufacturing allowed increased switching speeds and more components to be included in the processors. Parallel techniques, such as *pipelining*, were also used to improve performance. Pipelining is a technique whereby an operation is split into a number of stages. Once an operation has proceeded past the first stage of the pipeline, a further operation can be initiated, with the result that a higher throughput of results can be achieved. Improvements in manufacturing techniques allowed processors to be designed with, for example, more than one unit capable of performing arithmetic operations.

A significant advance was the introduction of vector processors, notably in the Cray series of the mid 1970s. These processors provided programmers with a target architecture incorporating parallel operations. In order to make efficient use of a vector processor, it was necessary for the programmer to organise data and to order operations so that operations could be carried out on vectors. In particular it was necessary to analyse the dependencies within program loops to ensure that the operations could be carried out independently. Many mathematical and scientific programs can exploit this style of programming, and machines employing vector processors became successful in the supercomputing market.

The advances in processor design with pipelining, multiple function units and vector units on processors, and corresponding advances in memory speed and disk technology, allowed high performance to be achieved by single microprocessors. However, there remains a fundamental limit to the performance of a solid state device: the speed at which the basic electronic components can work. The physical processes which govern the electronic signals within a microprocessor must ultimately limit the speed at which such a processor can be made to operate. A principal method of overcoming this limit is the use of multiple processors in one machine. It is interesting to note that in 1996, the processor manufacturers have yet to reach this performance 'ceiling'; faster processors are constantly emerging, although this is, in part, due to the use of parallelism within the processors.

In the late 1970s, it became clear that building parallel computers would be feasible; it was possible to use many commodity processors (for example, Intel 80486, DEC Alpha or Sun SPARC) to produce a machine with the same or better performance as one constructed using specialised processors. The relative simplicity and size of non-specialist processors allows them to be mass-produced, thereby giving them a price-performance advantage over their high performance specialised cousins. The growth in scientific computing, which had driven the development of vector processors, was set to continue and the computational demands would soon exceed the capability of conventional supercomputers. Software had also become more mature since the 1960s, with many of the principles of parallel computing having been investigated through the development of multi-user operating systems and time-sharing systems. These factors combined to make parallel processing a feasible solution to the demand for computing power.

2.2 A Classification of Parallel Architectures

The most widely used classification of computers is Flynn's Taxonomy (Flynn, 1972). Flynn classified computers according to whether they process single or multiple data items, and whether the same or different operations are applied at each stage. The resulting four classifications are Single Instruction Single Data (SISD), Single Instruction Multiple Data (SIMD), Multiple Instruction Single Data (MISD) and Multiple Instruction Multiple Data (MIMD).

SISD machines follow the von Neumann model. SIMD machines include the so-called 'array-processors' such as the Connection Machine CM-2 and CM-200 and the ICL DAP. SIMD machines consist of large numbers of simple processing elements which are sent instructions by a controlling processor. Within MIMD computers, each processor (which is normally much more powerful than a single SIMD processing element) executes its own program. Examples of early MIMD machines are the Sequent Balance and Meiko CS-1, and more recently the Cray T3D. MISD computers have not been built - no application seems to require this architecture at present.

2.2.1 Classification of memory models for parallel computers

Flynn's Taxonomy classifies computers according to their behaviour in terms of data and instructions. There is a further classification which is based on how the processors are able to access memory. With a single processor, all memory has to be addressed by that processor; with multiple processors there is a choice between allowing each processor access to the whole of the memory and allowing each to access only a certain part of memory. The former model is referred to as *shared memory*, the latter as *distributed memory*. Both techniques offer advantages and suffer disadvantages.

2.2.1.1 Shared Memory

In a shared memory environment, every processor in the computer can access every memory location. This offers a simple and fast mechanism for communication between processors, although mechanisms must be found to deal with contention for the memory hardware, which arises when several processors attempt to access the same location in memory simultaneously. The effects of contention have historically called into question the ability of shared memory architectures to scale well, i.e. to show a continuing improvement of performance as more processors are added.

In addition there are problems with controlling the way in which processors share variables to ensure consistency. This is usually resolved by protecting the shared variable using critical sections and locks as described in the next chapter.

There is a similar problem with ensuring that cached data is coherent. A cache is a local memory to which a processor has very fast access, and which holds copies of data from main memory which is in the process of being worked on, or which has been recently used or may be used in the near future. With shared memory multiprocessor systems there may be multiple copies of a single data word, at any one time, since each processor has its own cache. A strategy must be devised such that when data is written to the cache of one of the processors, the other processors use the updated value, rather than stale values. This can be done in a number of ways, including invalidating all of the data in the other caches, or by broadcasting the update values to all other cached copies.

2.2.1.2 Distributed Memory

The problems of contention and coherency encountered in shared memory systems are avoided by taking the distributed memory approach; each processor has exclusive access to a local memory.

In order for the processors to communicate, there must be a mechanism for data to be transferred between the memory associated with each of the processors. This is achieved by providing the processors with a network of links over which the data can be transferred. Special hardware and software are incorporated to drive the network and route information between one particular processor and another (known as *messages*). The result is that there is generally a greater delay in transferring data between processors than with shared memory architectures, which can have an effect on the performance.

2.2.1.3 Shared Virtual Memory

With advances in technology the distinction between shared and distributed memory is becoming blurred. A third memory architecture, described as 'shared virtual memory', has emerged. In this architecture, the processors have access to the entire memory of the computer, but not all parts of it can be accessed at the same speed due to a hierarchical organisation. For this reason the term non-uniform memory access (NUMA) is also used to describe this architecture. The advantages of this type of architecture are that the ease of programming of shared memory is retained, without the hardware memory contention problems which ultimately affect the scalability of the system. There is also the possibility that a program can be run which requires more memory than is physically present on one processor, allowing memory resources to be used more efficiently in a multi-user environment.

In a shared virtual memory system there is a single address space, although the memory is physically distributed across a number of processors. When a processor generates an address for data stored in its local memory, the access takes place without any inter-processor communication. When access to data which is not stored locally is required, the data is fetched by a memory management process from the local memory of the remote processor. Shared virtual memory is fully described in Nitzberg & Lo (1991).

The issue of cache coherency remains since a processor may be writing to a location which has been copied to several other processors. In this situation all copies of the data must be invalidated on the processors not involved in the write.

The shared virtual memory model offers the programmer flexibility in writing programs, although the efficient use of a virtual memory system will rely on the programmers being able to exploit the locality of the data. The memory to which a processor has fastest access should be used most frequently by that processor, and memory to which it has slower access should be used less frequently. The memory can be thought of as close to, or far from, a processor.

2.2.1.4 Shared vs. Distributed Memory

In practice, both distributed, shared memory and virtual shared memory systems have been implemented. Manufacturers of shared memory machines (often referred to as symmetric multiprocessors (SMPs) tend to use smaller numbers of powerful processors. SMPs are produced by many of the leading computer vendors, including Digital, Silicon Graphics, Bull, Sun Microsystems and Intergraph, and shared memory is the architecture adopted by many vendors of vector systems. Distributed memory machines, with thousands of processors, have been built. The arguments which have been posed against the ability of shared memory machines to scale effectively to large numbers of processors have been tempered by the advances in memory and bus technology, and novel techniques for maintaining cache-coherency, such as bus-snooping. It also worth noting that many real applications will not themselves scale to large numbers of processors due to parts of the algorithm being inherently sequential, and thus there comes a point at which there is little benefit in running them on more processors.

It is important to realise that clusters of workstations, connected using standard network technology, are a distributed memory parallel architecture. A distinction is usually drawn between these systems and distributed memory systems which incorporate special hardware to connect the components. The latter are often referred to as *tightly coupled* and the former as *loosely coupled*. The main difference will be in the performance of the communications.

SMPs have made a significant impact in the high performance computing market. They can be easily used either as parallel machines or as servers achieving a high throughput for sequential jobs (a multi-streamed workload). Existing software can be run, and low cost entry level systems are usually available, making them an attractive, low risk and highly commercial option. Distributed memory systems require more sophisticated configuration software, and the networking hardware required to implement the interprocessor communication can make small systems unattractive.

2.3 Interconnection Topologies

In order to facilitate the transfer of data within a parallel computer, there must be a network of connections linking the components.

There are two basic types of network; those which consist of fixed links between nodes, referred to as *static* networks, and those in which the nodes are not connected directly but communicate through switches, referred to as *dynamic* networks. In addition networks can be classified as *symmetric* if they appear identical when viewed from any node, and *regular* if each node has the same number of connections. Most common interconnection networks are regular and symmetric. Among other things this allows large networks to be built up from modular components, and allows routing algorithms to be easily implemented.

For large numbers of processors it is not feasible to provide a dedicated link to each processor, since the total number of links would rise with the square of the number of processors, and the cost would consequently be prohibitive. To use a 'star' configuration in which each processor is attached to a single central switch would lead to the central switch becoming a 'hot-spot'. The approach commonly taken is to provide each processor with links to a small number of processors, and to provide routing mechanisms whereby data can be moved between processors not directly connected to each other.

There are a number of ways to measure the effectiveness of a network. The *network diameter* is defined as the maximum of the shortest distance between all possible pairs, and represents the worst case distance between a pair of nodes in the network. The *network bisection width* is the number of links required to partition the nodes into two unconnected sets of equal size. This is a measure of the density of the connection, and gives an idea of the cost of increasing the size of the network. There are also metrics which measure the traffic density on each link which can be used to analyse the contention of links.

2.3.1 Static Networks

2.3.1.1 The Mesh and Torus

Due to their simplicity, two and three dimensional grids of processors are an attractive topology. It is usual for the connections to 'wrap-around' at the edges of the mesh, resulting in the torus shown in Figure 2.1. Communication is clearly most efficient for near neighbours. Many applications fall into this category, for example image processing and some matrix operations.

2.3.1.2 Trees and Fat Trees

A tree is a network in which each node is connected to one parent node, and to a number of children nodes. There is a single node which has no parent, referred to

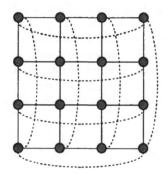

Figure 2.1: Torus

as the root, and a number which have no children, the leaves. The special case of a binary tree in which each parent node has 2 children is shown in Figure 2.2.

The tree is a useful structure for communications which are recursive, and for algorithms which use a 'divide and conquer' principle. However, it can be seen that the links closer to the root will be more heavily used if there is uniform message traffic. For this reason trees are not suitable for general communications.

An alternative is to increase the number of links connecting the nodes higher up the tree (closer to the root). This results in the so-called *fat tree* shown in Figure 2.3. Fat trees have been used in commercially available parallel machines, notably the Thinking Machines CM-5 and the Meiko CS-2.

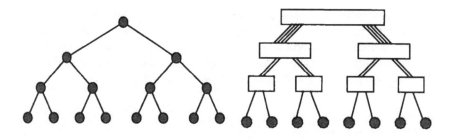

Figure 2.2: Binary tree   *Figure 2.3: Fat tree*

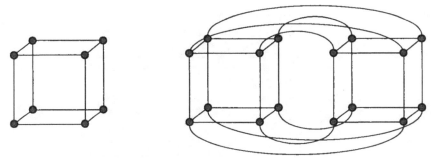

Figure 2.4: Hypercubes: 3-D (left) and 4-D (right)

2.3.1.3 The Hypercube

A frequently encountered interconnection topology is the hypercube. In this configuration, there are 2^n processors, each of which has n connections (see Figure 2.4). The particular properties of this topology are that doubling the number of processors requires only one additional link per processor, and that each processor is a maximum of n links away from every other processor.

Routing algorithms, which can exploit the fact that there are alternative routes between processors, are well known for the hypercube. The hypercube has a bisection width of $N/2$ for a network of N processors, compared to $\sqrt{N}$ for a torus.

2.3.1.4 Dynamic Networks

There are three basic types of dynamic network. In increasing order of cost and performance these are *bus systems*, *multistage networks*, and *crossbar networks*. Bus systems are based on processors having connections to a central communications bus. Only one transaction can be made at any one time, and thus multiple requests for transactions are dealt with by contention or time multiplexing. Bus systems do not exhibit good scalability because of the problem of contention.

A multistage network consists of a number of switching units which can send input data to a number of outputs. Practical networks consist of 2^n inputs and outputs with a number of 'stages' connecting them. The stages are linked by fixed links, and the setting of the switches changes dynamically to route data to subsequent stages. Figure 2.5 shows a multistage network connected in a commonly used configuration referred to as a *butterfly*. The butterfly network with 2^n inputs and outputs has n - 1 stages; a path of length n exists between each input and output node. The butterfly has a recursive structure, similar to the hypercube, and routing algorithms are well known.

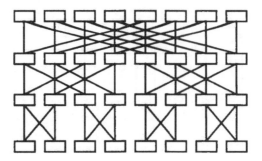

Figure 2.5: Butterfly multi-stage network

A crossbar network is essentially a single stage switched network which is capable of connecting any pair of components in the same way as a telephone switchboard might work, and is illustrated in Figure 2.6. The performance of the crossbar switch is high since there is great flexibility in the number of connections available. However, the cost is high, due to the large number of components; a crossbar network connecting N inputs and outputs requires N^2 switches, compared to $N \log_2 N$ for the corresponding butterfly. Crossbars are thus feasible only for connecting small numbers of units.

2.4 Balancing Computation and I/O in the Parallel Environment

For many applications, the additional computing power of a multiprocessor system can be fully exploited only if the input and output capability of the system also increases. The basic problem is that no matter how fast the processors and memory can operate, there will be a bottleneck caused by the speed and bandwidth of the I/O system.

The dominant technology in secondary storage is magnetic disk. There are inherent difficulties in improving the performance of devices based on this technology because of the limitations of the mechanical components involving moving parts. Alternatives such as magnetic bubble memory and optical media have been suggested as successors to conventional magnetic media, but have so far made little impact. It seems that magnetic disks will remain the dominant technology for a considerable length of time, although some solid state devices are now being used, for example as file caches (the SSD on the T3D).

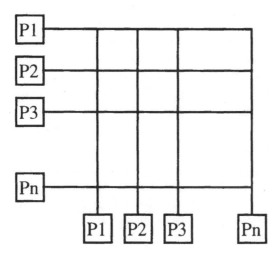

Figure 2.6: Crossbar network

Given the difficulties in improving the performance of individual disks, the alternative of using many disks is being adopted. Simply spreading the data across an array of disks can improve the throughput, and also allows the opportunity to build in fault tolerance by adding redundant disks to the array. This approach has led to the term RAID or Redundant Array of Inexpensive Disks. This approach has been taken with the Connection Machine CM-200 in the form of a *data vault* to which the processors have direct access without going through the single processor which controls processor execution (the so-called front end).

The disk array improves the throughput to the disk and is therefore suitable for an array processor running a single program. In addition, some MIMD systems are now being configured with disks attached to the processing nodes, rather than to a front end processor. In order to gain the full benefit from this configuration, it is necessary to develop high-level tools to manage file systems spread across multiple disks.

While it is possible to provide parallel computers with a number of disks which should allow the I/O bandwidth to scale-up as the number of processors is increased, there is a lack of software to support parallel file systems. Arguably, the closest thing to a standard file system interface is that implemented in the UNIX operating system. However, this does not lend itself very well to a parallel or distributed environment. There is much research going on into the field of parallel file systems, which is discussed further in Chapter 4.

2.5 Examples of Parallel Computers

In this section some examples are given of parallel computers which have been
developed since the early 1980s. The systems are chosen to represent the various
architectures described earlier in the chapter. Trew & Wilson (1990) present a
more comprehensive survey of parallel computers.

2.5.1 SIMD Examples

The restriction of executing the same instruction on every processor limits the
applications which can be run efficiently on a SIMD computer. Conversely there
are several applications which can very effectively exploit the SIMD architecture,
notably many scientific and mathematical problems. Driven by this niche market, a
number of SIMD machines (sometimes referred to *array processors*) have reached
the market place.

2.5.1.1 *Thinking Machines Corporation: CM-1, CM-2 and CM-200*

Thinking Machines Corporation (TMC) produced the first Connection Machine,
the CM-1, in 1984 (Hillis, 1985). The basic architecture is of a large array of
single-bit processors (16,384 in the first model, and up to 65,535 in later models)
controlled by an industry standard workstation, referred to as the front end.
Initially the machine was designed with symbolic computation in mind, but the later
models, the CM-2 and CM-200, can be upgraded by floating point accelerator units
which are shared by 32 of the single bit processors.

The single bit processing elements are grouped into blocks of 16. These blocks are
connected using a hypercube network controlled by a routing processor on each
block. The routing allows for point-to-point (i.e. any processor is able to
communicate with any other), and for nearest neighbour communication on a two
dimensional grid. The latter mode of communication uses a network called the
NEWS grid, deriving its name from the points of the compass. This grid was
implemented as a physically separate entity in the CM-1, but is implemented using
the hypercube in the later models and extended to more than two dimensions,
configurable in software. NEWS communications is highly efficient, and its use is
specified by the programmer when writing source code.

The Connection Machines are programmed in a *data-parallel* style using CM
FORTRAN, parallel versions of LISP (Steele, 1986) and C (*LISP and C*). CM
FORTRAN implements the array syntax defined in FORTRAN 90 (see the next
chapter for a description of data-parallel programming and languages).

All compilation is done on the front end machine, which runs a standard UNIX operating system. A program development environment called PRISM has been developed. This gives debugging facilities using a graphical user interface, and represents a recognition that software to aid software development is a vital supporting part of the system. In addition to the development software there is an extensive library of mathematical and scientific routines. These have been highly optimised to exploit the machine's characteristics and their availability is very attractive to software developers.

2.5.1.2 MasPar

The MasPar Computer Corporation is closely associated with Digital Equipment Corporation. The company produces the MP-1 range of SIMD machines.

The basic architecture is a 2-D grid of processors in configurations of 1,024 to 16,384 processors. The processors are custom designed with a RISC-like (Reduced Instruction Set Computer) architecture. Communications can use a network referred to as the X-Net which connects each processor with its 8 nearest neighbours. The X-Net is most efficient for regular communications. More general communications are implemented using a switched mechanism. This is a somewhat different approach to that adopted by Thinking Machines with the CM-200, and the use of the switch can result in contention for links and consequently processors may have to wait.

The interface to the MP-1 systems is through a DEC Vaxstation running a variant of UNIX.

A software development environment called MPPE (MasPar Parallel Programming Environment) is supplied with the system which provides debugging and visualisation capabilities.

2.5.1.3 The Distributed Array Processor (DAP)

Originally developed by ICL, the DAP was produced and marketed by a spin-off company, Active Memory Technology (AMT).

The basic architecture is a two dimensional grid of custom single bit processors. The interconnection network connects the 4 nearest neighbours in 2-D, with an additional 2 connections per processor to a bus like network. It was originally well suited for applications such as image processing, rather than those demanding high floating point performance. Later versions of the DAP were available with co-processors to improve performance in the latter regime.

2.5.2 Distributed Memory MIMD Examples

2.5.2.1 *Meiko CS-1 and CS-2*

Meiko Scientific was formed as a spin-off from Inmos, the company which developed the transputer. Meiko's products are the distributed memory MIMD Computing Surfaces CS-1 and CS-2. The CS-1 was first demonstrated in 1985 and was available as a commercial product a year later.

The basic architecture of the CS-1 is a switch connecting a number of processors, originally transputers, although later Intel i860 and SPARCs, each with its own private memory. Additional functionality (for example, graphics) could be added to a machine by the incorporation of special purpose processing boards.

The switch is effectively a crossbar implemented by custom processors which link the processors to a backplane. Communications between processors which are on different boards are routed through the backplane. It is possible to use software to configure the physical processors into a logical structure (such as a hypercube) by using the backplane routing. In addition, the CS-1 can be partitioned into a number of domains which can run independently, allowing a multi-user service. The physical limitations of the backplane restrict the number of processors which can be connected, although this limit is very high; a system of over 400 processors was in service at the University of Edinburgh from the late 1980s and until the early 90s.

The primary programming language available with the CS-1 was Occam supported by the Occam Programming System (OPS). However, Meiko quickly realised the importance of standard languages, and introduced FORTRAN and C compilers, with parallelism supported by the Communicating Sequential Tools (CSTools) message passing library.

The successor to the CS-1, the CS-2, was launched in 1992, with the aim of achieving scalability in every aspect of the architecture. The processing elements consisted of either SPARC Superscalar processors or vector processors. The data network uses a fat tree with constant bandwidth per stage. Scalability in I/O is achieved by enabling each processor to manage independent I/O devices. The operating system is based on Sun Microsystems Solaris, with extensions for resource management, parallel file-system and inter-processor communication. The programming environment includes the languages FORTRAN 77, C, FORTRAN 90, High Performance FORTRAN (HPF, discussed further in Chapter 3) and tools to allow debugging and performance monitoring of program codes. Many commercial applications have been ported to the CS-2.

Some 300 Computing Surfaces have been installed, in both industry and research establishments, notably at CERN in Switzerland and the Lawrence Livermore National Laboratory in the USA.

2.5.2.2 Cray Research T3D

In 1993, Cray Research introduced a massively parallel processor, the T3D, to complement the successful range of vector machines. The machine derives its name from the three dimensional torus which is the underlying interconnection network. The processing nodes are standard DEC Alpha processors.

The machine is targeted at the high-end of the market. This is reflected by the minimum configuration of 64 processing nodes with a peak performance of around 10 GFLOPS. The choice of a high specification Cray YMP vector processor as the front end also made an 'entry level' platform unfeasible. The machine can be partitioned into pools allowing space sharing of the system.

The T3D offers both a distributed memory and shared memory programming interface. The memory is physically distributed, but processors may read and write to the memory of other processors. Access to memory on a remote processor is much slower than to local memory, placing the machine in the NUMA category. The programming interface requires that references to memory on a remote processor are explicit.

Communication is via the six bi-directional connections each processor has to its neighbours, with longer range communications being implemented by routing along several links.

A variety of programming modes are available. Message passing using PVM and MPI (discussed in Chapter 3) are available, together with explicit shared memory and synchronisation calls. In addition, there is a *data-parallel* FORTRAN language, CRAFT, and a High Performance FORTRAN (HPF) compiler (see Chapter 3 for an explanation of data-parallel and HPF). Programming tools include a debugger called TotalView, and performance analysis tool called Apprentice.

The successor to the T3D, the T3E, was announced in 1995. The basic architecture remains the same, improvements have been made in caches, and there is better hardware for memory to processor transfers.

2.5.3 Shared Memory Multiprocessor Examples

Several computer hardware manufacturers have developed shared memory multi-processors, because of the commercial appeal of gaining additional throughput by plugging together existing and relatively low-cost processor technology. These machines were often sold as top-end extensions to a range of single processor machines, and their parallel processing (as against multi-processing) capabilities were often not significantly marketed. Particularly notable are the Digital Equipment Corporation (DEC) with TOPS-10, developed in the late 1970s (Deitel,

1984), followed by VAX-based SMP machines in the mid-1980s, Sequent using Intel 80386 then 80486 processors, and Sun with its SPARC-based SMP servers.

Silicon Graphics is worthy of particular consideration, because of its acquisition of Cray, and how that may influence its position in the high performance computing market-place.

2.5.3.1 Silicon Graphics Inc.

In recent years, SGI has extended its range from high specification graphics processing workstations to high performance servers and supercomputers.

The Power Challenge system, released in 1993, is based on powerful RISC processors and a shared memory, with uniform memory access. The systems are available with between 2 and 36 processors.

The machines offer the attraction of the dual purpose of a multiprocessing and a parallel processing system. The operating system (IRIX, a UNIX variant) allows programs with single or multiple threads (threads are discussed fully in Chapter 3). This versatility has attractions for users, since existing software for Silicon Graphics workstations can be run without modification.

The standard programming languages FORTRAN 77, FORTRAN 90, C and C++ are available. Tools are available to assist with parallelisation of C and FORTRAN codes.

The I/O subsystem uses striped disks and RAID technology.

2.6 Industry Trends in Parallel Processing

A number of trends have emerged since parallel computers became commercially available in the 1980s. These have helped to shape the still relatively young parallel and high-performance computing market.

2.6.1 Technology and Architectures

The SIMD model proved successful for certain types of problem (the so-called data-parallel problems), but this specialised range of problems left a gap in the market which began to be filled by MIMDs with both shared and distributed memory. Significantly, MIMD machines have become available from vendors that had previously specialised in SIMD architectures, notably with the CM-5 from

Thinking Machines. There has been a trend towards the use of commodity processors in these machines, and standard operating systems such as UNIX and its derivatives.

There has been a significant growth in Symmetric Multiprocessor systems. The attraction of these machines lies partly in the familiar environment presented to users, and allowing existing programs to run unmodified. The ability of SMPs to function as both multiprocessors and parallel platforms is an additional attraction, as it lowers the risk for users making their first entry into parallel computing.

The move away from centralised computing to distributed computing (i.e. from mainframes to workstations) has in part led to the notion of *clustered computing* whereby a number of workstations are connected by a network to provide a MIMD platform. The network technology restricts the performance of this type of architecture; an interconnect such as Ethernet will provide communications far inferior to the capabilities of a 'closely coupled' MIMD machine. Network technology is improving, with Asynchronous Transfer Mode (ATM) technology now capable of data transfer rates in excess of 155 Mbytes per second - this opens up the potential for parallel processing across multiple sites.

Clustered computing has a particular attraction for industry because it can make use of a resource which may otherwise spend a considerable amount of time idle. A typical industry scenario is for a department to have a number of workstations which represent a fixed cost; additional computing services such as time on a central mainframe or MPP cost additional money. Therefore if the idle time of the workstations can be put to good use, a saving can be made. Since workstations are frequently used in a standalone mode during the day and are idle at night, the cluster can be run as a MIMD platform overnight. Software to support such configurations is reaching maturity. Some vendors market workstation clusters as a dedicated parallel platform, and the definition of what constitutes a MIMD machine and the differences between the traditional MIMD architecture and a workstation cluster are becoming increasingly blurred.

2.6.2 Commercialisation

The early offerings in parallel computing were mainly from start-up companies, such as TMC, or spin-off companies from larger organisations, such as Meiko (originating with Inmos), AMT (ICL). The giants such as IBM, DEC and their Japanese counterparts offered multiprocessor systems combining a small number of very powerful processors. This has changed to a certain extent with the development by IBM of a range of distributed memory MIMD computers (the SP-1 and SP-2), and Digital's approach of connecting clusters of SMPs using high-performance proprietary, in addition to non-proprietary, interconnect hardware. In addition to this, the dominant force in supercomputers, Cray Research, has extended its range of machines to include the T3D, described above.

The parallel computing market has also proved turbulent for some of the players, with some failures (such as Kendall Square Research), some companies forced to restructure their businesses (such as Thinking Machines, which between 1995 and 1996 ceased to make the Connection Machine series, and focused on clusters of standard workstations), and some mergers and acquisitions (such as the buy-out of Cray Research by Silicon Graphics Inc. in 1996).

2.6.3 Software and Standards

In the field of software there have been initiatives to standardise on the programming interfaces which have for so long impeded the development of parallel software. These are discussed further in the next chapter. The emergence of standards has taken some of the risk out of developing parallel software, and many independent software vendors are now developing parallel applications, often with the assistance of hardware vendors. Funding from schemes such as ESPRIT in Europe has been available to companies to help 'kick start' the effort of porting.

2.7 Conclusions

Over the decade from 1985 to 1995, parallel computing has matured from a research technology to commercial viability. Evidence of this can be seen in the number of parallel programming systems in existence, in the effort which has been put in to producing parallel software, the activity in international standards and the entry of the large computer manufacturers into the marketplace.

The fundamental argument in favour of parallel processing is that the improvements in technology have enabled a growing range of applications to run sufficiently faster on multiple processors that the complexity and additional investment in development effort are worthwhile. Geographical processing and analysis are examples of complex applications where the potential of parallel processing is beginning to be recognised and exploited. The particular data-rich nature of GIS holds a set of additional challenges for achieving high performance and real-time analysis.

The question of whether one of the existing parallel technologies will become dominant to the exclusion of all others is less clear.

2.8 References

Burkhardt, H., 1992, Technical summary of KSR-1, Technical Summary, Kendall Square Research Corporation.

Dasgupta, S., 1989, *Computer Architecture: a Modern Synthesis*, Vol. 2, John Wiley & Sons.

Deitel, H.M., 1984, *An Introduction to Operating Systems*, Addison-Wesley, Reading, MA.

Flynn, M.J., 1972, Some computer organisations and their effectiveness. *IEEE Trans. Computers*, 21 (9), 948-960.

Goldstine, H.H. & Goldstine, A., 1946, The electronic numeric integrator and computer (ENIAC), *Mathematical Tables and Aids to Computation*, 2 (15), 197-110. Reprinted in Rendell, B. (Ed.), 1975, *The Origins of Digital Computing*, Springer, Berlin, 333-47.

Hillis, W.D., 1985, *The Connection Machine*, MIT Press, Cambridge, MA.

Morrison, P. & Morrison, E., 1961, *Charles Babbage and his Calculating Engines*, Dover, New York.

Nitzberg, B. & Lo, V., 1991, Distributed shared memory: a survey of issues and algorithms. *IEEE Computer*, 24 (8), 52-60.

Steele, G.L. Jr., 1986, CM-LISP, Technical Report, Thinking Machines Corporation.

Trew, A.S. & Wilson, G.V. (Eds), 1990, *Past, Present, Parallel*, Springer Verlag.

3

The Software Environment and Standardisation Initiatives

M. Sawyer

The previous chapter dealt with the issues of building hardware systems on which to run parallel programs. This chapter deals with the software issues of parallel programming. The emphasis here is on the basic concepts rather that at the level of how to design a parallel program. The latter issue is examined in Chapter 4. In this chapter, the issues of performance prediction and granularity are introduced, followed by a discussion of the various programming paradigms which have been developed. The chapter concludes with a discussion of the software standards which are emerging, and their importance.

3.1 Basic Concepts in Parallel Programming

The fundamental difference between a parallel and a sequential program is that in the parallel program several operations can occur at the same time on different data. When designing a parallel program, careful consideration must be given to the decomposition of operations and data, and to how the processors will communicate. A complex set of interrelated problems have to be dealt with. These include:

- the degree of parallelism i.e. the number of processing activities which can be simultaneously executed,

- data management, such as the partitioning or replication of data,

- mapping of processes to processing nodes,

- scheduling of tasks,

- communication between tasks which is not available at the machine level (i.e. not available through shared memory).

The goal of designing a parallel program is to balance the solutions to these problems in such a way as to optimise performance criteria, which include:

- the time taken to execute the program (*program latency*)

- the *bandwidth* (i.e. the throughput of repeated similar tasks)

- *speed-up*: the ratio of the performance of the parallel program to the sequential program

- *efficiency*: the ratio of the speed-up to the degree of parallelism.

3.1.1 Limitations to performance: Amdahl's Law

All but the simplest of processes consist of some tasks which can be performed in parallel and some which must be performed sequentially. The effect of the sequential component of a process is discussed by Amdahl (1967), and the expression for the maximum speed-up is often referred to as Amdahl's Law.

The speed-up which can be achieved when running on a parallel computer can be found by analysing the proportion of the program which can be performed in parallel, and that which must be performed sequentially. It is given by the expression:

$$S_N = \frac{T_s + T_p}{T_s + {T_p}/{N}}$$

Where S_N is the speed-up with N processors, T_S is the fraction of operations which must be performed sequentially, and T_P is the fraction of operations which can be done in parallel. The expression shows that in the limit of an infinite number of processors used to implement the parallel section, the speed-up is determined by the sequential component.

Amdahl's Law can be illustrated with an example: if a program has a serial component of 10 per cent, it can be speeded up by a maximum of ten. Additionally, once ten processors have been dedicated to this problem, the speed-up will be greater than nine and using further processors produces very little benefit. Figure 3.1 shows the effect of sequential processing when the sequential component is only 5 per cent.

3.1.2 Scalability

The way in which the speed-up of a program increases with the number of processors used (the degree of parallelism) is referred to as the *scalability*. A

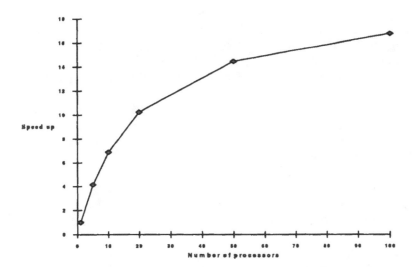

Figure 3.1: Graph showing the effect of sequential component on the speed-up of an algorithm. The graph shows the speed-up when the algorithm is 95% parallelisable

problem has good scalability if there is a roughly linear relationship between the speed-up obtained and the number of processors used up to a large number of processors. Good scalability implies an efficient use of processors, since they spend little time idle while waiting for sequential tasks to complete.

It is difficult to define what constitutes a large number of processors, but for practical purposes if a near-linear relationship between speed-up and number of processors can be maintained up to the largest size of machine on which the problem will be run (which may not yet have been built), then good scalability can be claimed. Amdahl's Law shows that good scalability can only be achieved when the sequential part of the algorithm is small compared to the part which can be undertaken in parallel.

3.1.3 Granularity

The *granularity* of a problem refers to the relative amount of computation in a task which can be performed in parallel. If the process can be broken down into many small parallel tasks it is described as *fine-grained*. If it consists of large tasks, a problem is said to be *coarse-grained*. Fine-grained problems can have a greater maximum degree of parallelism than coarse-grained problems, but it is perfectly possible for coarse-grained problems to be implemented efficiently.

As an example, some image processing algorithms involve operations which can be undertaken in parallel on individual pixels. This is a fine-grained example, as one could use a number of processors equal to the number of pixels in the image. Contrast this with the multi-block method used in many engineering calculations. In such a calculation, a structure such as a bridge or aircraft body is represented as a number of components. Each of these could be processed in parallel by allocating each block to a different processor, but the number of processors which can be employed is limited by the number of blocks available. This limitation on the degree of parallelism in itself need not be a problem, since it may still be possible to achieve an efficient parallel implementation.

3.2 Programming Paradigms

Designing a parallel program by solving the problems listed in Section 3.1 is a complex task for real-world applications. For reasons of portability, and to account for differences in machine characteristics, programmers need higher level methodologies for designing parallel programs. As yet, there are no general purpose tools available which can automatically parallelise an application effectively, although there are some which can recognise implied parallelism in certain cases. Thus in the overwhelming majority of situations, the skill of the designer is paramount.

With the rise of parallel computing have emerged parallel programming paradigms. The development of these paradigms has been led, to some extent, both by the facilities provided by hardware and by the ease with which the parallel nature of a program can be expressed. The *data-parallel* paradigm, for example, reflects the way in which a SIMD machine operates. MIMD machines more closely follow the *message-passing* paradigm. The data-parallel and message passing paradigms described below are relatively low level; higher level abstractions such as the *task farm* are described in Chapter 4.

3.3 The Data-Parallel Paradigm

The data-parallel paradigm involves the application of a particular function to each element within a dataset at the same time, and is characterised by:

* Global address space

* Single thread of control

* Loosely synchronous.

The concept of a global address space means that the programmer can regard any processor as being able to access any part of memory. The single thread of control reflects the idea that similar operations are performed for all data. This can include a conditional operation, in which a condition is evaluated for each data item, and appropriate action taken for the local data. The term *loosely synchronous* describes the behaviour of the program after operations on local data and conditional operations; the processors synchronise after each simple operation or after conditional operation. For processors where the conditional operation requires more operations, the remaining processors will wait until the completion of those operations before continuing.

Because data-parallel applications are those in which similar operations are applied uniformly to the dataset, they are well suited for SIMD computers. It is also possible to implement data-parallel programs on MIMD computers if the same instructions are used on each processor. Data-parallel programs are normally written using special languages which define operations on collections of data. In most practical situations, the parallel operations are expressed using extensions to standard languages such as FORTRAN or C. Communication of data is implied by the language extensions.

The data-parallel paradigm can be illustrated with an example, such as the contrast stretching algorithm used in image processing. This algorithm is used to enhance the quality of an image which has low contrast. In a low contrast image, the pixels have brightness values which are relatively close together. In order to use the full range of brightness, it is necessary to scale the value of each pixel. The algorithm has three stages:

- The highest and lowest values of brightness are found.

- The lowest brightness value is subtracted from each pixel.

- The new brightness value at each pixel is multiplied by a constant.

Clearly stages 2 and 3 consist of performing the same operation at each point independently, and hence this fits the data-parallel model. Notice that finding the highest and lowest values requires a global operation performed over all the data. Such operations are referred to as *reduction* operations, and are implemented by special functions in a data-parallel language.

The data-parallel paradigm can also be used effectively for problems which have regular communications.

Another example, drawn from image processing, which illustrates data-parallelism is edge detection. This technique attempts to find edges in an image based on whether there is a difference in brightness between adjacent cells. The

communications pattern is shown in Figure 3.2; each pixel requires data from the
north, south, east and west neighbours. The pattern of communication is the same
at all points, and thus the single thread of control can be used to express the
calculation.

Recently there has been an effort to standardise the language extensions necessary
to implement data-parallelism, notably through FORTRAN 90 and High
Performance Fortran (HPF). These are described in greater detail later in this
chapter.

The data-parallel paradigm is restrictive in the types of algorithms which can be
expressed. Generally, problems are either well suited to the data-parallel paradigm
or they are poorly suited, with little ground in between. This is due to the single
thread of control which, while allowing ease of programming, is restrictive because
different functionality cannot be assigned to each processor in an efficient way. It
is possible to perform different operations based on local conditions, but the loose
synchronisation would lead to very inefficient implementations if the functionality
were very much different.

3.4 The Message Passing Paradigm

The basic restriction of the data-parallel paradigm is the single thread of control
throughout the program. In order to allow the programmer more flexibility with
assigning different functionality to processors, a method of parallel programming
has been developed which allows different tasks to be performed on different

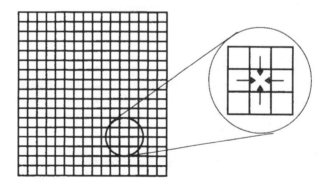

Figure 3.2: The value at the central cell depends on the four neighbours shown

processors. Data is sent to and received from other processors in an explicit fashion, by means of a set of library functions which are called from within each task. This paradigm is referred to as *message passing*. The message passing paradigm is general purpose and for this reason the use of MIMD machines, supporting message passing applications, is an attractive commercial configuration.

A basic message passing system needs to provide the programmer with the following types of operation:

- Point-to-point communications

- Collective communications

- Process management

- Synchronisation primitives.

Point-to-point communications are the most basic type of operation supported by a message passing system; one processor sends a message to another (Figure 3.3).

Collective communications involve larger groups of processors (Figure 3.4). Examples of collective operations are *broadcast*, in which one processor sends data to all other processors involved in the operation, and *reduction*, in which each processor contributes a value to one processor and which then returns an associative value, such as the maximum or sum of the values it has received.

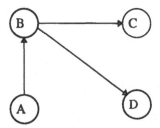

Figure 3.3: Point-to-point messaging: Process A sends a message to processor B. Processor B sends messages to C and D.

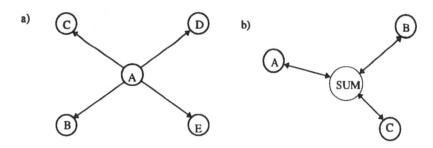

*Figure 3.4: Collective communications. a) Broadcast: processor A sends to B, C,
D and E, and b) reduction: each processor supplies a value and is returned a
value, in this case the sum.*

Process management governs the creation and termination of processes, while
synchronisation primitives are used to indicate when processors have reached a
certain point in an algorithm, but do not need explicitly to exchange data.

There are various ways in which the programmer can specify how the processes
involved in a message passing operation behave. The terms *blocking* and *non-
blocking* are used to describe alternative types of behaviour. In a blocking
operation, the calling process does not proceed until the operation has completed.
In the case of a 'send', the sender does not continue execution of the program until
the message has been received by the receiving process. In the case of non-
blocking calls, the program will proceed immediately after the send or receive has
been set up; the program must check later that the operation has completed. Non-
blocking communications can be used in order to continue processing while waiting
for messages to be sent or received. This can lead to more efficient programs,
since there can be less idle time, but the programs can be more complex.

In addition to the basic communication operations, some message passing systems
allow the programmer to group processors according to certain attributes, for
example to reflect the different functionality of processors. This allows message
passing operations to be confined to particular groups of processors, a feature
which is useful in the design of library software, since programmers can use
message passing within an application without interfering with communications in
the library. The development of library software to support higher level models of
parallelism is discussed further in Chapter 4. The industry standard for message
passing, the Message Passing Interface (MPI), goes further, defining *contexts* (see
section 3.9.3.2).

It is worth noting that although the message passing paradigm most closely follows the distributed memory MIMD architecture, there is nothing to prevent its implementation on other architectures, since it is apparent to the programmer simply as an interface to a library. Many shared memory architectures incorporate message passing libraries.

3.5 Shared Memory Programming

There are several ways in which shared memory systems may be programmed. The simplest way is to use the computer to run simultaneously a number of sequential jobs. Programs can be written exactly as for a single processor machine, and existing routines can be used without modification. The operating system schedules the execution of each process, which may migrate between different processors. Higher level software may be used to queue jobs, allowing the overall load on the machine to be managed. This mode of operation is often referred to as *multi-processing*. A higher throughput may be achieved in this way, but an individual program will not run more quickly than on a single processor machine.

Alternatively it is possible to implement a message passing library, enabling processes to communicate, and thus permitting parallel processing; programs written for the same message passing interface will be able to run on either a shared or a distributed memory machine. Again the operating system schedules the individual processes which make up the message passing application.

A third option is to exploit the fact that the memory is physically shared, and to write programs which exploit this directly. There are two ways in which this can be done: through the use of directives in the source code which instruct the compiler that sections of the program should be performed in parallel, or by directly specifying several *threads* of execution.

3.5.1 Threads

A process running under an operating system such as UNIX normally has a single thread of execution - there is a single program counter and context for the process, and the programmer could think of the program as being at a single point in the source code at any given time. Some operating systems allow more than one thread to be in existence at a time (e.g. Sun's Solaris, Digital's VMS with DECthreads). This can be thought of as having several program counters, and each thread being at a particular point in the source code, with the possibility of several threads running concurrently on separate processors. Threads within the same process have the important attribute that they share the same address space. If one thread modifies a data structure, all other threads in that process see the change immediately, and there is no need for explicit message passing. This has the attraction that it provides a simple programming model, and the overhead of communication becomes very small. However, there is a trade-off in that special programming techniques must be applied in order to guarantee that the data is consistent, and debugging multiple-thread applications can be extremely difficult.

3.5.1.1 Shared Data Structures

Because the threads within a process share the same address space, the programmer must exercise extreme caution to ensure that one thread cannot leave data structures in an inconsistent state for another thread. An example of how this could happen is in the management of a linked list of data objects. The removal of an object from a linked list requires a number of operations: the forward and reverse pointers must be adjusted in the sequence shown in Figure 3.5.

It can be seen that when only one pointer has been updated, the list is in an inconsistent state; the thread which is updating the list must complete the operation before any other thread may access it. Sections of a program which must be allowed to complete before other threads may access a data structure are referred to as *critical sections*, and require special programming techniques.

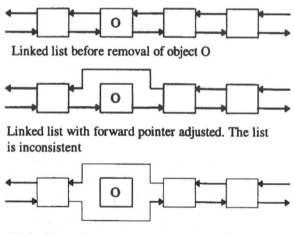

Linked list before removal of object O

Linked list with forward pointer adjusted. The list
is inconsistent

Linked list with both pointers adjusted. The list is
now consistent

Figure 3.5: Updating a linked list

3.5.1.2   Critical Sections

Programmers must ensure that critical sections (CS) have the following
characteristics:

- *Mutual exclusion*: there must be at most one thread executing the CS at a time.

- *No deadlock in waiting*: if two or more threads contend to enter the CS, one will
  succeed.

- *No pre-emption*: it must be impossible to interrupt the thread.

- *Eventual entry*: a thread contending to enter the CS will eventually gain entry.

Mutual exclusion can be implemented using *locks* or *semaphores*. Locks use a
variable known as the *gate variable* to determine whether a thread may enter the
CS. A thread which wishes to enter the CS tests the gate variable; if it is 'clear'
(i.e. its value indicates that the thread may enter the CS) then the gate variable is
set, and the thread executes the critical section. At the end of the CS, the thread
clears the gate variable enabling other threads to enter.

Since the gate variable itself is shared, care must be taken when it is accessed. In
order to ensure that two threads do not test the gate variable, find it clear and then
set it and enter the CS, the test and set must be done as a single operation. Such an
operation is said to be *atomic*.

When a thread finds that it cannot enter a CS due to the lock being set, it may
continuously poll the gate variable until it is clear, in which case it is referred to as
a *spin lock*. This is a simple approach, but results in processor time being wasted.

3.5.1.3   Semaphores

Semaphores are a synchronisation tool proposed by Dijkstra (1968) as a means of
implementing mutual exclusion based on a sleep and wait protocol to avoid wasting
processor cycles. A semaphore is an integer on which two atomic operations are
defined:

```
P(S): if S > 0

    then :  Decrement S

    else :  Block the process and place it in the 'wait queue'
            for S

            Select another process to run
```

```
V(S): Increment S

        if there are processes waiting for S

        then place such a process in the ready to run queue
```

Processes can then call P(S) before attempting to enter a critical section, and then call V(S) when the critical section has been completed. The advantage over using a spin lock is that the processes suspend themselves if the critical section cannot be entered, and are 'woken up' when it becomes available.

3.5.2 Programming with Threads

There are two main advantages of using threads. Firstly, their use allows a shared memory multi-processor to schedule the threads to run concurrently. Secondly, there is a lower overhead in switching between threads than in switching between processes. Thus, even on a single processor machine, there are advantages in writing an application as multiple threads within a single process, rather than as separate programs.

Threads can be created explicitly by the programmer through language extensions. It also possible for compilers to generate a multithreaded program, either automatically by analysing the data dependencies, or according to compiler directives.

User-defined threads must be written taking account of critical sections and shared data structures. This requires considerable skill on the part of the programmer, and such software is generally much more difficult to debug and test. If a critical section is incorrectly programmed, and it is possible for two threads to interfere with the processing of a shared data structure, the error may go undetected for a long time. Even when the error is detected, it may be hard to reproduce the conditions which led to it. For these reasons, software development using threads is generally a more expensive exercise than using separate processes.

3.6 Object-Oriented Languages

Programming languages based on the concept of objects have gained wide acceptance. The underlying principles of object-oriented programming are objects, classes, and inheritance.

Objects are a data abstraction which *encapsulate* data and operations. Each object has an interface which is visible externally and which defines the operations which may be applied to the object. The basic model of an object-oriented system is a number of objects, communicating by sending messages. The specification of objects is achieved using *classes* which define the attributes and interface; an object is an instance of a class. Inheritance is a mechanism whereby classes derive characteristics from other classes. This is a key attraction of object-oriented systems, as it enables software re-usability. Programming using the object-oriented method is described in detail in Meyer (1988).

The object-oriented system implies parallelism; the mechanism of communication between objects corresponds closely to message passing.

There is also the possibility of parallelism within an object; some of the operations within the object could be implemented using several processes of threads, communicating via message passing or shared memory. Such parallelism would be invisible externally, following the general principle of encapsulation, but clearly beneficial and, critically, re-usable.

3.7 Linda

Linda (Gelernter *et al.*, 1985) offers a shared virtual memory system with access functions. The rationale for the development of the Linda system was to allow a simple mechanism for communication between processes. The system can be likened to a bulletin board system in which people may add, read or remove messages.

Linda is based on a *tuple space* which is shared by all concurrent processes. Tuples are sequences of typed data, for example a tuple might contain a string of characters followed by an integer. Data items which have assigned values are referred to as *actuals*. Data items may instead be given formal pattern matches, and are referred to as *formals*.

The tuple can contain actual values, or formal pattern matching values.

Processes have asynchronous access to the tuple space, using the following six operations:

- **out** whereby a process outputs a tuple to the tuple space. Actuals are evaluated before the tuple is added to the space.

- **in** whereby a process withdraws a tuple matching the template supplied from the tuple space. If there is no matching tuple, the process suspends.

- **rd** is similar to in, except that instead of removing the tuple, a copy is made with the formals evaluated.

- **inp** and **rdp** are non-blocking versions of in and rd; if no matching tuple exists, they return immediately.

- **eval** allows tuples to be added to the tuple space, and is similar to in, except that the values are evaluated after it is added to the tuple space rather than before.

It can be seen that the basic operations required for message passing systems are available in Linda. Messages can be sent, point-to-point, by specifying a destination field in the tuple. Broadcasts to all, or groups of, processors can be implemented using the non-destructive read (rd).

One of the goals of Linda is to be simple enough that it can be implemented on a number of existing programming languages. There are commercially available implementations of C and FORTRAN which incorporate Linda extension.

A second goal was to obtain portability. This is achieved through the abstraction of the tuple space, which hides from the user the architectural details of the computer, and the communications method used.

3.8 The Parallel Random Access Machine (PRAM)

The PRAM (Fortune & Wylie, 1978) is an idealised model of a parallel computer which assumes zero synchronisation cost and memory access overhead. The model can be used for development of parallel algorithms, and for scalability and complexity analysis.

The PRAM consists of a number of processors with a globally addressable memory. The memory may be located centrally or distributed between the processors. The processors operate on a synchronised read, execute and write cycle. If the memory is central, the model must specify how reads and writes to the same memory location in the same cycle are handled. There are four memory update strategies:

- Exclusive read (ER): This model allows only one processor to read from a memory location.

- Exclusive write (EW): This allows only one processor to write to a memory location.

- Concurrent read (CR): This allows more than one processor to read from the same memory location.

- Concurrent write (CW): This allows more than one processor to write to the same memory location.

Various combinations of these read and write strategies can be combined to form four types of PRAM:

- EREW: In this model, only one processor may read or write to any location.

- CREW: This model permits multiple reads from a location, but only one processor may write to a location.

- ERCW: Only one processor may read from a memory location, but many may write to the same location.

- CRCW: Reads or writes may be made to the same memory locations by more than one processor.

Since the model is an idealised one, the issue of contention for the memory resource does not arise, and so there is no difficulty with concurrent reads; all processors will read the same value. However, in the case of concurrent writes some policy must be adopted to define which value is held in the memory location after the write. Four policies are proposed by Fortune & Wylie (1978):

- Common: The restriction is made that all processors which write concurrently to the same location write the same value.

- Arbitrary: Any of the values in a concurrent write are written.

- Minimum: The value written by the processor with the lowest index is stored in the location.

- Priority: Some kind of associative operation (e.g. sum, maximum) is performed on the values being written.

The assumptions about the cost of memory references, and synchronisation, clearly do not correspond to the real architectures. The SIMD is the closest real architecture to the PRAM model, provided the PRAM executes the same instructions on each processor. An alternative view could be of a synchronised MIMD with shared memory.

The PRAM models are useful tools for the design and analysis of algorithms. The CRCW and EREW variants are most commonly used in analysis. The CREW PRAM is an example of a MISD computer.

3.9 New Developments in Standards for Parallel Processing

3.9.1 FORTRAN 90

In 1966, the American National Standards Institute (ANSI) completed the first definition of a programming language, FORTRAN 66. During the following ten years, FORTRAN became the language of choice for programmers in scientific and engineering fields, and compilers became widely available. Unfortunately, along with the wide availability of compilers came a proliferation of extensions to the standards, as compiler writers gave programmers features to exploit machine performance or ease the task of programming. The result was that codes were not portable across machines. The decision to draw together and standardise the existing extensions was taken, and FORTRAN 77 appeared as an ANSI standard in 1978. At the same time ANSI set up a working group on FORTRAN standards as an ongoing activity to define a FORTRAN 8x standard. Unfortunately there were many delays in the publication of the standard, and the 'x' threatened to become greater than 9. In 1988, the International Standards Organisation (ISO) proposed FORTRAN 88 in response to ANSI delays. ANSI voted to accept the principle of the ISO plan, and in 1991 the final draft was published as a standard.

The FORTRAN 90 standard (ISO, 1991) represents the first major attempt to develop the language, allowing new programming constructs and the ability to express the data-parallel nature of a problem. The new features available to the programmer cover data-parallel syntax, additional support for structured programming, such as the ability to define data structures and the introduction of pointers.

FORTRAN 90 contains all the FORTRAN 77 features, although a mechanism has been introduced to remove features from previous standards, as future standards are defined. The FORTRAN 90 standard also categorises some FORTRAN 77 features as obsolescent (generally because they are regarded as bad programming practice, for example arithmetic IF), and redundant features for which there are better alternatives in the new standard.

The most significant feature, from a parallel programming point of view, is that the programmer may now program in a data-parallel manner using explicitly parallel constructs. FORTRAN 90 bears a great similarity to data-parallel languages such as CM FORTRAN.

3.9.2 High Performance Fortran

The High Performance Fortran Forum (HPFF) was founded by a group of users, academics and vendors with the aim of defining a new standard for FORTRAN which would provide high performance across a range of machines and architectures. The Forum has no official status, unlike ANSI and ISO.

The first meeting was held in 1992, and the goals of HPFF were set:

- to support the data-parallel programming paradigm

- to achieve top performance on MIMD and SIMD machines with non-uniform memory access

- to support code tuning on various architectures.

There were secondary objectives; notably to maintain compatibility with FORTRAN and other existing standards, and to define interfaces to other languages and paradigms. Since several members of HPFF also sit on the ANSI committee which deals formally with the FORTRAN standard, there is informal feedback between the two groups.

The array syntax of HPF follows that of FORTRAN 90. An important feature of HPF, which allows the programmer to write efficient parallel applications, is that the distribution of data can be specified.

3.9.2.1 *Data Distribution*

HPF defines a mechanism for describing the layout of arrays on the processors and the relative location of array elements. These features are implemented through compiler directives.

The distribution of data across processors is achieved using the DISTRIBUTE directive:

```
        REAL, DIMENSION(16):: A
!HPF$ DISTRIBUTE (BLOCK) :: A
```

In this example, with four processors, the elements A(1) to A(4) would be on the first processor, A(5) to A(8) on the second processor and so on.

An alternative is to distribute the data in a cyclic manner:

```
      REAL, DIMENSION(16):: A
!HPF$ DISTRIBUTE (CYCLIC) :: A
```

Again taking a four-processor example, the first processor would have elements A(1), A(5), A(9) and A(13), the second processor would have A(2), A(6), A(10) and A(14) and so on.

Arrays of higher dimension can be distributed by combinations of BLOCK and CYCLIC, or by specifying that a certain dimension is not distributed. Data in one array may be aligned with data in another array in a particular way. For example, elements with the same indices could be aligned on with each other, with indices permuted and offset.

For example:

```
      REAL A(16,16), B(14,14)
!HPF$ ALIGN B(I,J) WITH A(I+1,J+1)
```

This aligns the square array B with the interior of array A, leaving a border one element wide all round.

The standard also allows for a processor topology to be defined, using the PROCESSORS compiler directive:

```
!HPF$ PROCESSORS PROCS(8,8)
```

This example declares an array of 64 processors, arranged as a square grid. HPF restricts such declarations to rectilinear configurations.

The HPF directives are designed to give the compiler more information about the program, in order to allow it to produce the most efficient code. Clearly specifications such as the one given above will not be met on systems with fewer than 36 processors. However, it may be possible to implement the algorithm with almost the same efficiency on, say a four by four grid. For this reason the HPF standards refers to 'abstract processors' when describing distributions and alignments.

3.9.2.2 The WHERE Construct

The WHERE construct can be described as a parallel IF statement; it consists of a conditional expression (similar to an IF statement, except that the arguments are array sections) and a body of instructions. The conditional expression is evaluated for each element of the arrays specified; at each location where it is true, the body of the WHERE statement is executed.

3.9.2.3 Availability of compilers HPF 1.0

HPF implementations are becoming more widely available, with announced products by 1996 from the following vendors:

> Applied Parallel Research, Digital Equipment Corporation, Hitachi, Intel, Kuck and Associates, Meiko Scientific, Motorola, NA Software, NEC, Pacific Sierra Research, The Portland Group Inc. (PGI), and SofTech.

However, not all implementations conform fully to the standard. Some vendors have chosen to leave out some features, or to add further features. This has led to some industry fears that HPF is a fragmenting standard.

3.9.2.4 The Future of HPF: HPF-2

The HPF standard is being developed by the HPFF. The current development effort is centred on:

*   Enhanced mapping to support arbitrary mappings

*   Computation control and task parallelism (computation mapping, multiple HPF process model)

*   Input and output (striping/distributing files, checkpointing)

*   Communication optimisations (reuse patterns, locality assertions)

*   Language processor environment (interoperability).

3.9.3 Message Passing Standards

3.9.3.1 *Parallel Virtual Machine*

Parallel Virtual Machine (PVM) was originally designed as a basic message
passing system for heterogeneous collections of networked computers (Geist *et al.*,
1994; PVM, 1997). The development began at the Oak Ridge National
Laboratory, University of Tennessee and Emory University. PVM has been
implemented on many tightly coupled parallel computers, and became the *de facto*
standard message passing system prior to MPI. PVM is implemented as a library
of functions, which the user calls from an application. PVM supports the basic
functionality of message passing systems defined in Section 3.4.

3.9.3.2 *Message Passing Interface (MPI)*

Most of the distributed memory MIMD machines, which began to appear from the
mid 1980s onwards, included proprietary message passing systems, such as
CSTools on the Meiko CS-1. Although these largely offered the same facilities,
without standards, writing software which was portable between machines was
practically impossible. There were attempts at standardisation of an interface,
through efforts such as the Edinburgh Parallel Computing Centre's Common High-
Level Interface to Message Passing (CHIMP) which became available across a
wide range of platforms. The growth of PVM eased some of the portability issues,
but there was still the lack of a real standard.

The Message Passing Interface Forum (MPIF) was convened in a similar manner to
the HPFF. Its goal was to define a standard for portable and efficient message
passing applications. A preliminary draft proposal was put forward in November
1992. This was very much an attempt so start the discussions about message
passing requirements, and dealt only with point to point communications, not any
collective communications. Following the adoption of more formal procedures and
organisation, a draft standard was presented at the Supercomputing '93 conference
(MPI Forum, 1993).

The standard covers the following areas:

- Point to point communications

- Collective communications

- Groups contexts and communicators

- Process topologies

- Environment management.

A unique feature of MPI is the ability to define a *context* for communications. All communications take place within a specific context. Processors can only communicate if they share a context, and messages cannot be received in the wrong context.

Contexts can be used to avoid conflicts when message passing takes place within different parts of the program. In particular, it is possible to create library software to support programming models by defining a context for the library. Any message passing which the applications developer uses in a program will not conflict with the operation of the library.

In addition, the user either can use basic data types, such as integers and real numbers, or can construct derived data types based on collections of the basic types and other derived data types. This allows the programmer to retain the same structures for communications that are used in the remainder of the program.

MPI is available across a number of platforms including most UNIX workstations, Intel iPSC, CM-5, IBM SP-2, Meiko CS-2 and Cray T3D.

MPI is being further developed by MPIF as MPI-2. The developments being covered are parallel I/O, dynamic process creation, bindings for C++ and FORTRAN.

3.9.4 The Parallel Tool Consortium

The parallel tools consortium was created November 1993 to foster reliable, portable tools for parallel program development. It consists of US federal, industrial and academic representatives. The consortium recognises three factors key to the development of parallel tools:

- Parallel tools are hard for non-computer scientists to use

- Tools vary across platforms

- There is a lack of specialised support for heterogeneous or scaleable applications, which deters the development of parallel libraries and applications.

The projects completed or under way include a lightweight core-file browser visualisation tool for distributed arrays, message queue manager, parallel UNIX commands and portable timing routines.

3.9.5   Parallel I/O

One of the biggest obstacles to the effective use of teraflop computers (computers capable of 10 million million floating point operations per second) is that of moving data into, and out of, such systems. There is a significant amount of research into scaleable parallel I/O and parallel file systems, but the development of standards lags behind the developments such as HPF and MPI.

3.9.5.1   *The Scaleable I/O Initiative*

The Scaleable I/O Initiative (Bagrodia *et al.*, 1994) is a collaborative effort among application developers, computer scientists and hardware vendors which aims to ensure the efficiency, scalability and portability of its results and software. It also aims for the adoption of these techniques by users and vendors. The vendors involved include IBM, Cray, Intel and Convex. The applications under consideration are drawn from the sciences, engineering, earth sciences and graphics.

The initiative aims to guide the development of parallel I/O, taking account of compilers, runtime libraries, parallel file systems, high-performance network interface and operating systems. The PABLO performance evaluation tool (Reed *et al.*, 1994) is being used to examine the I/O characteristics of the applications using test-bed systems.

3.9.5.2   *Parallel File Systems*

In addition to the proprietary parallel systems offered by some vendors, there are a number of research projects dealing with parallel files systems. Brief details are given here:

PIOUS is a PVM-based parallel file interface developed at Emory University (Moyer, 1994). In the same way that PVM implements a virtual multi-computer on top of a heterogeneous network of computing resources, PIOUS implements a fully functional file system on top of PVM. Applications obtain transparent access to shared permanent storage using PIOUS library calls. The system is aimed at supporting high-performance applications and at use as a research tool.

PPFS is a portable parallel file system developed at the University of Illinois at Urbana-Champaign (Huber *et al.*, 1995). It is an ongoing project with the goal of developing a flexible application programming environments which allows users to specify hints of access patterns and to explore a variety of data distribution, distributed caching and pre-fetching strategies.

The Parallel and Scaleable Software for Input and Output (**PASSION**) project at Syracuse University is aimed at providing I/O support for data-parallel programs (Choudhary *et al.*, 1995). A runtime library is provided which can be used either from an explicit message passing language or from data-parallel languages such as HPF. A novel technique termed *data-sieving* is used to improve the performance of disk accesses which involve reading regular sections of data. Instead of using multiple reads for strided data, the technique involves performing a few larger disk reads, and then extracting the data required within main memory.

The **Hurricane File System** (HFS) is being developed at the University of Toronto for the Hector NUMA processor (Krieger & Stumm, 1993). The systems uses *hierarchical clustering* to obtain parallel scalability. The basic unit within a hierarchical clustering is the cluster which provides the functionality of a small scale tightly coupled SMP operating system. Larger scale systems are viewed as a collection of clusters.

3.9.5.3 MPI-I/O

A proposal for MPI-I/O has been drafted jointly by IBM Research and NASA Ames Laboratory to address the issues of portable parallel I/O. The general idea is to model I/O requests as message passing calls.

MPI itself does not address I/O, and not all parallel machines support the same interface. The closest thing to a standard is the UNIX file system interface which is ill suited to parallelism. The MPI I/O initiative, part of the general evolution of MPI, is addressing this problem.

The proposal is aimed at:

- scientific applications;

- common usage patterns rather than obscure ones;

- real world requirements;

- allowing programmers to specify high-level information about I/O to the system rather than low-level system dependent information; and

- favouring performance over functionality.

For clarity of purpose, some specific 'non-goals' of MPI/IO have been identified:

- Support for message passing other than MPI

- UNIX compatibility

- Transaction processing

- FORTRAN record I/O.

MPI I/O is MPI like, with file access is defined by a communicator. Access can be independent or collective, and derived data-types are used to express the layout of file and the partitioning, and how the data is laid out in the processor's memory.

3.10 Conclusion

Writing software for parallel computers is more difficult than for sequential computers. The considerable investment normally required to produce a parallel version of an application has been compounded by non-standard software development environments and proprietary programming interfaces, making portability difficult. This situation has been improved by standardisation efforts such as MPI and HPF. Many vendors also supply easy-to-use development tools with windowing interfaces.

There remain maintenance issues for the developers of message passing parallel software, since different versions of the source code are required for serial and parallel platforms. Applications written in languages which use compiler directives, or data-parallel languages, hold an advantage in this respect.

The difficulties of designing parallel applications have been eased by the development of some high level parallel programming models. These are discussed in the next chapter.

3.11 References

Amdahl, G., 1967, The validity of the single processor approach to achieving large scale computing capabilities. *AFIPS Conference Proceedings*, Spring Joint Computer Conference, **30**, 483-485.

Bagrodia, R., Chien, A., Hsu, Y. & Reed, D., 1994, Input/Output: Instrumentation, Characterisation, Modelling and Policy. *Scaleable I/O Initiative Working Paper No.2.*, http://www.cacr.caltech.edu/SIO/SIO.html.

Choudhary, A., Bordawaker, R., Harry, M., Krishnaiyer, R., Ponnusamy, R., Singh, T. & Thakur, R., 1995, PASSION: Parallel Scaleable Software for I/O. *NPAC*

Technical Report, SCCS-636, Syracuse University, http://www.cat.syr.edu/passion.html.

Dijkstra, E.W., 1968, Cooperating sequential processes. In Genuys, F. (Ed.), *Programming Languages*, Academic Press, New York.

Fortune, S. & Wylie, J., 1978, Parallelism in random access machines. *Proceedings ACM Symp. Theory of Computing*, 114-118.

Geist, A., Beguelin, A., Dongarra, J., Jiang, W., Manchek, R. & Sunderam, V., 1994, *PVM: Parallel Virtual Machine: A Users' Guide and Tutorial for Networked Parallel Computing*, MIT Press, Cambridge, MA.

Gelernter, D., Carriero, N., Chandran, S. & Chang, S. 1985, Parallel Programming in Linda. *Proceedings of the International Conference on Parallel Processing*.

Huber, J., Elford, C., Reed, D., Chien, A. & Blumenthal, D., 1995, PPFS: A high performance portable parallel file system. *Proceedings of the International Conference on Supercomputing (ICS) '95*, http://www-pablo.cs.uiuc.edu/Projects/PPFS/ppfs.html.

ISO, 1991, *Information technology - Programming languages - FORTRAN*, ISO/IEC 1539:1991, International Standards Organisation, Geneva.

Krieger, O. & Stumm, M., 1993, HFS: A flexible file system for large-scale multiprocessors. *Proceedings of the 1993 DASG/PC Symposium*, http://www.eecg.totonto.edu/parallel/hector.html.

Meyer, B., 1988, *Object-Oriented Software Construction*, Prentice Hall International Series in Computer Science.

Moyer, S., 1994, Parallel I/O as a Parallel Application, CSTR-941011, Department of Computer Science, Emory University, Atlanta, GA, http://www.mathcs.emory.edu/Research/Pious.html.

MPI Forum, 1993, MPI: A message passing interface. *Proceedings of Supercomputing '93*, November 1993, 878-883.

PVM, 1997, Parallel Virtual Machine, Oak Ridge National Laboratory, Oak Ridge, TN, http://www.epm.ornl.gov/pvm/pvm_home.html.

Reed, D., Ayth, R., Madhtastha, T., Noe, R., Shields, K. & Schwartz, B., 1994, An overview of the Pablo Performance Analysis Environment, Department of Computer Science, University of Illinois, Urbana-Champaign, http://www.hensa.ac.uk/parallel/performance/tools/pablo/index.html.

4

High-Level Support for Parallel Programming

S.M. Trewin

The parallel software environments described in the previous chapter provide the basic tools necessary for software development on parallel platforms. In particular, the development of the Message Passing Interface (MPI) provides a base upon which portable message passing applications can be built.

Despite this support, a great deal of code in such applications is still required to manage the interprocess communications. The structure of these communications is dictated by the structure of the parallel application as a whole: the way the problem has been decomposed to run in parallel. There are a number of common techniques for decomposing problems so that they can be distributed among a set of processes and solved in parallel.

For each technique, or paradigm, the interprocess communications have a specific structure, which any application using that paradigm will require. Code written to manage these communications is largely independent of the application task, and could be re-used between applications using the same paradigm.

This is the motivation behind the development of parallel utility libraries. Such libraries provide support for parallel programming at a higher level than the message passing system. They do this by providing routines which capture the communication patterns of common parallel paradigms in a flexible, general way. By using these routines, the programmer is freed from implementing a large volume of communication management code, and can concentrate more effort on the application-specific details. Furthermore, if time and effort is invested in optimising these parallel libraries, then the benefits will be enjoyed by a suite of applications, and the costs can be spread between them.

Parallel libraries also have a place in data parallel programming (introduced in Chapter 3). Under this paradigm, calls to parallel run-time libraries are inserted by a parallelising compiler, rather than an application programmer. However, this chapter focuses on the use of libraries in message passing applications.

This chapter describes several common parallel programming paradigms relevant to GIS applications, including those related to input and output of data to and from a parallel application - an important area which is too often neglected. The advantages of providing high-level support for these paradigms will be described and discussed, primarily using the example of the Parallel Utilities Library (PUL)

developed by the Edinburgh Parallel Computing Centre. One of the utilities within
this library - PUL-PF - was developed to meet the particular I/O requirements of
GIS and other similar applications. The development of PUL-PF will be described
and used to illustrate the difficult issues involved in developing genuinely useful
and reusable high-level support. Despite the difficulties, many successful parallel
libraries exist. These are an important extension of the support provided by
message passing systems, and make a valuable contribution to the usability of
parallel platforms.

4.1 Parallel Program Decomposition

For an application to run efficiently in parallel there must be some way of dividing
up the work so as to allow a set of processes running on the processors of a parallel
machine to operate simultaneously on the problem. There are two aspects of any
problem that can be decomposed: the operations and the data. See Foster (1995)
for a good discussion of these techniques and when they are appropriate.

Functional decomposition techniques are those which distribute the task operations
among the available processes. A common functional decomposition technique is
the *pipeline*, described more fully in section 4.2. In a pipeline, each operation or
set of operations is performed by a single process, and data is moved between the
processes as dictated by the order of operations required.

Data decomposition techniques distribute the data among the available processes.
Each process performs all of the necessary operations on its own part of the data,
and the results from each process combine to form the result for the whole data set.
Common data decomposition techniques include regular and scattered domain
decomposition, and irregular mesh and multiblock methods. These are described in
section 4.3.

The distinction between functional and data decomposition is not always clear, as
data and operations are often strongly related. Classification aside, the aim of any
decomposition technique is to distribute the work fairly among the available
processes. This is the concept of *load balancing*. Because the speed of the whole
application will be limited by the time taken by the slowest process, the better the
load is balanced, the better the performance from the parallel application as a
whole.

On some parallel platforms, it may also be necessary to take into account
differences between the individual processors on which the parallel processes are to
run. However, in general and throughout this chapter, the processors are (assumed
to be) homogeneous.

In order to achieve a good load balance, it is essential to choose a decomposition
technique appropriate to the characteristics of the problem to be parallelised. For
example, a problem where a large number of operations are to be performed on a

small vector of data is best parallelised by a functional decomposition technique, while a problem with a very large data set is often better suited to data decomposition. For complex applications, combinations of different techniques can be applied. In the following sections, the essential characteristics of appropriate problems will be described as each different decomposition technique is introduced.

GIS data can take on a wide variety of forms, so no single technique is universally applicable. The characteristics of GIS vector data are detailed more fully in Chapter 6, while Chapter 7 provides an overview of parallel issues for GIS applications.

4.2 Functional Decomposition Techniques

Functional decomposition is a technique best suited to true MIMD machines, where a different process can be run on each processor. SIMD machines require that the same executable be run on each processor. Although it is possible to implement functional decomposition codes in this way by branching into the different functionalities at the start of the code, it is inefficient, since the parallel hardware has to load all of the code and allocate memory for all of the different roles a process may take on. This can severely limit the memory available for each process.

Functional decomposition is one of the simplest ways to parallelise a program. Where the number of functional parts in a program is much smaller than the available number of processes, each part could be implemented by a set of processes, perhaps performing a data decomposition over the set. Functional decomposition is often the first step in parallelisation of large complex problems.

The *pipeline* is probably the most basic functional decomposition strategy. Given a set of N operations to perform on a set of data items, the operations are distributed over N processes. The first process reads in the data, one item at a time, performs the required operation, and then sends the data on to the next process. The last process in the pipeline is responsible for recording the final results for each data item.

On a small scale, this is the way that vector processors operate. However, at the application level, the operations to be performed must be more substantial than single arithmetic operations. If the operations take longer than the time it takes to transmit data from one process to the next, then once the pipeline is running, communication time can be overlapped with computation time for all but the slowest stage. If each stage of the pipeline processes data at the same rate, and there are a large number of data items, then the total speed-up will be proportional to the number of processes used.

A pipeline could also contain branches and loops. This might occur in reading NTF data (British Standards Institute, 1992a, 1992b), for example, where records describing different objects can appear in the same file, and need to be processed differently. A description of the NTF format is given in Chapter 6.

Another very common technique is the *task farm* (Foster, 1995). A task farm is applicable to problems which can be broken up into independent subproblems, where generation of the subproblems and processing of the results are trivial computations in comparison to the processing of a task. Such problems are sometimes referred to as *embarrassingly parallel* and arise in a wide range of application areas such as ray tracing or tree search problems. Wilson & MacDonald (1995) quote an example of a task farm used in an electromagnetic modelling application.

A classic task farm consists of a source process, a number of worker processes and a sink process. The source process generates tasks and gives them to the workers. The workers request tasks from the source, service them, and send the results to the sink. The sink process collates the results to produce the solution to the original problem. Even if the tasks are of varying sizes, the load balance remains good, since each worker need only request a new task when it has finished the previous one.

A distinction can be made between the size of a task and the amount of work associated with that task. For example, in a ray tracing application, the task size could be the number of rays to be traced. There can be enormous variations in the work required to trace individual rays, depending on whether the ray passes directly through the scene or hits one or more objects. So the actual time taken to compute a task is not always strongly related to the original task size.

Using a large number of tasks can improve the load balance. As the last tasks are being computed, if one of the final tasks is time consuming, and remains outstanding after most workers have become idle, there is a danger that the time spent on this task may dominate the run time of the application. The more tasks there are in total, the less impact this imbalance will have on the performance of the application as a whole.

However, if a problem is broken into too many separate tasks, then the performance may be dominated by the time spent communicating these tasks and results between the source, workers and sink.

Many variations on the classic task farm model are possible. Sometimes the source and sink are combined into one process. Where generation of subtasks is time consuming or there are a large number of worker processes to be serviced, the source process may become a bottleneck. It may be possible to use several source processes to balance the load. Similarly, sometimes multiple sinks are required. Tasks need not always produce results. In a search problem, for example, results may only be generated when the search is successful.

The task farm approach does not scale well to large numbers of workers. More powerful is the *divide-and-conquer* paradigm (Quinn, 1994). In this technique, the original problem is decomposed into independent subproblems and the subproblems are distributed. Each subproblem may then be solved immediately or further divided and distributed. The computation forms a tree, and as each branch is completed, the results are combined. The result appearing at the root of the tree will be the solution to the whole problem. Many search problems are of this type. Again, the load balancing properties of this approach can be very good, as any available process can be applied to any task.

4.3 Data Decomposition Techniques

Many applications, including GIS applications, require computation to be performed over very large data sets. The operations may be relatively simple, but because the data set is so large a serial solution is impractically slow. For such applications, data decomposition techniques are usually used. Appropriate techniques are dictated by the application data structures used. The following sections concentrate on techniques used to decompose two of the most commonly used data structures: the *grid* and the *mesh*, illustrated in Figure 4.1. The term 'grid' refers to a regular, rectangular grid of points, which may have many dimensions, and is stored in the computer as an array. Spatial relationships between points are recorded implicitly in their positions within the array.

A 'mesh' is an irregular network of elements covering an area of interest. A mesh will have a large number of small elements in complex or interesting areas, and fewer, large elements elsewhere.

4.3.1 Grid Based Data Structures

Many applications store data in a multidimensional array and perform identical operations on each data item. Important classes of application such as computational fluid dynamics and seismic data processing often fall into this category. The rectangular grid (or elevation matrix) structures used in digital terrain modelling are also examples (Weibel & Heller, 1991). The calculation of a new value for a given data point often depends not only on the previous value at that point, but also on the values at neighbouring positions in the array. Sometimes only nearest-neighbour values are required; other applications also look at values several positions away in the array. The set of points required to update a given point will be referred to as the *stencil* associated with the operation. Some applications use different stencils for different phases of operation. A stencil specifies a set of *local* points, that is points stored at adjacent or near-adjacent positions in the multidimensional data array.

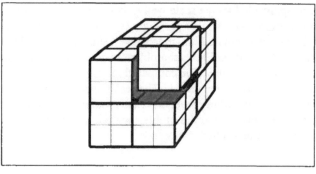

A Grid

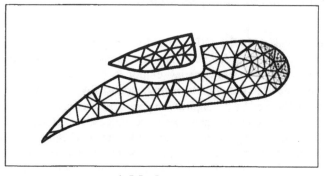

A Mesh

Figure 4.1: Grid and mesh data structures

Applications of the form described above can be effectively parallelised using *regular domain decomposition*. Figure 4.2 illustrates the parallelisation of an application based on a two-dimensional data array. The array is divided into evenly sized blocks, each of which is placed on a different process. The processes are arranged into a logical grid which maps onto the original data array, so that neighbouring processes in the logical grid own neighbouring blocks of data.

Depending on the stencil, the processes may require values held by other processes in order to be able to update their own data points. This is achieved by *boundary update* communications in which each process sends to its neighbours all values they require, and receives all values it requires in return. These communications must be managed carefully in order to avoid incorrect update of the values at the edges, or *deadlock* - a situation in which all processes are blocked, waiting for communications from other processes. Silberschatz and Peterson (1988) provide a general definition of deadlock. In the context of boundary updates, a naive boundary update procedure might have each process send information to the left, then receive information from the right. However, if the data set is cyclic, and blocking communications are used, deadlock will occur: each process cannot finish

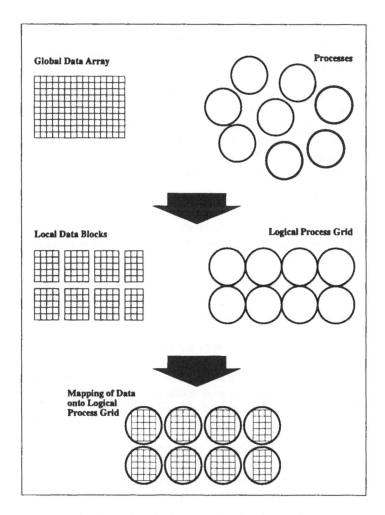

Figure 4.2: Parallelisation of an application based on a two-dimenstional array

sending until the data is received by its neighbour, but each neighbour cannot finish its own send, so no receive operations will be initiated. Several parallel libraries provide support for regular domain decomposition, avoiding this kind of pitfall. This support is described in section 4.6.4.

In some applications, there are large variations in the amount of work required to update the values at different data points. For example, special calculations are sometimes required at the edges of the data array. In other applications there are regions of heavy work which move as the calculation progresses, or cannot be predicted until the calculation has been initiated. In ice sheet modelling, for example, the ice sheet may exist only in a part of the grid. If regular decomposition is used, then the application may be very badly load-balanced; one process may have a block of the array containing all of the complex calculations while all of the other processes have only light work. The application performance will be dominated by the one process with the heavy load.

Scattered spatial decomposition is one useful approach to this problem, even when it is not known which areas of the data set are highly loaded. This paradigm is similar to regular decomposition, but here the data set is divided into many more blocks than there are processes, and each process is given blocks from scattered areas of the original array. The intention is that areas of heavy work will be broken up and spread more evenly over the available processes.

In some cases it may be possible to predict where the areas of heavy work lie within the data set, and adjust the decomposition accordingly, balancing the workload by giving the processes differently sized blocks of data. This approach would be appropriate, for example, for a geographically based data set in which it was known that the calculations over oceans were trivial while those over land were complex.

4.3.2 More Complex Data Structures

Some unevenly loaded spatial problems are based on an irregular mesh of elements or points covering the area of interest. The term 'irregular' is here used to imply that there is no regularity in the structure. One type of mesh is the triangulated irregular network (TIN) structure, used in digital terrain modelling (Peucker *et al.*, 1978). TIN structures are composed of triangular elements, and can incorporate structural features more easily than a grid.

In general, a mesh could be composed of elements of a number of different shapes. For example, the quadtree data structures used in GIS for compression of raster data (Aronoff, 1989) can be thought of as meshes in which all elements are rectangular, although meshes have no equivalent to a quadtree *level*.

Mesh structures are useful for problems in which the required, or available, resolution of data elements varies greatly across the data area. If a regular array data structure were to be used, then the finest resolution would have to be applied throughout the data area, resulting in perhaps tens of millions of data elements. The use of a mesh reduces the number of elements to a manageable size, while retaining the necessary accuracy in complex areas.

Decomposition of mesh based problems requires the mesh to be divided such that each process has a roughly equal number of mesh elements. Once the mesh is decomposed, boundary update communications similar to those employed in regular and scattered decomposition are required. Applications using mesh data structures include finite element or finite difference calculations, often modelling complex dynamic processes such as airflow around an aeroplane.

Mesh based applications tend to be more complex than array-based applications - in an array, the location of neighbouring elements can be found by a simple arithmetic calculation, whereas in a mesh, connectivity must be explicitly recorded or established by exhaustive search methods. An alternative technique for problems with unevenly loaded data sets is to base the calculation on a *multiblock* data structure. A multiblock is a set of separate regular grids which together cover the whole data area. Grids over complex areas will have a fine resolution and many elements, while those in less interesting areas will be coarse-grained. This approach retains much of the arithmetic simplicity of basic grid-based methods, but shares with the mesh approaches the advantage that the calculations in uninteresting areas can be minimised, so improving the load balance. An example of a multiblock is shown in Figure 4.3. Multiblock approaches are used in applications such as large scale climate modelling (Anon., 1993) and land cover dynamics (Townshend *et al.*, 1991).

4.4 Parallel I/O

As mentioned previously, many applications are parallelised because they have to deal with large volumes of data. Often, these data is initially stored in a file or files on disk, and sometimes a similar large data file must be written out at the end of computation. As the performance of parallel processors and intercommunication networks improves, efficient I/O is becoming a crucial requirement for achieving good parallel performance (Kotz & Ellis, 1993; Foster, 1995). Two issues are important here: the arbitration of access by a group of processes to a shared global file, and the provision of scaleable I/O performance through the distribution of a file over multiple disks.

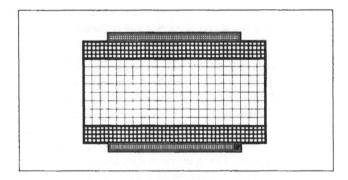

A Multiblock

Figure 4.3: Multiblock data structures using variable grid resolution

Unfortunately, although the newer parallel computing systems do tend to provide parallel file systems (including, for example, Thinking Machines Corporation's CM-5, Meiko's Computing Surface 2 and Fujitsu's AP1000), software support for these systems still lags behind the hardware capabilities (Choudhary, 1993). The reading and writing of large data files is the single largest bottleneck in many parallel applications (Crockett, 1989).

The I/O requirements of an application are strongly influenced by the parallel paradigm being employed, since that dictates the way in which the data stored in files is to be distributed across the parallel processors, or data is produced for writing to disk. For example, only the head process of a pipeline may need to read data from disk, whereas all of the processes in a regular domain decomposition require a block of values from a data array file. Another important consideration is whether the application is required to read and write data in a format compatible with other, perhaps serial, programs. Finally, the parallel platform in use will also impose certain restrictions on the I/O behaviour of applications.

On platforms where only a single I/O stream is available, the key to achieving the best possible performance is in making efficient use of the disk. This usually means minimising the number of independent read, write and seek operations performed on data files. This is often achieved by designating one process as a *server* responsible for co-ordinating and implementing the application's file access. This process may be dedicated to I/O operations, or the I/O may be performed in addition to the application operations. Dedicated I/O processes may use *cacheing* and *prefetching* of data in order to further improve the I/O performance. These are useful techniques which can provide significant performance improvements for applications where the data access pattern is predictable (Kotz & Ellis, 1993). A cache is a local store of data that is expected to be accessed in the near future. Cacheing is acknowledged to be "the architectural feature that contributes most to performance in a distributed file system" (Satyanarayanan, 1993). Prefetching is a technique used to fill a cache with data prior to the actual data requests being made (Foster, 1995). Cacheing and prefetching potentially allow all of the time taken by communications between the server processes and application processes, as well as the file access time, to be overlapped with useful computation on each process.

Where several disks are available, the I/O can be distributed among these disks, giving much faster data access. Files may be distributed over disks in a number of ways. Crockett (1989) gives one useful taxonomy similar to that used here. A *multi-instance* parallel file is replicated in its entirety on every disk. This scheme is identical to RAID (Katz & Gibson, 1987) and allows for maximum parallelism when reading data. The other important distribution techniques use *disk striping* (Kim, 1986; Kim & Tantawi, 1987), in which files are distributed by cycling through the available disks allocating the next block from the file until the end of the file is reached. At one extreme, a *partitioned* parallel file uses a block size of `(the_file_size / the_number_of_disks)`, so that each disk has only one block from the file. This technique is appropriate where application processes are independently accessing different sections of the file. Alternatively, in an *interleaved* file, the data is striped across the available disks such that each item x in a file appears on disk (x mod the_number_of_disks). An item could be a byte, or a record viewed by the application as a single element. Interleaving is a good approach for applications in which groups of processes access adjacent elements of the file, or when the processes all share a file pointer and read or write the file sequentially. Between these two extremes, *partitioned-interleaved* files apply striping to sets of contiguous elements.

Distribution of files across disks should spread the I/O such that all of the disks operate in parallel. This requires that the distribution technique should take into account the data access patterns of the application. A number of common access patterns can be identified:

* *Single* mode: All processes access the same data elements at the same time. This mode is common for files containing global data to be accessed by all processes.

* *Multi* mode: All processes simultaneously access a set of contiguous data elements, such that process x reads element x, or element set x, in the sequence. Each process accesses the same number of contiguous elements. This mode is common for files containing large data arrays and is often used by numerical applications employing a regular or scattered decomposition scheme. It would also be suitable for raster data files distributed among a set of synchronised processes.

* *Independent* mode: Processes independently access elements from any position within the file, without synchronisation. This style of access is typically required for parallel database applications.

* *Random* mode: Processes access elements from the file, without synchronisation, but share the same file pointer so that operations performed by one process will affect the other processes. This mode is useful for unordered files, such as vector data, since any process may write a new record, or read the next record, and each record is accessed only by one process. Files accessed in this way could also be log files to which all processes are contributing, or files of independent tasks read by workers in a task farm.

Given the large number of file distribution strategies and access modes, and the importance of efficient file access for parallel applications, any high-level support for file I/O needs to be extremely flexible and powerful. Examples of utilities supporting parallel I/O are discussed in section 4.6.6.

4.5 Portable, Reusable Parallel Utilities

As mentioned in the introduction to this chapter, parallel libraries can encompass the communication structure of many popular programming paradigms. In addition to the benefits described earlier, the use of parallel libraries reduces the knowledge and effort required to exploit parallel systems: an application programmer using the library need only understand the underlying parallelisation strategy, and need not be concerned with detailed implementation issues. This allows the possibility of prototyping a parallel version of an existing serial application quickly using the library routines, giving a profile of the behaviour of the application in parallel, so that the bottlenecks can be identified and development effort can be concentrated where it is needed. Foster (1995, Chapter 4) discusses further the advantages of reusing code in this way.

To be truly reusable, such libraries should be available on a variety of parallel platforms. In the past, every parallel platform had its own message passing interface, and porting the message passing code of an application was often difficult and time consuming. With the advent of MPI, a portable message passing standard which explicitly provides support for parallel libraries, message passing code written using this interface can easily be ported between platforms supporting MPI.

Parallel libraries, then, are developed with several goals in mind: code reuse, portability, performance, and improving the ease of use of parallel systems. Predictably, these goals conflict, and compromises must be made (Fletcher & MacDonald, 1994). Ideally, a code module should perform well for any application, across any platform. Unfortunately, differences between the characteristics of both platforms and applications mean that this is not the case. In order to achieve good performance, it is often necessary to provide several alternative implementations of a given module, each designed for a particular platform, or type of application. The same utility code cannot always be successfully reused. Importantly, though, the same interface can be used, so that applications are portable even where the parallel library is implemented specifically for a given platform.

There is also a conflict between the ease of use of a code module and its reusability. The most easily reused modules are those which provide a low level of abstraction for a very general level of functionality, for example a function to swap messages between two processes. Modules at higher levels of abstraction provide a high level of support to a specific class of application and so are less easily reused. One example is a routine for performing a boundary update in a regular domain decomposition.

4.6 Existing Utility Libraries

Many parallel programmers have been motivated by the arguments presented in the previous section to develop reusable utilities supporting common operations and paradigms. For example, the Edinburgh Parallel Computing Centre has developed a suite of such utilities, collectively called the Parallel Utilities Library (PUL) (Bruce *et al.*, 1995).

In order to illustrate the feasibility of producing general purpose, *useful* parallel utilities, this section describes existing utility development efforts. Since PUL provides a particularly broad range of utilities, and includes utilities developed with GIS applications in mind, the discussion is structured around the development of PUL.

4.6.1 Overview of PUL

The main aim of PUL is to provide a library of utilities which promote code reuse in parallel applications and ease the process of code migration from serial to parallel systems. The PUL interface can be used by both ANSI C and FORTRAN 77 message passing applications.

Complex applications use a variety of forms of parallelism. To accommodate these, PUL is designed so that any number or combination of the utilities can be used within a single application without conflict or confusion among the messages sent by the utility routines. This is possible through the process grouping facilities provided by MPI, on which communications within PUL are based. Each instance of a utility has a unique communication context on which messages are sent.

In recognition of the trade-off between ease of use and reusability, the utilities within the library fall into three distinct categories:

- *Non-specific modules* implement generic functionality for use in a wide range of applications. An example is PUL-GF, which provides basic file I/O facilities.

- *Paradigm-specific modules* provide support for stereotypical parallel paradigms such as task farms (PUL-TF) and regular domain decomposition (PUL-RD). Within these modules, a range of levels of support are available to applications. A *procedural* interface provides low-level flexible routines to manage the data flow within the paradigm, while a high-level *skeletal* interface takes over management of both data and control flow. The skeletal level of support owes much to Cole's "algorithmic skeletons" (Cole, 1989), and is covered more fully using examples for PUL-TF and PUL-RD later in this section.

- *Domain-specific modules* provide high-level support for particular classes of application. For example, PUL-SM supports applications based on irregular mesh data structures, such as finite element solvers.

The domain-specific modules are the most application friendly, while the non-specific modules have the greatest potential for reuse, and indeed can be used within the paradigm and domain specific modules.

The following sections describe some of the areas in which reusable parallel utilities have proved useful. The sections are ordered from the lowest to the highest levels of complexity. In each area, utilities that have been developed are described and contrasted.

4.6.2 Utilities Supporting Extended Message Passing

Before the advent of MPI, with its collective communication facilities, applications were written using message passing interfaces that provided only basic send and receive facilities. Common higher level operations, such as parallel gather and scatter, or pair-wise exchange operations, usually had to be hand written. Such operations are relatively easy to write in a general, reusable way. As a result, libraries providing this type of extension to the basic message passing systems available were developed in several centres.

Parasoft Express (Parasoft Corporation, 1988) and Intel NX/2 (Pierce, 1988) are early run time environments for parallel machines, both of which provide collective communication facilities in addition to basic message passing and process control.

When the PUL project was initiated, it used the Common High-Level Interface to Message Passing (CHIMP) developed at EPCC. To complement the basic send, receive and broadcast operations provided by CHIMP, PUL-EM (Extended Messaging) (Trewin & Chapple, 1993) was developed. This was a non-specific utility providing common collective communication operations. Similarly, the London Parallel Applications Centre (LPAC) developed a library of global communication routines for a range of different architectures (London Parallel Applications Centre, 1993).

These and similar libraries are becoming redundant, as their functionality is provided by the collective communication routines now available in MPI. This is a good example of functionality which started out as a set of useful extensions to existing message passing systems, and developed into a standard part of message passing provision. Ultimately, support for higher level parallel message passing paradigms should also become standardised and more widely available.

4.6.3 Utilities Supporting Task Farming

The task farming paradigm is one of the simplest decomposition strategies, and as such is a prime candidate for inclusion in any parallel utility library.

The second PUL utility to be developed was based on a classical task farm with single source and sink processes and a number of workers. The roles of source, worker and sink are assigned to particular processes, and can also be combined. The utility - PUL-TF - was initially developed for an automatic fingerprint recognition application (Trewin *et al.*, 1993). It has since been extended and applied to many different applications, including weather prediction and ray tracing.

PUL-TF provides both a procedural and skeletal interface, as introduced in section 4.6.1. The procedural interface has routines responsible for data flow within the application. The routines provided, `tf_put` and `tf_get`, will get tasks from the source or results from workers, and give tasks to workers or results to the sink. Example pseudocode for source, worker and sink follow.

- The SOURCE program

```
taskLength = 1;
while (taskLength > 0) do {
    taskLength = /* create a task into taskBuffer and return its size */;
    /* put the created task onto any ready worker */
    if (taskLength > 0) tf_put(farm,taskBuffer,taskLength);
}
tf_put(farm,NULL,0)    /*put out a zero length task to flush the workers */
```

- The WORKER program

```
taskLength = 1;
while (taskLength > 0) do {
    /* get a task from the source */   .
    taskLength = tf_get(farm,taskBuffer,maxTaskSize);
    if (taskLength > 0) {
        resultLength = /* execute task, leave result in resBuffer,
                        return its length */;
    }
    else if (taskLength == 0) resultLength = 0;                /* a
flush*/
    else break;
    tf_put(farm,resBuffer,resultLength); /* send the result to the sink */
}
```

- The SINK program

```
resultLength = 1;
while (resultLength > 0) do {
    /* get a task result from any worker */
    resultLength = tf_get(farm,resultBuffer, maxResultSize);
    if (/*it was not a flush */) /* process the result */;
}
```

When using the procedural level functions, the application has full control over, and responsibility for, the way in which tasks are generated, tasks are processed, and results are collated. However, many task farm applications use the same

control structure. The skeletal interface provides a simple function - tf_operate
- which encapsulates this control structure. The C prototype of tf_operate is:

```
int tf_operate(TF farm,

        void (*sourcefunction)(void *task, int *taskSize, int *returnValue),

        void (*workerfunction)(void *task, int *taskSize, void *result,

                            int *resultSize, int *returnValue),

        void (*sinkfunction)(void *result, int *resultSize, int *returnValue))
```

Application processes simply provide a function to generate tasks, a function to
process tasks, and a function to collate results, and the rest is taken care of by the
utility. This reduces the application programming to three simple functions. The
procedural and skeletal levels are also interoperable, so that while the workers may
be using the skeletal level, the source could be using the procedural level functions,
and performing some more complex operation than that allowed by the skeletal
framework.

The PUL-TF utility uses *task prefetching* to improve the efficiency of task
distribution. This means that when a worker starts one task, it immediately sends a
request for its next task. The hope is that the next task will arrive before the current
task is finished, so the worker can move straight on to this new task without waiting
for communications to complete. The task prefetching in PUL-TF is an example of
a performance optimisation that has been implemented without requiring any
changes in the applications using the utility.

A similar farm library was developed by Distributed Software Limited, for use with
the Helios Operating System (Perihelion Software Ltd). This library (Distributed
Software Ltd, 1992) is related to the skeletal level of PUL-TF in that the
application provides three routines - producer, consumer and result - which are
called by a Helios procedure. The farm uses a combined source/sink process,
where the roles are implemented as two separate threads. The use of threads for
combined source/sink functionality is simple and effective, but limits the farm's
portability, whereas PUL-TF processes can perform multiple functions without
requiring threads. The Helios Farm Library also incorporates some fault tolerance,
so that if a worker process crashes, the farm will continue to operate. PUL-TF is
not fault tolerant, since this functionality is difficult to implement in a portable way.
Limiting the portability of the software to platforms supporting Helios has allowed
some powerful features to be incorporated into this library. PUL-TF has sacrificed
some of this power for greater portability between parallel platforms. This trade-
off between functionality and portability will persist while the system software
provided on parallel platforms remains non-standard.

4.6.4 Utilities Supporting Regular Domain Decomposition

Like task farming, regular domain decomposition is a popular and important
parallel paradigm. Many groups and organisations have developed support for
problems based on regular grids, including Parasoft Corporation (1988), which

incorporated a *gridmap* component into Express. Gridmap provides functions to assist the programmer in mapping a multidimensional array onto a logical processor grid, translating grid coordinates to processor IDs, and exchanging update messages with neighbouring processors.

LPAC have developed a kernel library with similar functionality, supporting distributed two- or three- dimensional arrays (London Parallel Applications Centre, 1993). Routines are provided to send boundary data in specified directions, and the application developer is responsible for ensuring that deadlock is avoided. The extent of each process' local block is specified by an integer representing the number of extra layers of elements required around the local data. Similarly, at each boundary swap a specified depth of boundary data is exchanged. Other groups which have developed support for regular decomposition include Argonne National Laboratories (Gropp, 1993), Germany's GMD (Hempel & Ritzdorf, 1991) and NOAA/FSL (Rodriguez *et al.*, 1995).

The PUL library supports regular decomposition with PUL-RD (Chapple *et al.*, 1995), which provides facilities for applications based on an N-dimensional array of data. PUL-RD will choose an appropriate decomposition so as to equalise the sizes of the individual blocks while minimising the sizes of the borders along which communications are required. Where desired, the application can override PUL-RD's default decomposition with one more suited to the data set calculations to be performed.

The PUL-RD interface provides a specialised I/O routine to read in sub-blocks efficiently from an array stored in a file, and provides procedural routines to update the boundaries of sub-blocks, according to the operation stencil currently in use. These procedural boundary updates are similar to the update facilities provided by LPAC's Kernel Library, but because they use an explicit representation of the operation stencil, they not only can swap data in all directions at once, but also can exchange only the data that is actually required, and the application programmer does not need to handle deadlock avoidance. Both blocking and non-blocking boundary updates are available. A non-blocking boundary update consists of two functions: one which initiates all communications and one which completes the communications and updates the local boundary. The advantage of non-blocking updates is that between initiating and completing the update, applications can compute local values not dependent on the boundary data. This allows computation to be overlapped with communication, hiding some of the communication time. The following pseudocode shows the main body of a simple application using the non-blocking, procedural interface:

```
for (i=0; i<NUM_ITERS; i++) {
    rd_startSwap(rdh,array1);
    for (/* all local values not dependent on the boundary data */) {
        update(/* local value in array1, placing new value into array2*/);
    }
    rd_endSwap(rdh,array1);
```

```
for (/* all local values dependent on the boundary data */)    (
    update(/* local value in array1, placing new value into array2*/);
}
/* swap local arrays    */
}
```

In PUL-RD, the skeletal level interface (- rd_operate -) requires the application
to provide a routine to update a sub-block of the local array, and the utility
performs a specified number of update-communicate iterations, overlapping the
communication with the computation as much as possible. For comparison with the
procedural level, here is the skeletal equivalent of the same application:

```
rd_operate(rdh,array1,array2,NUM_ITERS,update)
```

Alternative skeletal level functions, perhaps with different termination criteria,
could easily be added.

Since its original design, PUL-RD has been used in many different applications,
and has been extended to support several separate distributed data sets of elements
of different types, so that applications can store data such as the pressure and
temperature at each point in separate arrays. If both pressure and temperature data
are to be updated, then instead of sending separate boundary update messages for
each array, the messages are concatenated to reduce communication overheads.

4.6.5   Utilities Supporting Irregular Data Structures

Management of irregular data structures tends to be more complex than that for
regular grids, and so utilities supporting such structures tend to be domain specific.

For applications based on irregular mesh structures, the PARTI/CHAOS (Moon &
Saltz, 1994), GRIDS (Geuder *et al.*, 1994) and PUL-SM (Trewin, 1995) libraries
all provide support for decomposition and boundary update, but they do so in very
different ways. PUL-SM is a run-time library for message-passing applications,
while PARTI/CHAOS is designed for use by parallelising compilers, and GRIDS is
a programming system which accepts a specification of the problem and compiles it
into a parallel executable.

One feature of such applications is that the workload associated with each element
in the mesh may be very different, and may also change during program execution.
To retain a good load balance it may be necessary to either remap the whole
application, or adjust the borders of the sub-meshes, expanding those domains in
which there is little work, and shrinking those containing heavy elements. PUL-SM
and PARTI/CHAOS both provide dynamic load balancing/remapping routines for
this purpose.

PUL-SM also provides specialised I/O facilities to load mesh data structures from
disk files. PUL-SM was developed for three-dimensional sub-sonic fluid dynamics,
and has since been applied to airflow modelling.

The PARTI library also provides primitives supporting multiblock codes (Agrawal *et al.*, 1994). Similar multiblock libraries include LPAR (Kohn and Baden, 1993), a Lattice Programming Model, and P++ (Lemke and Quinlan, 1992), a set of C++ libraries.

4.6.6 Utilities Supporting File I/O

In the absence of standards for parallel file I/O, several groups have developed libraries supporting I/O for parallel applications. UNIX-like support for parallel access to shared serial files is provided by PUL-GF (Global File) (Chapple *et al.*, 1994) and NOAA's SRS (Hart *et al.*, 1995). They are based on the functionality of the C `stdio` library, and files are viewed as streams of bytes. Both libraries provide formatted and unformatted I/O, together with specialised operations for reading and writing the sub-blocks of multidimensional files. They employ a client-server architecture, providing a dedicated server process through which all I/O is routed. This allows the utilities to control the file access so as to make the most efficient use of the disk.

PUL-GF supports the four file access modes described in Section 0: single, multi, independent and random, while SRS supports single, multi and random modes. The mode determines the behaviour of the I/O operations performed. A read operation in single mode, for example, will read and distribute the same data to every process, while in multi mode a set of data items will be read and each process will receive one item according to its position in the process ordering. Both utilities ensure that read and write operations are atomic, preventing two processes from writing to the same file position concurrently.

PUL-GF provides both blocking and non-blocking file access routines, enabling applications to overlap I/O with computation. However, the architecture of the utility is limited by the single server design. Where a large number of application processes are to be served, the PUL-GF process will become a bottleneck. SRS, which is targeted at numerical weather prediction models, relieves this problem by using a set of cache processes, which store data in blocks for sequential output.

A more scaleable solution is to use a parallel file system to multiply the number of streams over which I/O can be performed. To take advantage of parallel file systems, libraries such as PUL-PF (Chapple, 1994), NASA Ames' MPI-IO (Corbett *et al.*, 1995) and Argonne National Laboratory's PIO routines (Galbreath *et al.*, 1993) have been developed or (in the case of MPI-IO) proposed.

PUL-PF and PIO provide similar coordination of I/O to that of PUL-GF and SRS, but use multiple server processes to provide an interface to parallel disks. Both libraries hide the parallelism of the I/O system from the application, and present a view of a file as a single logical unit. The file distribution is not visible to the application processes. This allows a cleaner, simpler programming model and a good separation between the functionality of the system, represented by the interface, and the performance, achieved through the file distribution.

While PIO retains the UNIX byte stream model of a file, PUL-PF abandoned this model in favour of a file viewed as a set of *atoms*. An atom is a user-defined record type, which can be of variable length. The motivation for this change is described in Section 4.7.1. One of the advantages of this approach is that the file distribution can be matched to the structure of the application data, and also to the data access pattern of the application, which in turn allows disk accesses to be minimised.

In PIO, the file model does not allow the application's view of the file's structure to be reflected in the file distribution, and the distribution is dictated by the library implementation. However, PUL-PF provides all of the distribution patterns described in Section 4.4: multi-instance, partitioned, interleaved, and partitioned-interleaved. It also allows user-defined decompositions to be specified, giving the user complete control over the allocation of atoms to file segments. Applications can also provide PUL-PF with *hints* about the predominant access patterns for each file, which are used to choose an appropriate distribution strategy.

Knowing the data access pattern also allows PUL-PF to predict what information each process will request next. The library makes use of this information by prefetching the appropriate data in advance of the request. The data is sent to the process and stored in a local cache, as described in section 0.

4.7 Developing and Using Libraries to Support GIS

Having given some of the reasons why parallel utility libraries can be a valuable and useful tool for the parallel application programmer, this section focuses on the practical issues which arise in the development of these libraries, particularly in relation to GIS applications.

In order to develop a useful, reusable library, developers need to be aware of the ways in which applications may diverge from the typical form represented by the utility. This is essential so that the appropriate flexibility can be built into the library interface. Ideally, the design process should be iterative, with the utility interface constantly being refined to meet the needs of new applications, or the changing needs of existing applications.

4.7.1 An Example of the Utility Development Process

As an example of the typical development process, consider the PUL-PF utility. PUL-PF was proposed as a successor to PUL-GF, providing structured I/O facilities for parallel applications. The major difference between PUL-PF and PUL-GF is that while PUL-GF reads and writes serial files, PUL-PF is intended to take advantage of modern parallel file systems, providing truly parallel file I/O.

PUL-PF was developed initially for use in a large GIS application, but was also intended to be used in other application areas in the future. The specific application was to provide parallel versions of some common GIS operations (specifically, conversion between vector and raster data formats, and polygon overlay facilities) which are discussed in the following chapters. As the design of the application and of PUL-PF proceeded simultaneously, the utility interface was influenced not only by the previous example set by PUL-GF, but also by the characteristics of the GIS data to be read and written. Many iterations of design and refinement were required to produce an interface which efficiently supported the I/O required by the GIS application, while retaining the general purpose character of the PUL utilities.

As mentioned previously, both PUL-GF and PUL-PF were implemented using a client-server architecture. PUL-PF was to have several servers, each controlling I/O on an independent disk. The parallel files would be spread across these disks, each disk storing one *parallel file segment* of each file. Ideally, the organisation and operation of the server processes and the parallel nature of the file would be invisible to the application processes.

Just as there are many ways to distribute a data set across a set of processes, there are several ways to decompose a file into segments. In order to exploit the advantages of a parallel I/O facility, it is important that the decomposition strategy matches the data access pattern of the application using the file. However, there is also a potentially conflicting requirement that the distribution strategy be mathematically simple and concisely describable. This makes it easy to determine which server process holds a particular section of the file.

Most supercomputer applications are numerical problems based on a large data set of constant sized items. GIS data is often very different in nature. The particular project described here was required to read and write both vector and raster data. These are described in greater detail in Chapter 6.

The vector data was to be imported in NTF level 4 format (British Standards Institute, 1992a, 1992b) from both ASCII and binary data files. The project also needed to retain the ability to read other formats in the future. The NTF files consisted of records of a number of different types; some of these are of fixed size and others are of variable size with no upper limit. The records themselves could be in any order within the file.

Similarly, the raster data could be ASCII or binary, and were expected to be run-length encoded scan-lines. This format results again in files consisting of variable sized records, each representing one scan-line of the data set. However, here the records are ordered. The project was also required to handle data files that were too large to be held in core memory, the only limitation being that each individual record should be able to fit in the core memory of the parallel machine.

The application viewed these files as consisting of a set of records, and required the ability to read and write whole records at a time. The parallelisation of the application was such that individual records would reside on a single process as far as was possible.

This record-based model of the file conflicted with the byte stream model used in PUL-GF. If files are parallelised as an unstructured byte stream, then records will be split across many disks, making efficient access difficult. Consequently, PUL-PF adopted a model of a file as a set of atoms (or records), which are the logical units of which the file is composed. File distribution, therefore, took place at the atom level, so that the structure of the data was preserved, and records were not split across disks.

Having decided to decompose files as sets of atoms rather than as a stream of bytes, a distribution strategy appropriate to the data access pattern of the application was required. The basic decomposition options are listed in section 4.4. Since the data files could be arbitrarily large, the use of multi-instance files was not feasible, so some form of striping or partitioning was required. The project was also designed to allow the possibility for data to be streamed in from some other (serial or parallel) operation, and similarly streamed out to other operations.

The raster input and output files were sorted. The requirement to support streaming restricted the application's file access pattern to sequential reading and writing of these files. Consequently, file access was concentrated on one area of the file at a time. Using a partitioned file distribution would therefore mean that all of the file access at a given time would be on the same disk, and so no benefit would be gained from the parallel nature of the file. However, a striped distribution would spread the access load evenly between the disks, and the cacheing and prefetching facilities provided by PUL-PF could be used to achieve a good level of throughput, keeping all of the servers occupied in supplying data, and making good use of all of the disks. Clearly, the striped distribution model was the most appropriate for the raster data files.

Since the vector data files were unsorted, it was not necessary to access the records in any particular order, so a sequential access pattern was not essential. When writing vector files it was not possible to predict how many records there would be in the file, and so under a partitioned distribution strategy it would have been difficult to divide the file equally among the disks available. Consequently, a striped distribution was also the most appropriate for the vector data files.

As the project progressed, it became clear that there was a conflict between the requirement to write raster files in sorted order and the need for each process to write out processed scan-lines as soon as possible in order to make room for new scan-lines. Ordered output was provided by the MULTI file access mode, but that mode required all of the processes to synchronise over each set of scan-lines to be written. While this was reasonable for numerical applications using regular domain decomposition, since the processes are already loosely synchronised by the boundary update communications, in the GIS application the processes were not already synchronised, and enforcing a synchronisation at this point could severely affect performance. To avoid this, a new file access mode was developed. The new mode - MULTIRANDOM - combined the asynchronous properties of RANDOM mode with the ordered output provided by MULTI mode. Processes made requests to write a particular scan-line, and the servers stored these requests until all the preceding scan-lines had been written. The only restriction was that each process send in scan-lines in increasing order of appearance in the file. The scan-lines sent by each process could then be merged and written to the file in the correct order. Adding this functionality to PUL-PF removed the need for synchronisation when writing sorted records to a file or a stream.

Having based the file distribution on atoms, further complications were introduced by the fact that some record types, particularly in vector data, could be arbitrarily large. While it was assumed that every atom could be read into the core memory available to the application, it could not be assumed that every atom could be stored on a single process. Large atoms would have to be distributed. This conflicted with PUL-PF's model of the file as a set of indivisible atoms. When reading large atoms, they would have to be split into manageable *particles* to be distributed to different application processes. Similarly, sets of application processes holding particles needed some mechanism for writing them so that a single atom appeared in the final data file. Merging of particles was achieved by introducing a file locking mechanism. The set of processes owning an atom passed a token between them to synchronise the writing of the particles to the server. When the server received and had written out the final part, the file was unlocked and other atoms could again be written. This flexibility would not have been introduced into PUL-PF without the GIS application having illustrated the requirement for it.

Splitting of atoms was achieved by introducing an application specific function at the server processes. This function would operate on an atom, dividing it up into several smaller particles which could then be distributed in the usual way by the server processes. This mechanism was also used to perform data format conversion, for example between binary and ASCII data, and to provide filtering of atoms for operations which required only a subset of the record types in the file. This meant that the application processes operated independently of the original data format, and different functions could be slotted into the servers to handle different data formats. Although it violated the goal that the server operation be invisible to the application, it did provide a neat and efficient way of operating on the data as it was read.

All these facilities were provided while retaining the features of PUL-GF necessary for efficient I/O for numerical, array-based applications. In order to achieve sufficient flexibility to handle such different types of I/O requirement, the development of the PUL-PF interface was very much an interactive process. The final product is an extremely flexible and powerful interface for performing efficient parallel I/O.

4.8 Conclusions

This chapter has described some common programming paradigms employed in parallel message passing applications. These paradigms can be broadly classified into functional decomposition techniques, which distribute the tasks to be performed, and data decomposition techniques, which distribute the application data, although the distinction between the two is not always clear.

Common techniques include pipeline, task farm, regular domain decomposition, scattered domain decomposition, mesh-based decomposition and multiblock approaches. Each of these is appropriate for different classes of application. Factors influencing the choice of paradigm for a particular application include the data structures to be used, the size of the data set, the number of available processors, the operations to be performed, dependencies between data items, and the requirement to achieve a good load balance across the application.

Aside from the choice of decomposition technique, the management of I/O is another very important issue for parallel applications, since there is often a large volume of data to be handled. The major issues in parallel I/O are the arbitration of access to shared files from a set of application processes, and the creation and use of parallel files, to improve I/O performance. The I/O requirements of an application are often dictated by the parallel paradigm employed and, for good performance, it is important to match the I/O facilities to the data access pattern of the application.

The parallel paradigm in use shapes not only the application's I/O requirements, but also its communication requirements. Many communication patterns are paradigm-specific, and occur in all applications using a given paradigm. This communications and I/O code, then, can be written in a general purpose way and packaged into libraries for reuse by different applications. This insulates the applications programmer from low level details of data flow management, and the message-passing interface being used. It also reduces the knowledge and effort required to produce an initial parallel prototype of an application, so that its behaviour can be analysed and development effort concentrated where it is needed. Finally, such libraries are reusable, so the cost of developing and optimising them can be spread over several applications, and the benefits of optimisation will be similarly spread.

Many parallel libraries have been developed, and some of those aimed at message passing applications have been described here. In particular, the Parallel Utilities Library developed by Edinburgh Parallel Computing Centre supports a wide range of different paradigms, and includes support for efficient parallel I/O.

Real applications are often far more complex than the paradigms by which they are decomposed, and for high-level support to be practically useful, it must be flexible enough to allow for this additional complexity. Developing a powerful yet flexible interface can be a difficult task, and developers need to be aware of the requirements of the applications they are targeting, in addition to the form of the paradigm being represented.

One of the I/O utilities within PUL, PUL-PF, was developed with a GIS application as its primary target, and has been used as an example of some issues that arise in the provision of library support. The GIS application had non-trivial data access requirements in comparison with the numerical applications towards which previous PUL support for I/O had been biased. This example illustrates that library design should be an iterative process, with the interface being gradually refined or extended to meet the needs of the target applications. Given sufficient attention to the needs of real applications, it is possible to develop genuinely useful, portable and reusable high-level support libraries.

4.9 References

Agrawal, G., Sussman, A. & Saltz, J., 1994, Efficient runtime support for parallelising block structured applications. *Scaleable High Performance Computing Conference* 1994, IEEE.

Anon., 1993, Survey of principal investigators of grand challenge applications. *Workshop on Grand Challenge Applications and Software Technology*, May 1993.

Aronoff, S., 1989, *Geographic Information Systems: A Management Perspective*, WDL Publications, Ontario, Canada.

British Standards Institute, 1992a, *Electronic Transfer of Geographic Information (NTF) Part 1. Specification for NTF structures*. BS7567.

British Standards Institute, 1992b, *Electronic Transfer of Geographic Information (NTF) Part 2. Specification for implementing plain NTF*. BS7567.

Bruce, R.A.A., Chapple, S., MacDonald, N.B., Trew, A.S. & Trewin, S., 1995, CHIMP and PUL: Support for portable parallel computing. *Future Generation Computer Systems*, 11, 211-219.

Chapple, S., 1994, *PUL-PF Reference Manual*. Technical report EPCC-KTP-PUL-PF-PROT-RM, Edinburgh Parallel Computing Centre.

Chapple, S., Fletcher, R. & Trewin, S., 1994, *PUL-GF-Prototype User Guide*. Technical report EPCC-KTP-PUL-GF-PROT-UG, Edinburgh Parallel Computing Centre.

Chapple, S., Trewin, S. & Clarke, L., 1995, *PUL-RD-Prototype User Guide*. Technical report EPCC-KTP-PUL-RD-PROT-UG, Edinburgh Parallel Computing Centre.

Choudhary, A., 1993, Parallel I/O Systems. *Journal of Parallel and Distributed Computing*, 17, 1-3.

Cole, M., 1989, Algorithmic Skeletons: Structured management of parallel computations, *Research Monographs in Parallel and Distributed Computing*, Pitman.

Corbett, P., Feitelson, D., Hsu, Y., Prost, J., Snir, M., Fineberg, S., Nitzberg, B., Traversat, B. & Wong, P., 1995, *MPI-IO: A Parallel File I/O Interface for MPI Version 0.3*, NAS-95-002, NASA Ames Research Center.

Crockett, T.W., 1989, File concepts for parallel I/O. *Proceedings of Supercomputing '89*, 574-579.

Distributed Software Ltd, 1992, *The Helios Farm Library*, Distributed Software Limited, The Maltings, Charlton Road, Shepton Mallet, Somerset BA4 5QE, UK.

Fletcher, R.A. & MacDonald, N.B., 1994, Portability and code reuse in parallel applications. In Joubert, G.R., Trystram, D., Peters, F.J. & Evans, D.J. (Eds), *Parallel Computing: Trends and Applications*, Elsevier Science B.V., 609-612.

Foster, I., 1995, *Designing and Building Parallel Programs*, Addison Wesley.

Galbreath, N., Gropp, W. & Levine, D., 1993, Applications-Driven Parallel I/O. *Proceedings of Supercomputing '93*, 462-471.

Geuder, U., Härdtner, M. & Zink, R., 1994, GRIDS - a programming system for grid-based technical and scientific applications on parallel systems. *Future Generation Computer Systems*, **10**, 285-289.

Gropp, W., 1993, *BlockComm for Fortran*. Technical Report, Mathematics and Computer Science Division, Argonne National Laboratory, 9700 South Cass Avenue, Argonne, IL 60439-4801.

Hart, L., Henderson, T. & Rodriguez, B., 1995, *An MPI Based Scaleable Runtime System: I/O Support for a Grid Library*. NOAA Forecast Systems Laboratory, R/E/FS5, 325 Broadway, Boulder, CO 80303, http://www-ad.fsl.noaa.gov/mvpab/hpcs/hartLocal/io.html.

Hempel, R. & Ritzdorf, H., 1991, *The GMD Communication Library for Grid-Oriented Problems*. Gesellschaft fur Mathematik und Datenverarbeitung, mbH, Institute F1/T, Postfach 1240, 5205 St Augustin, Germany.

Katz, R. & Gibson, G., 1987, *Case for Redundant Arrays of Inexpensive Disks*, University of California at Berkely.

Kim, M., 1986, Synchronized Disk Interleaving. *IEEE Transactions on Computers*, **C-35**, 11, 978-988.

Kim, D. & Tantawi, A., 1987, *Asynchronous Disk Interleaving*. RC 12497, IBM T.J. Watson Research Center, Yorktown Heights, NY.

Kohn, S. & Baden, S., 1993, An implementation of the LPAR parallel programming model for scientific computations. *Proceedings of the Sixth SIAM Conference on Parallel Processing for Scientific Computing*, SIAM, 759-766.

Kotz, D. & Ellis, C.S., 1993, Cacheing and writeback policies in parallel file systems. *Journal of Parallel and Distributed Computing*, **17**, 140-145.

Lemke, M. & Quinlan, D., 1992, *P++, a C++ Virtual Shared Grids Based Programming Environment for Architecture-Independent Development of Structured Grid Applications*. Technical Report 611, GMD.

London Parallel Applications Centre, 1993, *A Kernel Library for Parallel Applications*. Computational Fluid Dynamics Project, London Parallel Applications Centre, Imperial College.

Moon, B. & Saltz, J., 1994, Adaptive runtime support for direct simulation monte carlo methods on distributed memory architectures. *Scaleable High Performance Computing Conference 1994*, IEEE.

Parasoft Corporation, 1988, *Express Version 1.0: A Communication Environment for Parallel Computers*. ParaSoft Corporation, 27415 Trabuco Circle, Mission Viejo, CA 92692.

Perihelion Software Limited, *The Helios Parallel Operating System*, Prentice Hall International, ISBN 0-13-381237-5.

Peucker, T.K., Fowler, R.J., Little, J.J. & Mark, D.M., 1978, The triangulated irregular network. *Proceedings of the ASP Digital Terrain Models (DTM) Symposium*, American Society of Photogrammetry, Falls Church, VA, 516-540.

Pierce, P., 1988, The NX/2 Operating System. *Proceedings of the 3rd Conference on Hypercube Concurrent Computers and Applications*, 384-391, ACM.

Quinn, M., 1994, *Parallel Computing: Theory and Practice*. 2nd Edition, McGraw-Hill Computer Science Series: Networks, Parallel and Distributed Computing.

Rodriguez, B., Hart, L. & Henderson, T., 1995, Programming Regular Grid-based Weather Simulation Models for Portable and Fast Execution, to appear in *Proceedings of the International Conference on Parallel Processing*, 14-18 August, CRC Press.

Satyanarayanan, M., 1993, Distributed file systems. In Mullender, S. (Ed.), *Distributed Systems*, 2nd Edition, ACM Press, NY.

Silberschatz, A. & Peterson, J., 1988, *Operating Systems Concepts* (Alternate Edition), Addison-Wesley.

Townshend, J., Justice, C., Li, W., Gurney, C. & McManus, J., 1991, Global land cover classification by remote sensing: present capabilities and future possibilities. *Remote Sensing of Environment*, **35**, 243-256.

Trewin, S., 1995, *PUL-SM Prototype User Guide*. Technical report EPCC-KTP-PUL-SM-PROT-UG, Edinburgh Parallel Computing Centre.

Trewin, S. & Chapple, S., 1993, *PUL-EM Prototype User Guide*. Technical report EPCC-KTP-PUL-EM-PROT-UG, Edinburgh Parallel Computing Centre.

Trewin, S., Chapple, S. & Clarke, L., 1993, *PUL-TF Prototype User Guide*. Technical report EPCC-KTP-PUL-TF-PROT-UG, Edinburgh Parallel Computing Centre.

Weibel, R. & Heller, M., 1991, Digital terrain modelling. In Maguire, D.J., Goodchild M.F. and Rhind, D.W. (Eds), *Geographical Information Systems: Principles and Applications*, Vol. 1, 269-297, London: Longman.

Wilson, A. & MacDonald, K., 1995, Electromagnetic Modelling on Parallel Computers. *International Symposium on 3D Electromagnetics, Schlumberger-Doll Research, Ridgefield, CT, October 1995.*

Part Two

Design Issues

5

Issues in the Design of Parallel Algorithms

S. Dowers, M.J. Mineter and R.G. Healey

5.1 General Context

The previous chapters have described the development of parallel processing in terms of the hardware, software and parallel libraries. They have also described some of the current initiatives to develop parallel processing from a very specialist tool that is difficult to use effectively to one facilitated by a set of standards and libraries. In order for parallel processing to be of general benefit in the area of geographical information systems, similar developments, specific to geographical problems, are required. This chapter introduces some of the distinguishing characteristics of geographical data processing which present a challenge for parallel implementation with particular emphasis on a subset relating to both vector and raster representation of areal data. Later chapters in Part 4 of this volume explore other application topics relating to image processing, terrain modelling and spatial analysis.

5.1.1 Implementation Framework

The range of algorithms for processing geographical data which can benefit from parallel processing is enormous. They cover a variety of applications areas with correspondingly diverse scope in terms of format and size of data sources. While some applications involve specialised data which is unique to the problem area, in general, data sources are reused in many application areas. The storage and management of data is therefore of major concern.

During the research and development phase, algorithms may be applied in an standalone environment, with specialised formats for input and output of data. However, as an application area develops the practical issues of collating and converting data sources, specifying parameters and analysing, displaying and visualising the results become increasingly important. The way in which the user interacts with the algorithm assumes greater significance as use of an algorithm spreads to less experienced and knowledgeable users. The standalone program can no longer be viewed in isolation and must be viewed in terms of inputs, outputs and user interface, i.e. as part of an information system. We would therefore argue that the design of all algorithms for processing geographical data needs to take into consideration the framework in which the algorithms will be used. For many algorithms this framework can be considered to be a geographical information system (GIS).

5.1.2 What is a GIS?

Aronoff (1989) provides the following description of a GIS:

'A GIS is designed for the collection, storage and analysis of objects and phenomena where geographic location is an important characteristic or critical to the analysis.'

Since this could describe a purely manual system, his more detailed definition is more important for our purposes:

'A GIS is a computer-based system that provides the following four sets of capabilities to handle geo-referenced data:

- input

- data management

- manipulation and analysis

- output.'

In any practical implementation the system extends beyond the hardware and software to include the people who operate the system including the technical staff, management and organisational structure (Maguire, 1991). While Chapter 15 considers some aspects of the role of a GIS within the IT framework of an organisation, this chapter is concerned with some aspects of the storage and management of the data and with the manipulation and analysis component. The analysis component in particular is targeted as the role where some operations are complex and computer-intensive and therefore ideal candidates for parallelisation. Some aspects of input and output, such as data transfer and interchange, may also be suitable for improvement.

5.1.3 Architecture of a Typical GIS

Commercial GIS are generally very complex systems that are the result of tens or hundreds of person-years of development. For many practical applications the complex nature of the system is partly concealed by a tailored user interface which allows the application expert to apply the tools of the GIS in a way that is natural to their discipline. This tailoring process is implemented in a development language which may range from a macro language to a full programming language. The analysis functions of the GIS may be accessed from the development language by invoking programs, by function calls or methods or by sending a message to an object, depending on the programming paradigm adopted. Most systems also have an interface mechanism which allows programs and modules that are *external* to the system to be invoked.

Recent developments in computing have seen the client/server model adopted in a wide range of application areas, including GIS. The user interface may therefore be presented on a desktop PC, while the complete system may include database services from corporate IT, and map data and analysis services from a set of GIS servers.

5.1.4 Integration of Parallel Components with Commercial GIS Software

From this background it can be seen that a complete parallel implementation of a GIS would be impractical and very inefficient. Parallelisation is only justified for those operations which are compute-intensive (Gittings *et al.*, 1994), where the requirement for real-time response cannot be met by serial implementations (Xiong & Marble, 1996) or where the volumes of data to be processed cannot be practically handled by other means. From the user's point of view the use of a parallel implementation of an analysis function should be transparent; that is, the parallel implementation should be encapsulated to hide all the internal details of implementation. This approach has the additional advantage of minimising the development effort of adding parallel components to an existing application. Parallelisation can therefore be seen as a way of 'supercharging' a GIS to give increased performance but controlled by familiar tools.

In order to develop parallel algorithms which will interoperate with the majority of commercial GIS packages at present in existence, the data model used in the algorithms must share common features with these packages. In addition, development effort is reduced by adopting a common transfer format between packages and the algorithms (van Roessel & Fosnight, 1984). A considerable amount of work has been carried out in recent years in developing storage and transfer formats which address many of these issues. This aspect of the design is explored in more detail in Chapter 6.

To integrate seamlessly with a GIS package, the parallel implementations of algorithms should be packaged in such a way as to substitute for the existing operations within the GIS package and provide a simple way to add new operations. The hardware and software environment in which a package may run also needs to be considered. The user may be using an interface on a workstation or PC which is connected by a network to the parallel system which will implement the algorithm; the algorithm may therefore be invoked through a remote call mechanism. The data to be used in the algorithm may be accessible through some shared network file space or alternatively may be provided by the invoking system through some form of network transfer or via a *pipe*. A pipe is a type of I/O device which one program writes to and another program reads from. It can therefore be viewed as a tool for feeding data from one program into another without the programs having to be concerned with the location of the intermediate data. Similar considerations may apply to the data which results from the application of the algorithm.

Operating with a pipe effectively constrains the GIS algorithm to sequential file access for initial input and final output. Obviously a sequential data transfer phase implies a limit to the scalability of the algorithm due to Amdahl's Law (Chapter 3). To reduce the effect of this the algorithms should be designed to allow the data input and output stages to be replaced by faster alternatives, such as a parallel file system, should the need arise. This implies a modular design with a clean well-defined interface between the modules. The modular aspects of the design described in this section are considered in more detail in Chapter 7.

A further advantage of adopting this type of approach is that enhancing the functionality of the GIS package by adding algorithms which are practical only in the parallel environment becomes relatively simple. Since the details of the parallel environment are concealed from the front-end GIS, alternative hardware and software may be introduced to improve performance. This approach mirrors the current developments in client/server database implementation where the server may be upgraded from a sequential to a parallel implementation with no changes required at the client end.

5.1.5 Alternative Frameworks

Although parallel geographical algorithms fit naturally in the general framework of a GIS package, they may also be implemented in other environments where the algorithm is more 'loosely-coupled', such as during the development phase, or as part of a larger information system. An example of the latter would be as part of a World Wide Web server. Although the issues of control and data flow may be less constrained in this type of environment, the issues of data model, representation, storage and transfer format are just as important as in a GIS package. Indeed, the data sources for the algorithm may well have been pre-processed and generated in a commercial GIS.

5.1.6 Exemplar Operations

To explore the problems in implementing parallel algorithms that interface with existing GIS, three operations dealing with areal data were chosen as examples which cover different combinations of input and output data models and type of algorithm:

- polygon overlay of two vector datasets

- conversion from vector to raster format of a single dataset

- conversion from raster to vector format of a single dataset.

Among the factors considered when choosing these algorithms was the need to concentrate on the way the algorithm uses and generates data, both in memory and in terms of I/O. More complex and compute-intensive algorithms may seem more appropriate to explore the benefits of parallelisation, but would have diverted design effort into implementation details which could not be reused in other algorithms. There are several other classes of data and corresponding operations which are considered in later chapters such as terrain models in Chapter 16, transport network modelling and spatial analysis in Chapter 17 and remotely sensed data analysis in Chapter 18.

The following additional requirements which were also imposed:

- arbitrarily large datasets, certainly larger then the memory available on the system

- reusable modules which could be used in the implementation of additional algorithms

- generic designs which would be portable across different parallel architectures

While it could be argued that the memory configurations of current systems is large enough to deal with practical problem sizes, the volume of data to be processed is also increasing as new remote-sensing platforms and large-scale digital map sources become available. The remainder of this chapter considers these requirements in the context of the hardware and software environments introduced in Chapters 2 to 4. Chapters 7 to 13 describe some of the design decisions made to address these issues.

5.2 Generic versus Architecture-Specific Design

Chapters 2 and 3 have described the variety of hardware and software architectures which are available to support parallel processing. Unfortunately, each variant requires different design and implementation techniques to obtain optimum performance. This results in implementations that are not portable. For example, SIMD architectures are suited to problems which are naturally data-parallel and in which the same operation may be carried out on many items of data simultaneously. While this type of process is common in processing raster or grid-based data, it is less common for the variable structure and size of vector-based representations. In addition, research on many vector-based processes has resulted in algorithms which give substantial improvements over the 'brute force' versions, by utilising the relationships between components in the vector representation or by utilising some form of spatial sorting or indexing (Franklin *et al.*, 1989 and Chapter 13). Using brute force implementation on a SIMD architecture is likely to prove much more expensive in compute time than an optimised design on a MIMD architecture. For these reasons the design issues explored in Chapters 7 to 13 concentrate on MIMD architectures. In addition, small-scale MIMD parallel environments are becoming common with the widespread introduction of SMP servers and workstations and the increasing use of groups of workstations connected by a high-speed network (Chapter 2).

5.2.1 Memory Considerations

Shared memory architectures (MIMD-SM) allow a relatively simple approach to parallel processing, since in theory all data in memory are available to all processes. Therefore the parallel implementation is concerned mainly with distributing the work to processes to ensure a balanced workload and in managing access to critical areas of memory using a variety of techniques such as spin locks, semaphores and barriers (see Chapter 3). The recent increase in the number of commercial systems which support SMP, plus the development of standard approaches to thread control, make such implementations fairly portable.

However, the increasing speed of processors and the increasingly complex cache architecture of SMP machines mean that considerable care is required in the use of shared memory. While processor speeds have been increasing geometrically for some time (approximately doubling every 18 months), the DRAM chips which are used for main memory on most systems have only doubled in speed in the past 10 years. The move to 64-bit processors has also doubled the bandwidth required from the memory subsystem. As a result, the pattern of memory access has a major impact on overall performance.

To address these problems, system manufacturers have taken several steps to provide high memory bandwidth including:

- increasing bus width from 32 bits up to 256 or 512 bits (Digital, SGI)

- page interleave, burst-mode and other techniques implemented either in the memory chips or in the memory subsystem aimed at increasing the bandwidth of block transfers (SGI)

- multi-segment or cross-bar switching memory buses to allow simultaneous access by more than one processor (Convex, Sun)

- multi-level caches either on chip or at the board level. In some cases there may be 2 levels of cache on the chip together with a third level on the board. In order to support multiple processors and ensure consistency between the caches and main memory, techniques such as bus-snooping must be used to ensure that the cache copy of a value is invalidated when the main memory copy is changed by another processor (Digital) (see Chapter 2)

- read-ahead on cache filling. This approach is designed to reduce the delay involved in a cache miss by anticipating that the next cache miss will be for the adjacent cache line (Cray T3D).

Since the performance of present day systems is crucially dependent on a high 'hit' rate in the cache, *invalidation* due to values being changed by another processor can have a serious impact on the speed. Invalidation occurs when the contents of stored in a cache are no longer consistent with the true value held in a memory location due to that location being updated by another processor. Designing for SMP must therefore ensure that this happens as seldom as possible.

Some distributed memory (MIMD-DM) architectures provide a global address space to allow access to non-local memory (Cray T3D). However, accessing non-local memory is normally slower than local memory by one or two orders of magnitude in terms of latency, as described in Chapter 2. The bandwidth for non-local access is also lower than for local access and may be subject to an overall limit on a system-wide basis. In addition, many of the issues relating to caches for MIMD-SM architectures apply equally to MIMD-DM, in particular the problem of cache-invalidation. For MIMD-DM architectures with no global address space, such as farms of workstations, any data-sharing must be handled by passing messages between processes.

It is possible to implement message passing on a MIMD-SM architecture by using shared memory buffers; this is the usual route for implementation of PVM and MPI libraries on MIMD-SM architectures. From this it should be clear that a design using message-passing should be portable across a range of architectures. The use of a standard such as MPI allows the details to be hidden in the library leaving the implementor of the algorithm to concentrate on issues such as decomposition strategy and load-balancing. Additional benefits arise from the use of message passing due to the ability to control the interaction between processes and thereby avoid conflict in the shared memory space.

5.2.2 I/O Considerations

There are several approaches to the implementation of input and output (I/O) in parallel architectures. For the purposes of GIS algorithms almost all I/O is concerned with reading and writing disk files or across network *pipes* (5.1.4). The use of pipes constrains the application to sequential I/O for the initial and final stages. Therefore the discussion here is concerned primarily with disk I/O. One common characteristic of I/O implementation is transfer of data between a disk device and a buffer in memory, through a controller. In most instances the mechanics of the transfer is independent of a processor since the controller can manage the transfer to/from memory, generating an interrupt to inform a process that the transfer is complete. Hence the controller must logically share access to memory with the processor(s). Some architectures may provide independent routes from I/O controllers to memory using a separate bus, dual-ported memory, or separate cross-bar connections to remove conflict between processors and I/O.

In a MIMD-SM architecture there may be several disk controllers but I/O is effectively shared in a logically similar way to memory. Therefore each process running on such a machine has equal access to disk storage but the total bandwidth available is shared. In addition, disk requests to one disk from two processes are likely to lead to reduced performance from the disk due to excessive seeks, unless the controller or the operating system optimises scheduling of operations on the disk to minimise seek time. This type of optimisation can have unpredictable effects on the latency of requests seen by individual processes.

MIMD-DM machines may have a variety of I/O architectures ranging from a single shared I/O system to a separate disk subsystem for each processor. The main issues which are important from the point of view of the implementor are the bandwidth available to the system as a whole, and to any individual processor, and the bandwidth and *latency* for each disk as seen from a processor. Latency is a measure of the elapsed time between a processor issuing an I/O request and the data starting to move between memory and the I/O device. In some architectures the access to disk storage from most processors is via the communication network which is also used for messages. The I/O scalability of a system also varies with the architecture. For example, an architecture with a disk subsystem for each processor will scale well for local disk access, while the scalability of an architecture with a shared I/O system depends on the scalability of the I/O system itself and the channels used to transfer data to processors.

GIS algorithms are concerned with the performance of I/O for loading and saving the data being processed but, since the data volumes are normally very large, also with temporary local storage during processing or between phases of processing. Chapter 7 explores in more detail the implications of I/O performance for GIS algorithms.

From the point of view of the GIS algorithm designer, the variety of approaches to

I/O makes generic implementation problematic. The Parallel Utility Libraries described in Chapter 4 (PUL-GF and PUL-PF) are designed to hide much of the complexity of implementation from the user. This encapsulation of I/O functionality supports portable code and allows optimisation of I/O, such as caching, to be hidden from the algorithm. However, the success of caching relies on the data in the cache being reused before it is flushed out thereby saving a disk access. For many GIS algorithms each data element is only accessed once or twice. A cache can therefore only be effective when blocks of data are buffered in the cache and successive reads from the application access the same block. This pattern of access implies close to sequential disk access. The penalty for random access to disk can be quite severe depending on the block size used. For example, for a RAID array with maximum transfer rate of 32 Mbytes/s in sequential mode, the corresponding transfer rate for random access in 16 Kbytes blocks would be about 1.6 Mbytes/s; a reduction of throughput by a factor of 20. To obtain maximum benefit from disk I/O systems algorithms should access data sequentially if possible or should attempt to block I/O in sequential "chunks".

5.2.3 Generic Design Approach

In order to minimise design and implementation effort it is desirable to produce a generic design which is capable of running on a wide range of machines including MIMD-SM and MIMD-DM. The discussion above on memory access patterns leads to the conclusion that a design based on message passing can be implemented on both architectures. In addition the use of standard message passing libraries such as MPI isolates the hardware architecture from the algorithm and therefore enhances portability. The task of optimisation to match a particular hardware environment is therefore carried out once for the message passing library. To extend the layered approach described in chapter 4, a generic approach to GIS algorithm design might look like that shown in Figure 5.1.

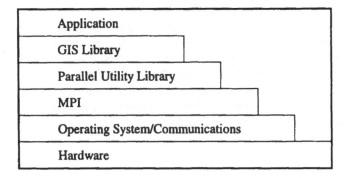

Figure 5.1: Model of parallel GIS library

In this idealised approach the application uses the GIS library to take advantage of the PUL and MPI layers, while the PUL layer isolates the application and the GIS library from some aspects of the memory and I/O implementation and the MPI layer conceals the implementation of communications. This approach facilitates the development of portable code, and simplifies the parallelisation of further algorithms through reusable software libraries.

5.3 Implications of Geographic Data for Parallel Processing

5.3.1 The Impact of Alternative GIS Data Models

Chapter 6 introduces some of the wide variety of data models used in GIS today. These fall into two main classes - vector and raster. Since vector data models use constructs built on points, lines and areas in two- and three-dimensional space to represent real world objects, data held in this format is naturally of variable length. The volume of data required to represent any one object also varies depending on the accuracy with which it is recorded. If topological relationships are stored explicitly in the model then the data representing these relationships may also be of variable length; for example, the list of edges making up the boundary of a area or the list of edges which join a node in a network. In contrast, raster-based data models use a rectangular array or grid of rectangular elements of a fixed size to represent a part of the real world. Each element in this array (grid cell) may hold values for several attributes, which are assumed to apply to the area represented by the cell, e.g. dominant land use or number of people. While the raster representation is very simple, there are a number of techniques used to reduce the amount of storage required to hold the data including run-length encoding, quad-tree encoding (Samet, 1984) and compression techniques borrowed from image processing. Worboys (1995) provides more detail on representation and data structures for both vector and raster data.

While the raster GIS data model is similar to the type of data associated with many applications to which parallel processing has been successfully applied (see Chapters 4, 16 and 18), the variable length data of the vector data model with its more complicated structures is quite different. Some of the issues relating to this have been discussed in Chapter 4 regarding the development of the parallel utility library for parallel files (PUL-PF). The issues for development of corresponding GIS parallel libraries are explored further in Chapter 7.

5.3.2 Decomposition Strategy

Chapter 4 identified two alternative decomposition strategies to distribute the workload across the available processors; functional decomposition and data decomposition. For most GIS algorithms the processing is relatively simple and can only be broken down into a limited number of functions. In contrast, the volume of data is normally very large and therefore a data decomposition strategy is appropriate. There remains the problem of partitioning the data into suitable units for distribution to the available processors.

5.3.3 Problems of Spatial Data Decomposition

A range of specific problems associated with the decomposition of vector data in a parallel environment need to be examined. These include irregular spatial distribution and complexity of data, which may make load balancing more difficult, and the variable length or extent of spatial elements. Also, the chosen algorithm may require spatial coherence in the data and therefore an ordering or sorting phase may be required.

With raster data there are a number of issues concerned with compression and encoding, as these affect the actual volume of data that needs to be processed. Appropriate methods of dividing rasters into manageable blocks for individual processors also have to be considered.

These issues are explored in more detail in Chapters 7 to 13.

5.3.4 Problems of Dataset Re-assembly

While attention is always devoted to data decomposition, in the GIS context the final phase of building one dataset from data held in many processes assumes equal if not greater importance, particularly for vector topological structures.

Some of the major points that have to be addressed under this heading are as follows:

- The vector output format may require a direct or indirect (via record identifiers) relationship between elements which may introduce synchronisation problems between processes, as is shown in Chapter 9.

- Data elements may have large spatial extents and may require reconstruction across processes, e.g. polygons whose component line segments are distributed across more than one data sub-division. Raster output formats will require ordering of rows or columns which may also introduce synchronisation problems.

- These synchronisation problems between processes may impose limits on possible scalability.

- Very large datasets will require temporary disk storage since the intermediate results will not fit into the total memory of the system. Hence parallel I/O capability will be required to support a scalable design.

5.4 Summary of Design Aims

The overall focus of this discussion has been towards an approach which is general purpose and which addresses the practical problems of making parallel software usable in a GIS context.

Therefore a portable generic design is sought, which can process both vector and raster datasets of arbitrary size. The resulting algorithms should be both fast and scalable, and the modules from which these algorithms are built should be reusable in the form of a library. In addition, it should be straightforward to bolt the algorithms onto existing commercial GIS packages, to act as a 'super-charger' as well as allowing implementation in a more loosely-coupled environment.

5.5 References

Aronoff, S.,1989, *Geographic Information Systems: A Management Perspective*, WDL Publications, Ottawa.

Cray Research, 1996, http://www.cray.com.

Digital Equipment Corp, 1997, http://www.digital.com/info/hpc/hpc.html.

Franklin, W.R., Narayanaswami, C., Kankanhalli, M., Sun, D., & Wu, P.Y., 1989, Uniform grids: a technique for intersection detection on serial and parallel machines. In Anderson E. (Ed.), *AutoCarto 9: Proceedings of the 9th International Symposium on Computer-Assisted Cartography* (April 1989, Baltimore, MD), The American Society for Photogrammetry and Remote Sensing and The American Congress on Surveying and Mapping, Falls Church, VA, 100-109.

Gittings, B.M., Sloan, T.M., Healey, R.G., Dowers, S. & Waugh, T.C., 1994, Meeting expectations: a review of GIS performance issues. In Mather, P.M. (Ed.), *Geographical Information Handling - Research and Applications*, Wiley, Chichester, 33-45.

Hewlett-Packard, 1997, http://www.convex.com/prod_serv/exemplar/sx-class/exemplar2.html.

Maguire, D.J., 1991, An overview and definition of GIS. In Maguire, D.J., Goodchild, M.F. & Rhind, D.W. (Eds), *Geographical Informations Systems: Principles and Applications*, Vol 1, Longman, London.

van Roessel, J.W. & Fosnight, E.A., 1984, A relational approach to vector data structure conversion. In Marble, D.F. *et al.* (Eds), *Proceedings of the International Symposium on Spatial Data Handling*, Zurich, Vol. 1, 78-95.

Samet, H., 1984, The quadtree and related hierarchical data structures. *ACM Computing Surveys*, **6** (2), 187-260.

Silicon Graphics Inc, 1996, http://www.sgi.com/Products/hardware/Power/challenge-xldata.html.

Sun Microsystems, 1997, http://www.sun.com/hpc/products/index.html.

Worboys, M.F., 1995, *GIS: A Computing Perspective*, Taylor & Francis, London.

Xiong, D. & Marble, D.F., 1996, Strategies for real-time spatial analysis using massively parallel SIMD computers: an application to urban traffic flow analysis. *International Journal of GIS*, **10** (6), 769-789.5.2

6

Data Models, Representation and Interchange Standards

S. Dowers

This chapter describes the data model and interchange formats which are utilised in the algorithms described in Chapters 7 to 13. To provide some indication of how these relate to other models the chapter opens with an overview of data models and of some current interchange standards.

6.1 Role of Data Models

In order to consider the techniques required to implement GIS algorithms one must first consider the ways in which geographic entities are represented; i.e. the data model. Commercial systems use a range of data models for this purpose but may be broadly classified according to whether they use some form of raster or vector representation. Since the aim of this part of the book is to present generic algorithms which are valid across a range of parallel architectures and which can interoperate with a variety of existing systems, the data models must not be based on any one implementation but should include characteristics present in most, if not all, systems in present use. A more complete review of data models is provided by Peuquet (1984), while Worboys (1995) explores models, representation and data structures from an implementation perspective.

Worboys describes the *field-based* and *object-based* approaches to modelling. The field-based approach starts from a set of spatial locations with one or more *functions* from the locations to the *attribute domains*. *Tessellation* models fall into this category and the raster model described later in this chapter is particular type of tessellation model. In contrast, the object-based approach starts from real-world identifiable features for which certain characteristics can be measured among which is *spatial extent*. The extent can be represented by a *spatial object* such as a point, a line or a polygon. The object-based may be seen to correspond with the traditional vector model used in many GIS packages. For the particular types of operations considered in this part of the book, the distinction is not quite so clear-cut, since the vector areal data can be considered as a field-based model where the tessellation is defined by irregular polygons. The interpretation is therefore dependent on whether the polygons are attributes of real-world features or are defined by the value and distribution of some attribute. Albrecht & Kemppainen (1996) review data models from the wider context of a *Geographic Open Systems Environment* (GOSE) and consider some current and draft standards from the viewpoint of their support for spatial operations.

Rather than design a new model to act as an interface between a parallel algorithm and a GIS, it is more practical to take advantage of a range of standards which are designed to promote the archival and interchange of spatial data. Considerable effort has already expended in the design of these standards to ensure that data may be exchanged between various systems with full semantic integrity. It follows that the data models used in these standards must be sufficiently general to represent the data. Therefore the models at the core of these standards cover the real-world entities which are to be represented such as a farm, the abstractions used to represent the entities such as land parcels and the geometric and topological representations.

Several of these standards are considered in this chapter. While the standards cover several forms of representing geographical data including raster data, they are reviewed largely from the perspective of the vector data model. This has been done largely due to the limited space available in a work such as this and to avoid distracting the reader with unnecessary detail. Those wishing to explore further are referred to the relevant standards documents.

The review is not exhaustive; it presents the trend of the current thinking in spatial data standards. They differ in the extent of the adoption of object-orientation and in the extent to which spatial and non-spatial characteristics are integrated or are represented separately. Although significant differences appear at the generic levels, at the topological and geometric levels there is a considerable amount of common ground. The terminology may be different but the basic concepts are common. Therefore despite the fact that a particular standard has been chosen in the designs described in this section, the principles will in general apply to other standards. The same issues are true in relation to direct interfacing to specific vendor systems. The very disaggregated model used in the designs has allowed exploration of the issues involved in processing the data. In general interfacing to a specific system will normally involve a simplification of the designs. Note that this chapter is concerned with the interface between parallel operations and GIS packages. The internal data model used by the operations is likely to be modified by the internal requirements of the algorithm. It is the job of the input and output modules to implement the mapping between the two.

The remainder of the chapter is concerned with a description of the formats used in the designs described in sections 6.2 and 6.3 as the interface mechanism to and from vendor systems. For vector data the chosen format was NTF Level 4 standard while a simple run-length encoded format was used for raster data.

6.2 Conceptual Framework

Before considering individual standards it is worth considering a framework of the ground which may covered by a standard. Each of the standards may then be considered in relation to this framework. There are several levels of which a framework may be composed depending on the degree of simplification.

6.2.1 Real World Objects

Real world objects are the entities or phenomena which we perceive in the world and with which we interact (Gatrell, 1991). They can range from simple things such as a borehole to complex aggregates such as farms or towns. Objects which are similar to one another can be grouped together and therefore share characteristics. In object-based modelling these groupings are called classes. Individual objects can then be seen as instances of a class. Classes may be further grouped together into superclasses which are more general in nature. The classification of objects may therefore be seen as a hierarchical relationship with more generalisation towards the root of the hierarchy and increasing specialisation towards the leaves of the hierarchy tree.

This object-based view of the world can be very useful in deciding just how things in the real world should be represented in a spatial database. The first stage in this process of representation is to consider the abstractions of reality which should be represented.

6.2.2 Abstractions

Abstraction can be seen as a way of simplifying the complexity of real world objects down to the level of detail necessary to meet the purposes of the application. It can therefore be seen as defining the mapping between the real world objects and the features used to represent them in the data model at the highest level of representation. This may for example involve considering a farm as a collection of buildings, fields and roads, while the buildings and fields can be considered as different types of area and roads as a type of line.

6.2.3 Representations

At the basic level the features used at the abstraction level are represented by basic components which describe the spatial and non-spatial attributes of the features. The range of types available varies between the various models depending on whether 2- or 3-dimensional coordinates are supported, whether there is explicit support for spatio-temporal coordinates and the way in which topological relationships are represented.

6.2.4 Vector versus raster models

For the purposes of this section we are concerned with two main differences in the way in which the spatial features may be represented.

- Raster Data Models

- Vector-Topological Data Models

A general description of raster models can be found in section 6.6; a more detailed explanation of the format on which the designs are based is presented in section 6.7. Section 6.3 explores the salient features of vector models and describes some of relevant characteristics of three national standards for the interchange of spatial data. Sections 6.4 and 6.5 provide more detail on the vector format which forms the basis for the work described in Chapters 7 to 13.

6.3 Vector Models

Vector data models represent features as points, lines, areas or volumes corresponding to zero-, one-, two- or three-dimensional objects. Lines are normally represented as sequences of points, areas as sequences of lines and volumes as sequences of areas. The precision with which the position and extent of each feature is defined in space is limited only by the resolution of the coordinate system in use.

Spaghetti is a term used to refer to vector data where each feature is not explicitly related to any other. In this case spatial relationships must be inferred by analysis of the coordinates of the features. Vector-topological data extends the data model to include explicit relationships between features such as the nodes at each end of a line or the areas on each side of a line. The representation normally also implies that each feature has an identifier to allow the topological relationship to be expressed.

6.3.1 Generic Characteristics

For the purposes of the designs presented in this section it is the lowest levels of the vector model which are most important; i.e. the geometric representation, the way in which topological relationships are recorded and the way in which attributes are attached to the units. This is true because the types of operations covered by these designs operate at the representation level and below. Therefore the data standards covered in this section are compared from the point of view of basic units which are used in operations such as vector/raster conversion, raster/vector conversion and polygon overlay.

6.3.2 Standards for Interchange

6.3.2.1 NTF (BS 7567)

Development of National Transfer Format (NTF) started in 1985 as a UK standard for the transfer of spatial data in digital form. It was released as a British Standard (BS 7567) in three parts in May 1992 when NTF became known as the Neutral Transfer Format (BSi, 1992). Part 1 is concerned with data model and structures used in NTF, Part 2 describes the implementation of plain NTF where the data is transferred in ASCII form while Part 3 describes the implementation of NTF using BS 6690 (equivalent to ISO 8211).

The NTF standard is not explicitly object-oriented and the most abstract form of representation is the feature which may be composed of simple features or collections of features. The standard supports five levels of modelling varying from simple spaghetti data at level 1 through to full topology at level 4. Level 5 allows the user to redefine existing formats and to define new record types. Data is transferred via records with a different record type for each different element to be represented.

Figure 6.1 describes the general structure of the data model used in NTF level 4 for full topology. This clearly shows three levels in the model:

• Features which correspond to real world objects

• Topology which describes the topological relationships between features and provides the linkage between features and the geometry

• Geometry which describes the spatial location and extent of features.

Simple features consist primarily of areas, lines and points which represent two-, one- and zero-dimensional entities while complex entities may be represented as collections of either simple or complex features. Attributes may be associated with features. Coordinates in NTF geometry records may be either two- or three-dimensional. A more complete description of the individual record types in NTF level 4 is considered in more detail in section 6.4.

NTF provides facilities for including information about the quality of the data being transferred at various levels from the transfer set right down to the individual point in a geometry record. The records for attribute description, feature classification and code lists are intended to allow a fairly complete description of the data within the transfer set. The data dictionary field in the database header provides another mechanism for describing the way in which the data are recorded.

6.3.2.2 SDTS

The Spatial Data Transfer Standard is a national data transfer mechanism for the United States of America. It was developed over the period 1982 to 1988 and a proposed Digital Cartographic Data Standard was published in 1988. Subsequent work by the Spatial Data Transfer Standard Technical Review Board has resulted in the current standard (United States Geological Survey, 1994).

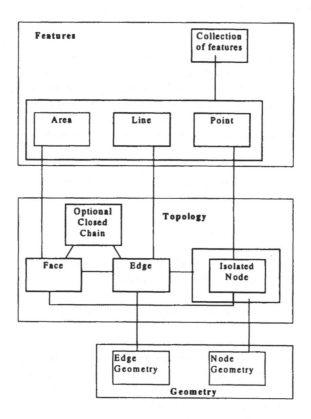

Figure 6.1: NTF Level 4 Data Model (after BS7567 Part 1)

SDTS consists of 3 major components covering the structure of the SDTS transfer, the definition of spatial features and attributes and the encoding method used for the transfer. It is designed to serve the spatial data transfer needs of the Federal agencies, especially the proposed National Geo-Data System, and the work of State and local governmental entities, the private sector, and research organisations.

The SDTS conceptual model is composed of three sub-models relating to spatial phenomena, the spatial objects used to represent phenomena and the spatial features which relate spatial objects and spatial phenomena. The model is based on real-world objects, occurrences or circumstances which are called *phenomena*. Similar phenomena may be assigned to a common class and therefore an individual phenomenon is an instance of its class. Classes may be generalised to form superclasses. For example Edinburgh Castle is an instance of the class castles which is part of the more general class buildings. *Aggregation* is the process of constructing more complex phenomena out of component phenomena. For example a farm is composed of houses, barns and fields. As an alternative to grouping phenomena by classification, *association* may be used to produce sets, using criteria different from those used for classification. A relationship is a special instance of an association.

The definition of each class specifies which characteristics a phenomenon must have in order to belong to the class and how the class differs from other classes. The classes are called entity types and individual phenomena are entity instances. Each entity instance has values for a set of attributes which are also specified in the class definition. A subset of these attributes are key attributes which define a unique identifier for each entity instance. Generalisations of entity types are called themes and therefore an entity type can be considered to be an instance of a theme.

Entity instances are represented in digital form by one or more spatial objects which may be an aggregation of other spatial objects. When there is a one-to-one correspondence between an entity instance and a spatial object, the spatial object is an entity object which may be classified into an entity object class. Entity objects have generalisations and associations.

SDTS defines a set of simple spatial objects. These simple spatial objects are either primitive objects, or are aggregated only from spatial objects belonging to different classes. The only exception is the composite object which may be aggregated from simple objects or from other composites. Spatial objects are classified into module types, one of the basic building blocks of the standard. Once defined, modules may be associated into sets by spatial domain, temporal domain, data quality, security requirements, topological relationships, or any other criteria.

The spatial objects which form the basis of the standard fall into three general classes; geometry only, topology only and both geometry and topology. In general the following sections are concerned with last of these categories and therefore objects such as GT-polygon, GT-ring, chain and node. Although the terminology differs from that used in NTF, SAIF and other standards the basic concepts are broadly comparable. The correspondence between similar classes in the various standards is summarised in Table 6.1.

A significant part of SDTS is concerned with the way in which the quality and provenance of data are represented and with a definition of terms and codes which are to be commonly adopted. This largely reflects the role of the standard in data interchange, particularly as part of a transaction between a data producer and a consumer in the US. These are not relevant for this discussion since we are concerned with the geometric and topological primitives and the way they relate to real world objects.

SDTS provides the facility to define a profile for particular types of transfer which in general describes a subset of the full standard with constraints on the type of modules which may be included and on the relationships between them. The Topological Vector Profile which forms part 4 of the standard is an apt example of this facility since it is concerned with the transfer of vector objects which comprise *2-dimensional manifolds*. A manifold is a complete partitioning of a 2-D planar space, i.e. a non-overlapping set of polygons.

6.3.2.3 SAIF

The Spatial Archive and Interchange Format (SAIF) was developed by the Surveys and Resource Mapping Branch of British Columbia Environment, Lands and Parks. It was accepted as a draft standard in 1991, has since been modified several times and is now a national standard in Canada (British Columbia Ministry of Environment, Lands and Parks, 1995). The current version, 3.2, has been modified to reflect the development of the ISO SQL Multimedia Spatial standard which is related to SAIF and the development of the Open Geodata Interoperability Standard (OGIS) under the auspices of the Open GIS Foundation. The standard includes specification of the modelling paradigm, a standard schema, and two notations to allow definition of the classes and representation of instances of the classes.

SAIF follows a multiple inheritance, object oriented paradigm and distinguishes between the description of information and the encoding. It includes explicit recognition of the temporal component by the availability of 6 different variants of coordinates to reflect one, two and three dimensions with or without a time value. It recognises three levels of information which approximately correspond to the levels described in section 6.2 above. The primitives available include abstract objects, list, sets and enumerations, which are used to build spatio-temporal geometric objects and then geographic objects. The model is unified in the sense that spatial and non-spatial characteristics are both seen as attributes of an object.

The principles of classification, generalisation, association and relationships described in 6.3.2.2 are all inherent in the SAIF model.

The SAIF standard schema includes over 300 generic constructs which are used in an application to define the specific classes used through object oriented inheritance. These classes can be expressed in the Class Syntax Notation (CSN) which allows each new class to defined in terms of the parent superclassess, the attributes specific to the class, default values for attributes, restrictions on domains and contraints which limit the objects which may be included in the class. In a practical application the different types of entities to be represented would be defined as classes in CSN inheriting from the standard schema as appropriate either by creating subclasses or defining attributes in terms of classes defined in the standard schema. Therefore the model of spatial representation or topological relationships can be extended as required by the application. While the set of basic primitives is very rich it does not enforce only one view of the representation of topology. Topological relationships can be defined in CSN as with any other characterstics of classes.

The transfer or storage of data in SAIF is carried out using Object Syntax Notation which allows a readable representation of object instances. Data represented in OSN may be stored in compressed form for more efficient use of storage space or transmission bandwidth. Alternative representations in binary may be used. The data for an object is generally represented as one unit except where identifier references are used to save storage. For example, several chains may reference one copy of an arc rather than including a copy of the coordinates. This form of representation differs from both NTF and SDTS where different spatial objects are represented in different record or module types. In the case of SDTS records of a particular module type are grouped together as a unit and must appear in numerical sequence while in NTF no ordering is implied and therefore there can be no assumptions about the records for an object being close in sequence within a transfer set.

SAIF was designed to support both interchange of data and storage of data. In this sense it differs from SDTS and NTF which were developed primarily for transfer of data between different systems. The Class Syntax Notation can be viewed as a data definition language for a spatial storage system.

6.3.3 Common Features

The various standards mentioned differ in the way they approach the mapping between the real world and the digital representation which we are primarily concerned with here, but are based on a common approach to representing vector data based on nodes, edges, chains and faces which form a 2-dimensional manifold.

Table 6.1: Comparison of spatial constructs

SDTS	SAIF	NTF
Point (NP)	Point	Node
Entity Point (EP)		
Label Point (NL)		
Area Point (NA)	Attribute of Polygon	
Line Segment	Vector	Geometry
String (LS)	Arc (CurveChoice = line)	Geometry
Arc (AC, AE, AU, AB)	Arc	
Link (LQ)		
Chain	PathDirected	Edge
Complete Chain (LE)	PathDirected + SpatialRelationships	Edge
Area Chain (LL)	PathDirected + SpatialRelationships	
Network Chain (LW, LY)	PathDirected + SpatialRelationships	
Ring		
G-ring (RS, RA, RM)	Ring	
GT-ring (RU)		Chain
Interior Area		
G-Polygon (PG)	Polygon with holes	
GT-Polygon (PR, PC)	Polygon with holes + ?	Face
Universe polygon (PU, PW)	U	
Void polygon (PV, PX)		
Pixel		
Grid Cell		
Digital Image (GI, GJ, GK, GM)	ImageGrid2D	
Grid (GI, GJ, GK, GM)	Grid	
Layer	Coverage	
Raster	Raster	
Graph	Graph	
Planar Graph	SingleLineNetwork	
Two-dimensional manifold		?

Table 6.1 provides some comparison between the standards. An additional characteristic shared by some of the standards is the separation of different types of vector object into different record types in the encoded format. In some cases the

record types are required to be stored in specific groups or modules and in a specific order as in SDTS. In other formats there is no specification of sequence or grouping and therefore no assumptions can be made about the order in which records will appear in any particular instance.

SAIF provides the most structured representation since all the attributes of an object (both spatial and non-spatial) are stored as one 'record'. NTF represents the most disaggregated form of representation since geometry is represented in a separate record from the topological relationship. This characteristic has a significant impact on the design of data input and output modules and the steps taken to deal with it are discussed in Chapters 8 and 9.

6.4 NTF Level 4

Although the description of some of the standards in current use in section 6.3 demonstrates substantial differences in modelling approach, the underlying geometric-topological models are basically similar. In addition, these models are also shared by many current commercial software packages. While the choice of NTF Level 4 was made to allow one common format to be used to interface parallel algorithms to a limited set of packages and to avoid duplication of input/output coding, the similarity of models means that the design presented later in this section and in the following section would be equally valid if an alternative transfer format were adopted.

The use of a model such as NTF means that the designs are not tied to any particular vendor's data model but are genuinely portable. In general the NTF model is more complicated and more disaggregated than vendor systems and therefore allows all the complexities of vector data input/output to be explored. Modifying the designs to interface to a particular GIS package is likely to be a case of simplifying the current designs due to some categories of data being stored in one record instead of many and therefore removing some of the stages of data sorting or pre-processing.

6.4.1 General Characteristics

NTF Level 4 contains records that allow the transfer of data in a rigorous full topological model which conveys the information necessary for operations involving vector data input or output.

Level 4 includes records for describing:

   Simple features   that is, areas, lines, points, text and external features.

Topology	that is, faces, edges, chains and nodes. These provide the topological relationships between simple features.
Geometry	there are two types of geometry record, for describing two- and three-dimensional features. These define the position of topological objects.
Text	text position and text representation.
Attributes	associates attributes with other record types.

Some of these records are not relevant to the algorithms (such as text related records) and therefore are not described in this chapter.

6.4.2 Description of Level 4 Topological Model

Figure 6.1 gives an overall view of the data model which underlies NTF level 4. A simple geographic feature is represented by an AREA, LINE or POINT record while more complex features can be built using the COLLECT record to define a set of AREA, LINE, POINT or COLLECT records.

6.4.3 Entity Relationship Description of Level 4

The relationships between features, topology and geometry in the NTF model are represented in a transfer set by the primary and foreign keys in each of the record types. Figure 6.2 presents this in the form of an entity relationship diagram for 2-dimensional features. At the highest level, features are represented by the area record which has references to one or more attribute records and to one or more face records. Since the face records in an NTF section form a non-overlapping partition of the coordinate space (see 2-D manifold in section 6.3.2.2) there is normally a one-to-one relationship between areas and faces in data used for the operations discussed in Chapters 11 to 13.

6.4.4 Topological Relationships and the Implications for Data Input and Output

The complexity of the relationships expressed in Figure 6.2 underlines the different items of data which must be brought together during the input phase prior to any operation. This sorting and joining of data into units appropriate for the operations is the subject of Chapter 8.

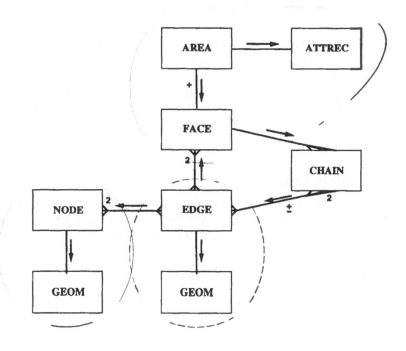

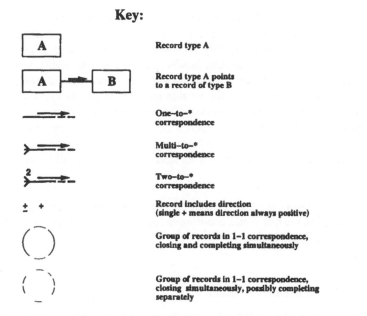

Figure 6.2: Entity-relationship diagram for NTF records

The disaggregated and unordered nature of the data model brings benefits to the output of vector data since it allows records of a particular type to be output as soon as all the information they contain is complete without regard to their position in the files relative to other types of records. For example, GEOMETRY1 records can be output as soon as all the coordinates and the ID for the record are known. This is likely to be well before the corresponding EDGE information is available since that depends on the availability of face IDs for completion. The issues are discussed in more detail in Chapters 7 and 9.

6.4.5 Attribute Linkages

A set of attribute values is represented by an ATTREC record. Each record has a numeric identifier which is unique within the transfer set and contains a variable number of attribute values. Features may contain references to an attribute record by identifier.

6.5 NTF Level 4 - Record Types and ASCII Representation

6.5.1 Logical File Structure

A *transfer set* using NTF contains a number of *databases* each of which is described by a *database header* and a set of records describing attributes, feature classification, codelists and quality statements. Each database may contain one or more *sections* containing a *section header*, section and data quality statements and then section data. A transfer set also has a *volume header* record identifying the transfer set and the NTF level and version number. In ASCII format each transfer also has a *volume termination* record. The database header describes common characteristics of all the sections such as default *data dictionary*, dates for the data dictionary, feature classification scheme and the quality report. It also defines the data model which is used ranging from spaghetti through to full topology and user-defined.

The section header describes the transformation between coordinates used in the transfer set and the real world coordinates together with information on the projection, grid and scale.

6.5.2 Logical Record Structure

Records are treated as independent entities in the logical file structure but are expected to appear in the appropriate sections as described above. The actual representation of the records in a file and the way they are delimited depends on the standard used for encoding. The descriptions which follow use the ASCII representation since it is more straightforward to explain, generate and interpret although the BSi standard also contains specification for transfers using BS 6690 (ISO 8211). Each record in an NTF transfer set has a numeric identifier which is unique within the record type; i.e. there may only be one record with identifier 10 of type AREA within a section but there may also be a record with identifier 10 of each type FACE, EDGE, GEOMETRY, etc.

Table 6.2: Logical file structure

	Full name	Mnemonic	Number of Records
Header records	Volume header record	VOLHDREC	1
	Database header record	DBHREC	1 per database
	Attribute description record	ATTDESC	Any number
	Comment	COMMENT	Any number
	Section header record	SECHREC	1 per section
Topological records	Area, attribute, face, chain, edge, node, 2D geometry	AREA, ATTREC, FACEREC, CHAIN, EDGEREC, NODEREC, GEOMETRY1	Any number in any order
Termination record	Volume termination	VOLTERM	1

When using ASCII representation each logical record may be split over several physical records. Each physical record is terminated by the appropriate carriage control characters for the operating system preceded by an NTF EOR character which is usually the percent (%) character. Immediately preceding the EOR character is a continuation character CONT_MARK which is set to '1' to indicate whether this logical record is continued on a physical continuation record. While the first two characters in a record are usually the NTF record type descriptor, in a continuation record the first two characters are '00'. Another feature of ASCII representation is the use of the DIVIDER character to terminate fields which are inherently variable length, for example A* text fields. The default value for the DIVIDER character is '\'. Alternative values for the CONT_MARK and DIVIDER characters may be specified in the Volume Header record. The distinction between physical and logical record structure has implications for the input and output modules of any operation, although the physical structure may be encapsulated within the I/O modules leaving the core algorithm to deal solely with logical records.

6.5.3 Individual Record Descriptions

This section lists alphabetically the format of each record type which is relevant to the description of the algorithms covered in chapters 8 to 13. NTF defines several other record types, for example TEXTREC, which we do not describe. Each table contains 4 columns which indicate:

- the field name; an asterisk (*) before the field name indicates that it is optional

- the format used in plain ASCII representation

- whether the field is a primary key, a foreign key referencing another record or has a particular domain of values

- a description of the contents of the field and any characteristics of special relevance for the algorithm design.

Vertical bars | next to a group of fields indicate that the *whole group* is repeated a number of times. Except for a few cases (specified as they occur) the format of the records and fields in a general NTF file (such as read by the vector input module described in Chapter 8) will be exactly as in this section.

Each record consists of fields and all records (except NULL) start with a REC_DESC field identifying the record type. Each field has one of the formats listed in Table 6.3.

Table 6.3: Definition of format types

Format	Description
In	Integer of width n
Rn,d	Real of total width n with d digits after the decimal point. The point is not included, so 2.75 in R4, 2 is 0275.
An	String of width n
A*	String of variable width. The width is either determined from another format descriptor or the string is terminated with {DIVIDER}
D6	Date of form yymmdd
D8	Date of form yyyymmdd

Table 6.4: Format of the AREA record

Field	Form	Key/Value	Notes
REC_DESC	A2	32	AREA Record Descriptor
AREA_ID	I6	Primary	Unique identifier for the AREA record
NUM_FACE	I4		Number of faces that form the area
FACE_ID	I6	Foreign ref FACE(FACE_ID)	Unique identifier of a FACE record
SIGN	A1	+ or -	Indicates whether the FACE_ID is added or subtracted to the area
NUM_ATT	I2		Number of ATTREC records attached to the AREA
ATT_ID	I6	Foreign ref ATTREC(ATT_ID)	

A two-dimensional feature as represented by the AREA record (Table 6.4) is composed of a number of FACEs which define the spatial extent of the feature. Note that it is possible to reference FACEs which are holes in an area by the use of a '-' value for the SIGN field. For the purposes of the algorithm designs it has been assumed that there is exactly one face for each area. The non-spatial attributes of the feature are represented by references to one or more ATTREC records via the ATT_ID field(s).

Table 6.5: Format of the ATTREC record

Field	Form	Key/Value	Notes
REC_DESC	A2	14	ATTREC record descriptor
ATT_ID	I6	Primary	Unique identifier for the ATTREC record
VAL_TYPE	A2	Foreign ref ATTDESC(VAL_TYPE)	A two character mnemonic used to identify an attribute which is defined in an ATTDESC record
VALUE	??		An attribute value associated with a VAL_TYPE

The ATTREC record (Table 6.5) records the valueof attributes for a feature with the different types of attribute values being described in the ATTDESC record (Table 6.6). Each attribute is identified by a two-letter mnemonic in the VAL_TYPE field. The way in which values of this type are recorded and the width of the values when they occur are specified in the FINTER and FWIDTH fields which are interpreted in much the same way as in FORTRAN.

Table 6.6: Format of the ATTDESC record. In general NTF, ATTDESC may have further trailing fields

Field	Form	Key/Value		Notes
REC_DESC	A2	40		ATTDESC record descriptor
VAL_TYPE	A2	Primary		Unique two character code for each type of identifier
FWIDTH	I3			Width of attribute value i.e. 4 if FINTER is I4; blank if FINTER is A*
FINTER	A5	In	Integer	Format descriptor for attribute value
		Rn,d	Real	
		An	Character	
		A*	Character of variable width	
		D6	Date 'yymmdd'	
		D8	Date 'yyyymmdd'	
ATT_NAME	A*			Name of the attribute
*FDESC	A*			Textual description of the attribute
*NO_DATA	A*			Contents of the field if no data is present
*RANGE_MIN	A*			Minimum allowable/possible value
*RANGE_MAX	A*			Maximum allowable/possible value
*UNITS	A*			A code representing the measurement units which may be defined in a code list or in external documentation

Chains are the mechanism by which the borders of a face are described in terms of edges. The list of edges is defined in the CHAIN record (Table 6.7) by the repetition of pairs of the EDGE_ID and DIR fields. The EDGE_ID identifies a particular edge while the DIR field specifies whether the GEOMETRY1 (Table 6.10) record corresponding to the edge is to be traversed in the direction in which it is stored (when DIR = 1) or in the reverse direction (when DIR = 2). From the list of edges in a chain it is possible to construct a closed sequence of points which bound a face. Note that each edge will appear in two and only two chains, and that since the DIR field is stored in the CHAIN record rather than the FACE record even in the simple case where a hole consists of a single face there must be a chain for the inside of the boundary of the hole and a separate chain for the outside.

Table 6.7: Format of the CHAIN record. Chains are always closed boundaries

Field	Form	Key/Value		Notes
REC_DESC	A2	24		CHAIN record descriptor
CHAIN_ID	I6	Primary		Unique identifier of the CHAIN record
NUM_PARTS	I4			Number of edges in the chain
EDGE_ID	I6	Foreign ref EDGEREC(EDGE_ID)		Unique identifier of an EDGEREC record
DIR	I1	0	Not relevant	Direction in which the geometry
		1	Forwards	associated with the edge is
		2	Backwards	traversed

Edges are the basic building blocks for one-dimensional features such are lines and for the boundaries of two-dimensional features. Table 6.8 shows that the EDGEREC record serves to define the relationship between the GEOMETRY record which defines the extent of the edge, the two NODERECs which define the start and end nodes of the edge and the two FACERECs correspondig to the faces on the left and right of the edge.

Faces are the basic two-dimensional building-blocks for areal coverages. The set of faces in a section do not overlap and completely cover the area of interest for the coverage. This implies the existence of a 'universal' polygon which fills the space not explicitly covered by all other faces (Figure 6.3). This is explicitly identified in SDTS via the Universe and Void polygons. For the purposes of the designs presented in later chapters we have assumed that the universal polygon can be identified by a particular attribute value.

Table 6.8: Format of the EDGEREC *record*

Field	Form	Key/Value	Notes
REC_DESC	A2	17	EDGEREC record descriptor
EDGE_ID	I6	Primary	Unique identifier for the EDGEREC record
SNODE_ID	I6	Foreign ref NODEREC(NODE_ID)	Identifier of the NODEREC record for the starting node
ENODE_ID	I6	Foreign ref NODEREC(NODE_ID)	Identifier of the NODEREC record ending node
LFACE_ID	I6	Foreign ref FACEREC(FACE_ID)	Identifier of the FACEREC record for the left-hand face
RFACE_ID	I6	Foreign ref FACEREC(FACE_ID)	Identifier of the FACEREC record for the right-hand face
GEOM_ID	I6	Foreign ref GEOMETRYn(GEOM _ID)	Identifier of a GEOMETRYn record which defines the extent of the edge

Table 6.9: Format of the FACEREC *record*

Field	Form	Key/Value	Notes
REC_DESC	A2	18	FACEREC record descriptor
FACE_ID	I6	Primary	Unique identifier for the FACEREC record
NODE_ID	I6	Foreign ref NODE(NODE_ID)	Identifier for the NODEREC record used as a representative point. May be 0 for none
NUM_CHAIN	I2		Number of chains which define the border of the face
CHAIN_ID	I6	Foreign ref CHAIN(CHAIN_ID)	Identifier of a CHAIN record which forms part or all of the border of a face

A face may have a representative point which is a convenient point to which to attach attributes or labels in some systems. The NODE_ID field in the FACEREC record (Table 6.9) is used to reference the NODEREC record for this point. The value for this field is zero if there is no representative point.

The border of a face is described by a list of chains each of which refers to a list of edges. Although not explicitly stated in the standard, it is assumed that there is one chain to describe the 'outer' extent of the face and one chain for each of the separate 'holes' in the face.

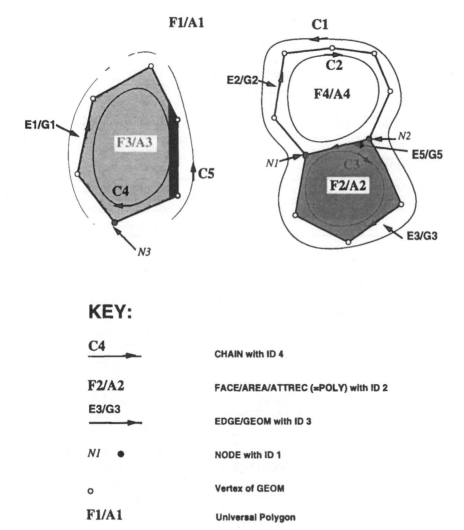

KEY:

C4 ⟶	CHAIN with ID 4
F2/A2	FACE/AREA/ATTREC (≡POLY) with ID 2
E3/G3 ⟶	EDGE/GEOM with ID 3
N1 ●	NODE with ID 1
○	Vertex of GEOM
F1/A1	Universal Polygon

Figure 6.3: Representation of the universal polygon and composite holes in the Vector Interface Format

Table 6.10: Format of the GEOMETRYn *record*

Field	Form	Key/Value		Notes
REC_DESC	A2	22		GEOMETRY2 record descriptor
GEOM_ID	I6	Primary		
GTYPE	A1	0	Disconnected	Only records of type 1 or 2 are considered in the following chapters
		1	Point	
		2	Line	
		3	Unspec spline	
		4	Specified spline	
		5	Arc of a circle	
		6	Circle (3 points)	
		7	Circle (2 points)	
NUM_COORD	I4			Number of coordinate pairs
X_COORD	I*			X-coordinate
Y_COORD	I*			Y-coordinate
Z_COORD	I*			Z-coordinate (height)
QPLAN	A1	SPC No data		Code used to indicate method of data capture

Table 6.11: Format of the NODEREC *record*

Field	Form	Key/Value	Notes
REC_DESC	A2	16	NODEREC record descriptor
NODE_ID	I6	Primary	Unique identifier for the NODEREC record
GEOM_ID	I6	Foreign ref GEOMETRYn(GEOM_ID)	Identifier of a GEOMETRY1 record which defines position of the node
*FACE_ID	I6	Foreign ref FACE(FACE_ID)	Identifier of a FACE record for which this node is a representative point

There are two different variants of the GEOMETRY record (Table 6.10) depending on whether a two- or three-dimensional coordinate system is in use. The only difference between the GEOMETRY1 and GEOMETRY2 records is the presence of the Z_COORD field in the latter. The GEOMETRYn record defines the locus of

Table 6.12: Format of the start of the SECHREC record

Field	Form	Key/Value		Notes
REC_DESC	A2	07		SECHREC record descriptor
SECT_REF	A10			Section reference
COORD_TYP	I1	1 2 3	geographic rectangular geocentric	Coordinate type
STRUC_TYP	I1	1	vector	Structure type
XYLEN	I5			Width of X_COORD, Y_COORD
XY_UNIT	I1	1 2 3	degrees metres feet	Unit for planimetric coordinates
XY_MULT	R10,3			Multiplier for xy coordinates
ZLEN	I5			Width of Z_COORD
Z_UNIT	I1	As XY_UNIT		Unit for height coordinates
Z_MULT	R10,3			Multiplier for height coordinates
X_ORIG	I10			Origin shift for X-coordinate
Y_ORIG	I10			Origin shift for Y-coordinate
Z_DATUM	I10			Vertical datum
XMIN	I10			Minimum X-coordinate
YMIN	I10			Minimum Y-coordinate
XMAX	I10			Maximum X-coordinate
YMAX	I10			Maximum Y-coordinate
XY_ACC	R5,2			Planimetric accuracy
Z_ACC	R5,2			Height accuracy
SURV_DATE	D8			Nominal date for survey
LAST_AMND	D8			Date of last amendment
COPYRIGHT	D8			Copyright date
+ other fields				

an edge or node in space in terms of a list of points recorded in the X-, Y- and Z-COORD fields. The interpretation of the points is specified by the GTYPE field. For GEOMETRY records which correspond to edges the options range from a simple series of line segments, through splines to part or all of the circumference of a circle. Note that the coordinates are recorded as integers with the number of digits specified in the section header record (Table 6.12). Therefore the coordinates must be transformed to and from the actual coordinate system for the data which is specified in the section header via the XY_MULT, X_ORIG and Y_ORIG fields.

The NODEREC record (Table 6.11) is used to define the link between geometry records and point locations such as the end points of edges, the representative points for faces and for point features.

The section header record defines the coordinate projection, transformation and format of the geometry records in the following data records. Fields are also present to describe the quality of the data in terms of accuracy, date of collection etc.

6.5.4 Implications of the Representation for Input/Output of GIS Operations

The vector-topological interface which forms the basis for the design described in later chapters is an implementation of Level 4 of the British Standards Institution neutral transfer format (NTF) (specifically, plain BSi NTF level 4 on formatted media using a full topology model).

This format is adopted for vector-topological data because it is a recognised standard and it provides all the information necessary for the Operations. However, being a general purpose transfer format it is not the most efficient interface between GIS packages and the Operations. For example, NTF Level 4 is an ASCII-based format and performance could be improved using a binary format; NTF Level 4 is also unstructured and relatively rich in terms of the number of record types, therefore requiring more processing on the input side than a typical commercial GIS package's data format.

The NTF standard for level 4 data places no restriction on the ordering or relative placement of records. Routines which deal with the input of data can therefore make no assumptions about whether all the records for a particular feature occur together or about the ordering of the records in spatial terms. As Chapter 7 shows, efficient data distribution and load balancing rely on discovering spatial coherence in the data and ensuring that data which is near each other is sent to the same process. Indeed operations such as polygon overlay and vector to raster conversion rely on ALL the data relating to the current region of interest being available. This applies not only to the geometry and topology records but also to the attribute records.

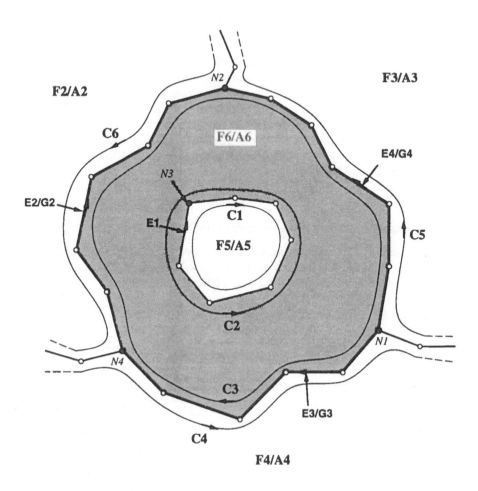

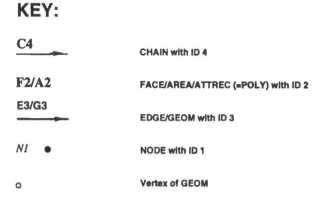

KEY:

C4 ⟶	CHAIN with ID 4
F2/A2	FACE/AREA/ATTREC (=POLY) with ID 2
E3/G3 ⟶	EDGE/GEOM with ID 3
N1 ●	NODE with ID 1
○	Vertex of GEOM

Figure 6.4: An example of the PAP GIS Vector Interface. F6/A6 is a polygon with a hole chain C2. The records to represent F6/A6 are listed in section 6.5.3.

6.5.5 Limitations of NTF Level 4

During the implementation stage of our work it became clear that the ASCII representation of NTF contains some limitations due to the fixed length ASCII fields allocated to item identifiers, the number of edges in a chain and the number chains in a face. In addition the R10,3 format used for XY_MULT could limit the precision with which coordinates are recorded when specified in units such as degrees. These problems only become apparent when dealing with large data sets in ASCII format and should not be seen as a limitation of the NTF data model itself or with transfers using BS6690.

6.5.6 Example of Topological Records in BSi NTF Level 4

As an example of the representation of vector-topological data in the PAP GIS Vector Interface Format, the polygon F6/A6 shown in Figure 6.4 would be represented by the following records (the ATTREC is not shown).

AREA record:

Field	Value
AREA_ID	6
NUM_FACE	1
|FACE_ID	6
|SIGN	+
NUM_ATT	1
|ATT_ID	6

FACEREC record:

Field	Value
FACE_ID	6
NUM_CHAIN	2
|CHAIN_ID	2
|CHAIN_ID	3

CHAIN records:

Field	Value
CHAIN_ID	2

NUM_PARTS	1
|EDGE_ID	1
|DIR	+

Field	Value
CHAIN_ID	3
NUM_PARTS	3
|EDGE_ID	3
|DIR	-
|EDGE_ID	4
|DIR	+
|EDGE_ID	2
|DIR	-

EDGEREC records (shared with other CHAINs):

Field	Value
EDGE_ID	1
SNODE_ID	3
ENODE_ID	3
LFACE_ID	6
RFACE_ID	5

GEOM_ID	1
Field	Value
EDGE_ID	3
SNODE_ID	1
ENODE_ID	4
LFACE_ID	4
RFACE_ID	6
GEOM_ID	3
Field	Value
EDGE_ID	4
SNODE_ID	1
ENODE_ID	2

LFACE_ID	6
RFACE_ID	3
GEOM_ID	4
Field	Value
EDGE_ID	2
SNODE_ID	4
ENODE_ID	2
LFACE_ID	2
RFACE_ID	6
GEOM_ID	2

The EDGE GEOM records (one per EDGE) have the following form:

Field	Value
GEOM_ID	1
GTYPE	2
NUM_COORD	7
X_COORD	...
Y_COORD	...
X_COORD	...
Y_COORD	...
X_COORD	...
Y_COORD	...

In a plain NTF file these records would appear as:

```
320000060001000006+010000060%
180000060000000020000020000030%
240000020001000001+0%
240000030003000003-00004+00002-0%
240000030003000003-00004+00002-0%
170000010000030000030000060000050000010%
170000030000010000040000040000060000030%
170000040000010000020000060000030000040%
170000020000040000020000020000060000020%
210000012 ...etc...
```

6.6 *Raster Data Models*

A raster is a regular tessellation of two-dimensional space into pixels or picture elements each of which is assigned a value for a single attribute. Very often the pixels are square and the value is restricted to a range of integer values. The resolution of this type of data is limited by the size of a pixel. The most common example of this type of data is remotely-sensed data from satellite and airborne sensors and scanned aerial photographs.

Rasters may be stored in full matrix format, or a compression algorithm may be applied such as run-length encoding, quadtree encoding, block codes and chain codes. Further details on raster formats and the ways in which they may be encoded may be found in Gatrell (1991).

6.6.1 Rectangular Array

Rasters in full matrix format are customarily stored row by row (as scanlines) starting from the top-left.

6.6.2 Run-length Encoding

Run-length encoding can be used to reduce storage requirements by effectively storing the locations in each scan-line where the value of the pixels change. Usually this is done by storing a series of *value-count* pairs where the *value* is to be repeated for *count* cells. This type of compression is therefore only effective where a limited number of values occur in groups as is often found in classifications of satellite imagery or raster representation of land-cover.

6.6.3 Quadtree and other Compressions

Quadtree encoding and block-encoding take advantage of regions of uniform pixel values to create a hierarchical description of the raster. Chain encoding is used to describe the boundaries of homogeneous pixel regions as a series of steps by counting the number of pixels edges and the direction of the edges. Scanned image formats such as TIFF (Tagged Image File Format) (Aldus Corporation, 1992) generally allow some form of sophisticated compression like Lempel-Zev-Welch (LZW) which attempts to replace sequences of bytes by codes from a dynamically constructed table (Welch, 1984). Each time the same sequence of bytes occurs it is replaced by the appropriate code. Since the set of codes must be able to represent single byte values as well a reasonable number of mutiple-byte codes to achieve reasonable compression, the number of bits needed to represent the code changes dynamically according to the number codes currently in the table. The encoded data is therefore best thought of as a bit stream and the boundaries between codes will frequently not lie on a byte boundary. Compression and expansion therefore involve a reasonable amount of computation.

6.7 Raster Interchange Formats

A wide variety of formats for the interchange of raster data exist, some of which are de-facto standards and some of which are proprietary. Three significant groupings of transfer formats are those associated with scanned images, those associated with remotely sensed data and those associated with digital elevation models (DEM).

6.7.1 Scanned Image Based Formats

The standards concerned with scanned images are usually concerned with some photographic or metric representation of the real world. They may also be concerned with a digital representation of a paper document, for example a fax or a scanned image of a map. The number of bands represented in the format ranges from one for monochrome images to three for RGB images and four for CMYK images. The range of values for each band can vary from two in the case of a fax to 256 for eight bit images and even up to 4096 for 12 bit images. Scanned images can use either a lossless compression scheme such as LZW or a lossy compression scheme such as discrete cosine transform (DCT), Joint Photographic Experts Group (JPEG) or fractals. TIFF is a commonly used format of this type although both GIF and JPEG formats are in common use for the display of images in the World Wide Web.

6.7.2 Remotely Sensed Image Based Formats

The formats used in remote sensing are largely vendor specific, for example ERDAS LAN files, or are associated with particular sources of data by satellite, by receiving station or by supplier depending on the level of processing applied to the data en route to the user. Most sensors measure the reflectance of the earth at several wavelengths, so each wavelength is represented by a separate band in the format. Bands may be recorded by interleaving at the pixel level, at the row level or at the image level. Formats used to transfer remotely sensed data do not generally use compression due to the variability between adjacent pixels owing to either 'noise' or small variations in surface characteristics.

When images are classified the resulting values for each pixel are usually drawn from a relatively limited set and therefore adjacent pixels have a higher probability

of having the same value. For classified images some form of compression is
practical especially if it takes accounts of spatial coherence.

6.7.3 Digital Elevation Model Based Formats

A Digital Elevation Model (DEM) is a particular specialisation of the raster format.
It records the height of the terrain at points on a regular grid. The characteristics
associated with DEMs are explored in Chapter 18 which examines the issues
involved in parallelising algorithms for processing models of the terrain. In general
the formats for DEMs do not involve compression.

6.8 Raster Format for the Parallel Algorithm Designs

Despite the number of formats currently in existence, it was not felt to be
appropriate to choose one vendor-specific format on which to base the designs.
The format used was based partly on the natural relationship between run-length
encoding and the algorithms used for the conversions to and from vector format.
Since the raster data to be transferred in the designs was largely in the form of
classified images, it was also felt that run-length encoding could offer benefits both
in reducing storage requirements and in reducing the bandwidth required for
network transfers. A further contributory factor was the ability of run-length
encoding to be used in a data-stream environment, i.e. when input and output are
pipes or streams of data which do not support rewinding.

The chosen format of raster file comprises a fixed length header of controlling
information, followed by the raster data, for each scan-line in turn, in a run-length
encoded format. (In the description below, cells and scan-lines are numbered from
1.) Coordinates are assumed to be Cartesian and measured in metres.

6.8.1 Raster File Field formats

The formats used in the description of the raster data file are as given in Table 6.13.

The V format is used to allow efficient storage of positive integers which may often
be small enough to fit in one byte, but may occasionally require two or more bytes.
Rather than waste space by defining a fixed size binary field, the V format
represents a byte array of variable length. Numbers smaller than 128 are held in

one byte. When the top bit of a byte is one, the remaining seven bits specify a multiple of 128, and the following byte the remainder.

Table 6.13: Raster format types

Format	Description
LONG	Four-byte binary integer
REAL	Four-byte binary real (IEEE format)
V	Variable length byte array (see below)
ATTR	Attribute value, formatted according to the header
Rn	Real number in ASCII, using an n-char field (left justified with trailing spaces e.g. "12345.12 " is an example of R12)
In	Integer in ASCII using an n-char field (left justified with trailing spaces)
An	n-char ASCII text, left-justified with trailing spaces
A*	NULL-terminated ASCII text, left-justified

Thus, for example (remembering that the top bit of a byte is 0x80)

Numbers from 0 to 127 require one byte

128 is held in two bytes as (0x80 +1, 0)

129 is held in two bytes as (0x80 +1, 1)

256 is held in two bytes as (0x80 +2, 0)

and (0x80+2, 0x80+5, 32) is decoded as $((2 \times 128) + 5) \times 128 + 32 = 33440$. A consequence of using V format is that the number of transitions in a scan-line is not necessarily a simple function of the number of bytes.

6.8.2 Raster File Header

The format of the raster file header is shown in Table 6.14.

Table 6.14: Raster file header

Field Name	Format	Description
RAS_IENDIAN	LONG	"Magic" integer number to allow recognition of mismatched endianism. Set to 123456789 at file creation.
RAS_RENDIAN	REAL	"Magic" real number (IEEE format) to allow recognition of mismatched endianism. Set to 123.456
RAS_HDR_LEN	LONG	Length of header in bytes
RAS_VERSION	R12	Version of raster interface used
RAS_BND_X1	R12	Real-world coordinates at *top-left*
RAS_BND_Y1	R12	of first cell of first scan-line
RAS_BND_X2	R12	Real-world coordinates at *bottom-right*
RAS_BND_Y2	R12	of last cell of last scan-line
RAS_N_COLS	I10	Number of cells per line
RAS_N_LIN	I10	Number of scan-lines in raster
RAS_ATTR_NAME	A*	Name of attribute Corresponds directly to BSi NTF ATT_NAME field in ATTDESC record
RAS_ATTR_TYP	A1	Method of storage of attribute (ATTR):
		I --- binary integer,
		R --- real (IEEE format),
		A --- ASCII,
		V --- variable format.
RAS_ATTR_SIZ	I1	Size of attribute in bytes (ATTR):
		For type R: 4 or 8,
		For type I: 1, 2, 4 or 8,
		For type A: in range 1 -- 8,
		For type V: unused.

Field Name	Format	Description
RAS_TITLE	A10	Name of dataset. Corresponds directly to BSi NTF SECT_REF field in SECHREC record
RAS_COMMENT	A*	Comment for log history etc. Corresponds directly to BSi NTF COMMENT record
RAS_PROJECTION	A*	Projection data corresponding to that in the NTF SECHREC record. This follows the exact ASCII format of the fields GRID_OR_X to PARAMETER_2 in the SECHREC (section 6.5.3)

6.8.3 Run-length Encoding

Following the header, each scan-line has data of the format shown in Table 6.15.

6.8.4 Implications of Run-length Encoding for Designs

The format of the run length encoded data has several implications for the design parallel algorithms as follows:

- The run-length encoding means that the storage occupied by a scan-line is variable.

- Since the scan-lines vary in length, the offset of the start of a scan-line within a file cannot be calculated.

- Since offsets cannot be calculated random access to a scan line is not possible and therefore the file must be read sequentially.

- Simple data distribution schemes such as those available in PUL-RD (Chapter 4) cannot be used since they assume uncompressed matrix storage of data.

- Relatively sophisticated load-balancing techniques must be adopted. The nature of the run-length encoding gives some indication of the variability of data along the scan-lines and therefore provides some estimate of the amount of work involved in processing the scan-line.

- Scan-lines must be written to the file sequentially which forces synchronisation
  constraints on the raster output module.

Table 6.15: Raster scan-line format

Field Name	Format	Description
`LIN_LENGTH`	LONG	Number of bytes used for this scan-line
`LIN_N_TRANS`	LONG	Number of transitions in this scan-line including those at the start and end
`LIN_NUMBER`	LONG	Scan-line number within image
For each of the transitions (i.e. for each new attribute value along the scan-line)		
| `TRN_ATTR`	ATTR	New attribute value (the format is as specified in the header)
| `TRN_CELLS`	V	The number of contiguous cells with this attribute value. (Although for the last transition of the scan-line this number is calculable, it is also included)

6.9 Conclusions

This chapter has reviewed some of the range of available interchange formats and
their characteristics to identify the common components and the problems they may
pose for interfacing to parallel algorithms. The vector formats share common
ground at the basic geometric and topological level despite differences in the
approach to modelling real-world objects. The choice of NTF level 4 has allowed
almost the worst-case scenario to be designed into the vector input module due to
the large number of record types. Similarly, the run-length encoded (RLE) raster

format presents problems for raster input which prevent some types of data decomposition strategies but offer potential storage savings. In fact the RLE could also be seen as a worst case since the more complex TIFF format allows a form of random access within a file by use of a strip index.

Most vendors or their agents have convertors to and from NTF at levels 1 to 3. Relatively few have convertors to and from level 4 but this largely reflects the availability of data in this format from the UK Ordnance Survey. Conversion to and from most remote-sensing formats is relatively straightforward.

6.10 References

Albrecht, J. & Kemppainen, H., 1996, *A Framework for Defining the new ISO Standard for Spatial Operators*, GISRUK'96.

Aldus Corporation, 1992, *TIFF^{TM} 6.0*, Aldus Developers Desk, Aldus Corporation, Seattle.

British Columbia Ministry of Environment, Lands and Parks, 1995, *Spatial Archive and Interchange Format: Formal Definition, Release 3.2*, Vancouver: Surveys and Resource Mapping Branch, British Columbia Ministry of Environment, Lands and Parks.

British Standards Institution, 1992, *Electronic Transfer of Geographic Information (NTF) Part 1. Specification for NTF Structures*, British Standards Institution, London.

British Standards Institution, 1992, *Electronic Transfer of Geographic Information (NTF) Part 2. Specification for Implementing Plain NTF*, British Standards Institution, London.

Gatrell, A.C., 1991, Concepts of space and geographical data. In Maquire, D.J., Goodchild, M.F. & Rhind, D.W. (Eds), *Geographical Information Systems: Principles and Applications*, Vol 1, Longman, Harlow, 119-134.

Peuquet, D.J., 1984, A conceptual framework and comparison of spatial data models. *Cartographica*, **21** (4), 66-113. (Also in Pequet, D.J. & Marble, D.F. (Eds), *Introductory Readings in Geographic Information Systems*, Taylor & Francis, London, 250-285.)

United States Geological Survey, 1994, *Spatial Data Transfer Standard. US Federal Information Processing Standard (FIPS) 173*, SDTS Maintenance Authority, United States Geological Survey, Reston.

Welch, T.A., 1984, A Technique for High Performance Data Compression, *IEEE Computer*, vol 17(6).

Worboys, M.F., 1995, *GIS: A Computing Perspective*, Taylor & Francis, London.

7

A Modular Approach to Parallel GIS Algorithm Design

S. Dowers, M.J. Mineter and R.G. Healey

7.1 Introduction

This chapter explores the extent to which the apparent modularity of the exemplar algorithms which were discussed in Chapter 5, and are shown in Figure 7.1, can be realised. The modules aim to separate the I/O and data distribution from the core processing of the algorithm in each case and comprise:

- Vector Input and data distribution
- Raster Input and data distribution
- Operation
- Vector collation and Output
- Raster collation and Output

In addition to common ground in input and output, the modularity requires that decomposition strategy, process synchronisation and process software structure be investigated. Following general comments in section 7.2, the next section explores each of the operations in turn. The functionality of each module is then described in sections 7.4 to 7.7.

7.2 Benefits of Modularity

7.2.1 Reusability

Modules provide a convenient tool to support the reuse of software in the implementation of different algorithms which share some common functional requirement. Reuse can be a relatively straightforward process provided the interface to the module is well defined and sufficiently generic. The ideal tool for reusable code is in the form of a library such as has been described in chapter 4.

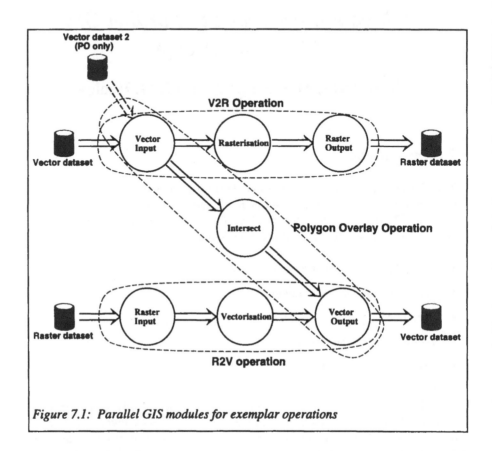

Figure 7.1: Parallel GIS modules for exemplar operations

7.2.2 Encapsulation

The use of modules allows the details of the implementation to be encapsulated within the module. The operation which uses the module need only be concerned with providing the appropriate input and output data to the module to take advantage of it. The module may therefore be optimised without affecting the other modules with which it interacts. Some modules may also act as converters between different data representations and therefore isolate the core operation code from details of the input or output data structures. The resultant effect is that the complexity of dealing with the parallel environment is largely hidden from the designer of an algorithm. This effect is essential in opening out the availability of high performance computing environments to developers who are expert in an application area, but not in parallel processing.

7.2.3 Platform Independence

Encapsulating the implementation details within modules also allows the overall algorithm and large parts of the implementation to be relatively independent of the platform. Much of the independence is provided by the lower layers of software shown in Figure 5.1, which provide message passing (MPI) or utility libraries such as PUL-PF (Chapter 4). The modular approach allows particular aspects to be tuned to suit a particular hardware environment without such changes having knock-on effects within other modules. The operations would be portable to platforms on which the lower layers have been implemented.

7.3 Design Approach

The design approach is constrained by the choice of data formats and the goals identified in Chapter 5 to support data streaming and arbitrarily large datasets. These goals constrain the use of I/O and the assumptions which can be made about occupancy of memory. The approach is also influenced by the choice of decomposition strategy for parallel implementation and by the desire to isolate the parallel infrastructure from the algorithms. Chapter 5 has already identified data decomposition as the appropriate choice and a natural consequence is the partitioning of the problem space into sub-areas: non-spatial partitioning is not appropriate for the classes of algorithms considered here, since all data pertaining to a region of space is required for the algorithms. To avoid unnecessary communication, this implies that each processor holds all the data for a region. The requirements of the chosen algorithm, with performance implications, will then determine the most appropriate shape of sub-areas.

One approach to isolating the algorithm from the parallel infrastructure is to allow the algorithm to operate on each sub-area in a sequential environment defined by the input and output interface to the algorithm. This implies an overall process structure similar to a task-farm (Chapter 4) with enhancements to allow communication between workers when required, for example, to stitch data from sub-areas together in some manner. This strategy also allows the legacy of serial algorithms to be used.

We now consider the extent to which these serial algorithms can be exploited in a parallel implementation with decomposition into sub-areas. The focus is on how the data requirements of alternative algorithms for each operation relate to the data models described in Chapter 6, and consider the corresponding requirements for memory, I/O and communications.

7.3.1 Vector-to-Raster Conversion

Vector to raster conversion involves assigning the attributes of each polygon in the input dataset to corresponding raster cells.

The conversion algorithms require the coordinates defining the boundary of the polygons and the associated attributes. NTF Level 4 allows these to be associated with each other either by linking from a polygon through a list of edges or by using attributes assigned to the polygons on each side of a boundary. The first approach uses the linkage between area, face, chain, edge and geometry records; the second uses the linkage between area, face, edge and geometry respectively.

The details of some possible algorithms and their implications are summarised briefly here and described in more detail in Chapter 11. Terminology is based on that used by Piwowar *et al.* (1990)

7.3.1.1 *Frame Buffer Algorithm*

The frame buffer algorithm needs all of the raster in memory at one time and operates in a similar manner to the software and hardware engines used to generate raster visual displays from vector data. The input vector data need not be sorted. Since the ordering of the input data is not known, output of the raster data may not commence until all of the vector input has been processed.

7.3.1.2 *Image Strip Algorithm*

The image strip algorithm needs vector data grouped into blocks of adjacent scan-lines known as strips. This is equivalent to bucket-sorting of the edges prior to the operation. Edges within a strip may be processed in any order and the allocation of attributes to pixels is carried out in the same manner as the frame buffer algorithm. Output of the raster data for a strip may start once all vector data for that strip has been processed. However the sequence of output of the strips may be constrained by the raster output format. If run-length encoding is used then the strips must be output in order, since the position of a particular record in the output data stream cannot be calculated.

7.3.1.3 *Scan-line Algorithm*

The scan-line algorithm is equivalent to the image strip algorithm when the strip size is reduced to one scan-line. The vector data must be sorted in one dimension to determine the scan-line on which it first appears. Since raster data, as used in remote-sensing, image-processing and other applications, is normally stored in scan-lines running horizontally and starting at the top, the most natural ordering for the input vector data is based on the maximum y-coordinate. A further refinement which simplifies processing involves splitting geoms at turning points to produce monotonic segments; i.e. the slope of the line always stays in the same quadrant within the segment. This simplifies the processing of pixels which are covered by more than one polygon. Since the scan-lines are processed in the correct order for writing to the output dataset, no special restrictions are imposed by the raster output format provided the load balancing between processes is even and run-length encoding is available naturally.

The constraints of memory utilisation, together with the poor handling of raster cells which overlap more than one polygon, preclude the use of the frame buffer algorithm. The use of bucket sort in the image strip algorithm implies a data partitioning scheme which is decided before the data is read and therefore cannot take into account variable data distribution unless the input data are fully sorted. In this case the requirements of the image strip and scan-line algorithm are virtually identical but the scan-line algorithm is simpler to implement. The sub-areas used for data-decomposition can be variable numbers of scan-lines, depending on the spatial distribution of the data across the area of interest.

The choice of the scan-line algorithm requires the following components from the vector data model as input:

- Edge records with left face, right face and geometry identifiers

- Geometry records

- Areas with attribute and face identifiers

- Attribute records.

Output from this operation is a raster data set in run-length encoded format.

7.3.2 Vector Polygon Overlay

Polygon overlay involves the intersection of the boundaries from one set of polygons with the boundaries of a second set of polygons to produce a third set. Each polygon in the output set is related to one polygon in each of the input sets. The attributes of the output polygons can therefore be derived directly from the attributes of the corresponding input polygons.

A summary of the algorithm and the implications are presented below and are described in more detail in Chapter 13.

7.3.2.1 Plane Sweep Algorithm

The plane sweep algorithm needs vector data representing the boundaries between pairs of polygons with the associated attributes for the polygons on each side of the boundary. The efficiency is improved if the boundaries are divided into monotonic segments sorted in one dimension. This property is used to optimise the line-intersection test which is at the core of the overlay operation by reducing the number of comparisons between line segments. Data decomposition may be based on rectangular sub-areas in either or both dimensions. More detailed analysis, outlined in Chapter 13, identifies one-dimensional decomposition orthogonal to the sweep line as the most satisfactory choice. Vector output data is derived independently for each sub-area and therefore some reconstruction of objects which span more than one sub-area is required.

Since the vector output format is heavily reliant on identifiers to relate the various types of records to each other, a parallel implementation must have a mechanism for the allocation of these ids which is consistent across all the processes involved. The vector output format implies no ordering in the records following the headers and therefore the algorithm is not constrained by sequencing within the output format *per se*. However, it is constrained by the time at which the various identifiers in each of the records become known.

This operation requires the following components from the vector data model as input:

- Edge records with left face, right face and geometry identifiers

- Geometry records

- Area records with attribute and face identifiers

- Attribute records.

Output from this operation comprises vector data with the following records:

- Edge records with left face, right face and geometry identifiers

- Geometry records

- Chain records with edge identifiers

- Face records with chain identifiers

- Area records with face and attribute identifiers

- Attribute records.

7.3.3 Raster-to-Vector Conversion

Raster-to-vector conversion involves the identification of regions of contiguous cells which have the same attribute value in order that a vector topological description of the corresponding polygons can be derived.

The details of the algorithm and its implications are described in Chapter 12. Issues associated with parallelising standard serial algorithms for the conversion are previewed below, with particular reference to the consequences of processing arbitrarily large datasets.

7.3.3.1 Polygon Cycling Algorithms

Polygon cycling algorithms operate by following the edges of regions of pixels of the same value. From this it follows that the complete raster must be available in memory at one time and that output is generated for each region as it is completed. Identification of edges between polygons is not readily available. The memory requirements alone rule out this type of algorithm.

7.3.3.2 Edge Stepping Algorithms

Edge stepping algorithms allow the implementation to make a single pass through the input raster data. The extraction of the boundary segments may thus be made on a small section of the raster. However the reconstruction of topology is performed by traversing the boundary segments in vector format that infers that the complete vector representation must be available in memory. Output of the vector cannot start until all raster data has been read and processed. The memory requirements also preclude this approach.

7.3.3.3 Boundary Linking Algorithms

Boundary linking algorithms allow processing of the raster data in scan-line order but allow boundary extraction, topology construction and line simplification to be carried out 'on the fly'. The algorithms also generate nodes, edges and chains. This provides a natural mapping to the vector data model described in Chapter 6, allowing vector data to be derived independently for each sub-area of the data set. 'Stitching' of the topology for each sub-area is therefore necessary.

The chosen boundary linking algorithm requires raster input data, preferably run-length encoded, in scan-line order for rectangular sub-areas of the data set. Output from this operation includes the following records:

- Edge records with left face, right face and geometry identifiers

- Geometry records

- Chain records with edge identifiers

- Face records with chain identifiers

- Area records with face and attribute identifiers

- Attribute records.

7.3.4 Raster to Raster Operations

Raster to raster operations have not been formally considered since parallel paradigms already exist for this type of operation, e.g. PUL-RD and PUL-TF. Chapter 10 provides an overview of the techniques available for managing raster data in a parallel environment.

7.4 Vector Input Module

7.4.1 Conversion Interface

The vector input module provides an interface between the external representation provided by another system and the internal representation used by operations. The data required by the vector-raster conversion and vector polygon overlay operations are relatively simple compared with the highly normalised and disaggregated data model used in BSi NTF Level 4. They require vector co-ordinate strings with left and right attributes attached, sorted spatially in one dimension. However, the superficial simplicity is misleading, not least because the input module can make no assumptions about ordering of the data in the input dataset.

This design is one instance of a general class of vector input modules which have a common output representation, but which may have alternative input representations. For example, there could be alternative designs to interface directly with particular proprietary GIS formats. In most cases these will be simplifications of the design presented here and in Chapter 8.

7.4.2 Vector Data Join

The nature of the NTF model implies that the vector input module must build the links between the various elements present in the dataset such as GEOMETRY and EDGE and between EDGE and FACE using the identifiers present in the corresponding records. These identifiers act as primary and foreign keys in database terms and the linking operation is known as a 'join' (Date, 1990). The complete dataset is involved in the joins and the data is undergoing conversion during the processing. The design therefore uses a sort-merge algorithm, a classic technique which is commonly used to carry out join operations in database implementations.

One of the goals specified in Chapter 5 was to deal with datasets which are too large to fit into the main memory of all the processors in the machine, and to provide graceful degradation as the dataset sizes increase. The input module could form a potential sequential bottleneck in the operation since it involves substantial disk I/O. The design has been chosen to optimise the use of the disks to obtain maximum throughput.

7.4.3 Use of External Storage Devices

Although the performance and capacity of disks have improved over recent years the rate of increase falls a long way short of the performance curve for processing power, which doubles approximately every 18 months. Disk I/O performance will always be constrained by the physical limitations of the mechanical components such as the rotating platter and the seek mechanism for the heads. Currently available disks are capable of sustaining transfer rates of 12 Mbytes per second while the burst transfer rate over the interface may be up to 40 Mbytes per second. In contrast, the average seek time of a high performance disk is around 8 milliseconds with an additional delay due to rotational latency of around 4 milliseconds. To put these figures into perspective, a geometry record consisting of 20 points could be transferred in approximately 20 microseconds while it could take 10 milliseconds on average to find the record; this is 500 times slower. Caching techniques can be used to improve apparent disk performance, but even a hit rate of 95% would only reduce the average access time to 500 microseconds. To obtain optimum performance it is necessary to ensure that almost all I/O to the disk is sequential rather than random access. This basic principle holds true when a parallel file system is used. Although the effective transfer rate is improved, the seek-time and latency are still constrained by physical devices.

7.4.4 Data Partitioning

The accumulation of statistics on the spatial distribution of the data is an important by-product of the sort/merge process. This is used to estimate the best way to decompose the data to achieve a balanced load across all the worker processes. The partitioning is carried out in the y-dimension of the dataset. Each task sent to a worker is a strip across the full width of the input dataset. The choice of partitioning in the x-dimension or the y-dimension is explored further in Chapter 13.

7.4.5 Minimising Communication between Workers

The Vector Input module is responsible for ensuring that each worker is sent only the data which is relevant to the area it is processing. It achieves this by defining the limits of the sub-area to be processed by a worker (Chapters 8 and 11). The geometry and attribute data for edges which span more than one sub-area will be replicated and sent to each of the corresponding workers. This approach minimises the amount of communication necessary between workers to deal with input data and processing since each worker should have all the data it needs: without replication of these data, workers would need to exchange sub-area border information.

7.5 Vector Output Module

7.5.1 Split Objects

This module encapsulates functions which build the parts of topological objects which are within a sub-area, functions to stitch the topology, and functions to collate and write out records. The abbreviation 'TSO' is used: Topology building, Stitching and Output. Due to decomposition into sub-areas, data for a single element may be held in several worker processes requiring 'stitching': geoms, edges, chains and faces may all be split across several processes, and the relationships between these must be derived, and data for the various records of the vector data format must be collated.

7.5.2 Identifiers

The normalised NTF Level 4 format requires unique identifiers as primary keys for each record to be output. Since these identifiers are also used as foreign keys in other records, there is a substantial level of interdependence between types of records. For example, an edge record cannot be written until the identifiers for the faces on the left and right, for the nodes at either end and for the corresponding geometry record are known.

Many processes may be requesting identifiers at the same time. The identifiers for each type of record should be drawn from a consecutive set of integers. A separate identifier generator process is required to ensure this requirement is met. Various optimisation techniques are available to avoid this generator process becoming a bottleneck.

7.5.3 TSO Framework

Stitching is accomplished by exchange of data between worker processes, collation of data for a record entails the cooperation of workers holding parts of that object, and identifier allocation is achieved by an exchange of messages with an identifier generator process. A worker can therefore receive several types of messages, so that these have to be prioritised to minimise delays in other processes. This leads to a multi-threaded approach using an event loop to respond to messages. As a result, the interaction with the operation code is more sophisticated than simple function calls and requires the vector output module (TSO, Chapter 9) to control the overall operation of a worker via the event loop.

The encapsulation of the complexity of the communication and associated processing below an interface to the functions which build the topology for a sub-area allows not only the parallelism but also many facets of the topology building to be hidden from the implementor of an algorithm.

7.6 Raster Input Module

Most remotely-sensed and scanned image formats structure data as scan-lines from the top of an image. The format chosen for these designs is similar except that each scan-line is compressed using run-length encoding into a variable length record. Data distribution is therefore naturally in blocks of scan-lines; i.e. one-dimensional partitioning.

Decomposition to reduce the perimeters of the sub-areas and to allow several processors to receive data concurrently can be achieved by two-dimensional partitioning; i.e. by splitting a block of scan-lines into a set of rectangles each the full height of the block (See Chapter 10 for details). The size of the rectangles can be estimated from a sub-set of scan-lines in the block buffered by the input process before distribution. The amount of work involved for a worker can be estimated from the number of *spans*, runs of pixels of the same value, in a scan-line. With run-length encoding as proposed in chapter 6, the number of bytes required to represent a scan-line gives an approximation to the number of spans.

7.7 Raster Output Module

The sequential, ordered nature of the raster format implies that records must be written in a strict order. This ordering may be ensured by several techniques, but normally implies a level of synchronisation between worker processes. A process cannot output its first scan-line until all scan-lines above it have been output by other workers. This strict constraint can be relaxed by using buffering of output, both at the worker process and in the sink or output server process. One of the design aims for the PUL-PF library (Chapter 4) was to provide transparent buffering and reordering facilities. Since the memory requirements of the workers are relatively low, output scan-lines can be queued for output using non-blocking communication until the queue runs out of buffer space. To prevent synchronisation limiting performance and scalability, the data distribution process must allocate tasks of relatively uniform size to each process.

7.8 Conclusions

This chapter has described some aspects of components of three GIS algorithms to identify generic modules which are suitable for reuse in the input and output of raster and vector-topological data. These modules provide for data distribution and data collation in a processing framework which can be used by a variety of operations. The decomposition into sub-areas allows appropriate, existing, sequential algorithms to be exploited in the processing of each sub-area, with necessary extensions to account for the data partitioning. In the case of vector output, the TSO module provides facilities for data set reassembly and an event driven structure which may be tuned to adapt to the constraints of different hardware environments. The input modules are designed to feed an operation module which is spatially coherent, by providing all the data an operation needs for a particular region of space but no more.

The range of operations which could be built upon these modules extends well beyond the scope of those considered in this book e.g. point-in-polygon, topology building from spaghetti. By co-ordinating the division into sub-areas, parallel implementations of algorithms which require both vector and raster data could be developed. An example of this type of operation is boundary assisted image classification.

7.9 References

Date, C.J., 1990, *An Introduction to Database Systems*, Addison-Wesley, Reading, MA.

Piwowar, J.M., LeDrew, E.F. & Dudycha, D.J., 1990, Integration of spatial data in vector and raster formats in a geographical information system environment. *International Journal of Geographical Information Systems*, 4 (4), 429-444.

8

Parallel Vector Data Input

T.M. Sloan and S. Dowers

8.1 Introduction

This chapter describes an algorithm for the input of vector-topological data to parallel GIS operations such as vector-polygon overlay and vector-to-raster conversion of non-overlapping classified polygons. The algorithm exploits parallel I/O facilities and traditional merge sort methods to process data efficiently into a format suitable for a GIS operation. It also gathers information on data distribution within datasets which is used to load balance the parallel GIS operation.

A geographical feature is generally represented within a GIS by one or more attributes describing the feature, or some aspect of it, and a spatial description of the extent and location of the feature in space. One of the principal characteristics of those GIS which use hybrid vector-topological data models is that the spatial description of a geographical feature is not stored with its associated attribute information. Instead, unique identifiers relate the different characteristics of a feature in a non-redundant manner. That is, the geographical feature is represented by a record containing the unique identifiers of other records which in turn contain or refer to records holding the spatial and attribute information. Models of this kind have been examined in detail in Chapter 6.

The result of this division of attribute and spatial information is that for any particular geographical feature a number of relationships between different records and record types must be followed to process the spatial and/or attribute information in a GIS operation. In addition, this information may have to be sorted into a format suitable for the algorithm used by the operation. This extraction and sorting of data is a considerable task, requiring significant I/O and processor time in any computing environment.

In parallel GIS operations, there is an additional overhead since the input data must also be distributed to parallel processes. Moreover, for such an operation to be performed efficiently, the data must be distributed to the parallel processes in packets of comparable workload. This is to prevent any one process being swamped with too much work and so holding up the operation (Chapter 4). Further, in a parallel environment where extra compute power increases the

likelihood of processing very large datasets, the potential performance impact of
the extraction and sorting phase is compounded.

Therefore, the issues concerning the input of vector-topological data to a GIS
operation are involved primarily with the effect of indirection on the retrieval of
data, and the sorting and pre-processing of data into a format and sequence suitable
for an algorithm. In parallel GIS operations, these issues are compounded by the
need to distribute data to parallel processes and to ensure equity of load across
these processes.

In section 8.2, some aspects of vector-topological data format BSi NTF Level 4 are
described and used to illustrate the relationships between spatial and attribute
information in such formats, with particular reference to the requirements for the
vector polygon overlay and vector-to-raster conversion operations. In section 8.3
the parallel sorting and merging of data are described. Section 8.4 provides an
overview of the algorithm, with section 8.5 detailing the process structure and
section 8.6 describing the various phases of the algorithm in more detail. Finally
section 8.7 discusses the limitations of the algorithm.

8.1.1 Vector Input Requirements

This section considers the vector input requirements for the vector polygon overlay
and vector to raster conversion operations.

The data requirements common to both operations are

- the spatial coordinates of the edges forming the polygon boundaries with the
  corresponding attributes of the polygons on their left and right sides. That is, in
  NTF Level 4 terms, the GEOMETRY1 records and the attribute values of the
  FACE records on their left and right sides.

- edges sorted by their uppermost y-coordinate.

The reasons for these requirements are explained below.

8.1.1.1 Geometry with Left/Right Attributes

In GIS operations such as polygon overlay (see Chapter 13), where attribute values
and the spatial description of geographical features are manipulated, it is the
coordinates of the polygon boundaries and the left and right attribute values at these
boundaries that are required by the underlying algorithms. In polygon overlay the
polygon boundary coordinates are required for the intersection tests which
determine the extent of the new polygons and it is the left and right attribute values

of the polygon boundaries which determine the attribute values of the newly created polygons.

In a similar way, in vector-to-raster conversion of non-overlapping polygonal areas (see Chapter 11) (which can be viewed as a polygon overlay where one dataset is a blank grid), it is the polygon boundaries and their left and right attribute values which are required to determine the attribute values of the raster cells.

Similarly, overlay operations involving linear datasets require the coordinates of the lines and the attribute values for each line. Operations involving point datasets require the coordinates of the points and the attribute values for each point.

8.1.1.2 Sorted Spatial Information

Line intersection tests are required in operations such as polygon overlay, topology creation and buffering. Some of the techniques used to improve line intersection algorithm performance require spatially sorted data. Other operations such as vector-to-raster conversion also require sorted spatial information if they utilize image strip or scan-line type algorithms described in Chapter 11 (Piwowar *et al.*, 1990; Harding *et al.*, 1993b). Sorted spatial data improves the efficiency of the algorithms. In general, any algorithm which processes set of objects which touch or interact spatially will benefit from sorting spatially in this way.

8.2 Input Data

In this chapter BSi NTF Level 4 is used to illustrate the issues concerning the retrieval and processing of spatial and attribute information in vector-topological data formats and to describe an algorithm for the input of GIS vector data into parallel GIS operations. The format is fully described in Chapter 6 but, by way of a reminder, its salient features are summarised in section 8.2.1.

For any particular GIS operation not all records are required. Sections 8.2.2 and 8.2.3 detail the records and fields which must be extracted and sorted from NTF Level 4 datasets for the polygon overlay and vector to raster conversion operations. Section 8.2.4 explains how the contents of the required fields are extracted and the initial processing that is necessary. Section 8.3.1 explains how the extracted information is sorted into a suitable format for these operations.

8.2.1 BSi NTF Level 4 Representation

A simple geographic feature is represented by an AREA, LINE or POINT record.

A COLLECT record is used to list a number of AREA, LINE or POINT records which form a more complex geographical feature.

The AREA, LINE and POINT records contain a list of identifiers for ATTREC records and a list of identifiers for topological records. The ATTREC records contain the values of particular attributes of the feature. An attribute and the format of its value are described by an ATTDESC record. The topological records define the relationships between the spatial components that form a geographic feature. The topological records are the FACEREC, EDGEREC, NODE and CHAIN records. In turn these topological records contain the identifiers of geometrical records which contain the coordinates of the spatial components and hence the location and extent of the feature.

8.2.2 Feature/Geometry/Attribute Relationships

The relationships between a geographic feature and the records containing its spatial location and attributes are best illustrated by use of the example shown in Figure 8.1. NTF Level 4 contains a greater deal of indirection between the spatial and attribute information than most vector-topological data models. However, it still usefully illustrates the point that the spatial and attribute information in many vector-topological data models is stored separately.

From Figure 8.1 it can be seen that the attribute values associated with an area can be found relatively directly through the ATT_ID field. However, the route between boundary of the area and the attribute value on either side is more convoluted, involving the links between faces and areas in the AREA record. From a face there are two possible routes to the GEOMETRY records defining the edges. The first involves following the link from the FACE record through the CHAIN record to the EDGE records and thence to the GEOMETRY records. The second possible route matches the FACE_IDs held in both the EDGE and AREA records to allow the indirect relationship between ATTREC and the GEOMETRY records to be established. (AREA records include both FACE and ATTREC IDs; EDGE records include both FACE and GEOMETRY IDs). The second route is the one which is used in this design.

The VALUE of an attribute associated with the LFACE_ID or the RFACE_ID of an EDGEREC is found via the AREA record. The ATT_IDs of ATTRECs (where attribute values are held) are stored in AREA records along with FACE_IDs. Where a LFACE_ID or RFACE_ID matches a FACE_ID in an AREA record and the sign of that FACE_ID is positive, (+), then the ATT_IDs in that AREA record refer to the ATTRECs containing the attribute values associated with that LFACE_ID or RFACE_ID. When the FACE_ID is negative (-), it represents a hole within the AREA and the attributes are ignored.

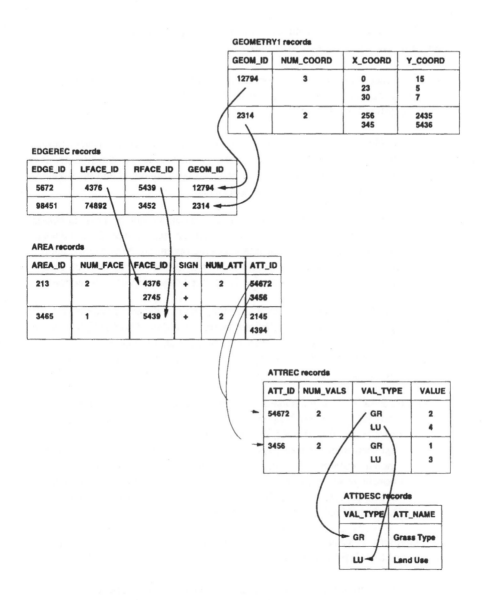

Figure 8.1: Relationship between relevant NTF records

8.2.3 Which Records and Fields are Required

A number of details are required from the header records of an NTF Level 4 dataset. Some of these, such as the NTF level and version in the Volume Header, are for checking the dataset is in the correct format, whilst others, such as the storage method of attribute values or the coordinate multiplication factor, are required for the operations to work correctly.

The Header records from which information is extracted, are listed here along with a brief explanation of some of the information extracted. Most of these records are more fully defined in Chapter 6.

> **Volume Header** An NTF transfer set starts with this record. It identifies the contents of the transfer set and the NTF level used for data storage. It is also examined to determine the end of record character and the divider character used for variable length fields.

> **Database Header** Each NTF transfer set can contain one or more databases which start with one of these records. In this record only the contents of the data model field are of interest to the operations to verify the data model. The algorithms described in this chapter assume the first database in an NTF transfer set contains the desired dataset in its first section.

> **Attribute Description** For each attribute type in a database in an NTF transfer set, there is an Attribute Description record (ATTDESC) which describes it and how its values are stored in an attribute record. Therefore, the ATTDESC records must be examined to determine how to extract the relevant attribute values from the attribute records.

> **Section Header** Each database contains one or more sections. Each section starts with a Section Header record. In the context of the algorithms described in this chapter, the dataset of interest is assumed to occur in the first section of the transfer set. The Section Header record is required since it contains the details of the bounding area of the dataset, the coordinate scale, projection and transformation. The end of the data in a section occurs when another Section Header record is read or when a Database Header or Volume Termination record is read.

In this chapter it is assumed that the header records described above all occur at the beginning of the NTF transfer set.

A number of different types of records can exist in a section. However, only the ones listed below are of interest to the operations. No ordering is specified in the standard for these records within a section. These record types are more fully defined in Chapter 6.

> **Area** The Area records (AREA) contain the identifiers of the Face and Attribute records (FACEREC and ATTREC) which describe the spatial location and attribute values of the polygonal geographical features in a section.

Attribute These records contain the attribute values of the geographical features in a section.

Edge Each face is bounded by a number of edges. Each edge is described by an Edge record (EDGEREC) which contains the identifiers of the FACEREC records on its left and right sides and the identifier of a 2D Geometry record (GEOMETRY1).

2D Geometry This record contains a string of coordinates. It is also known as the GEOMETRY1 record. Only these records which exist wholly or partly within the area of interest to an operation are required.

The remaining record types in a section, including FACEREC, CHAIN, and NODEREC records, can be discarded since all the necessary information for the operations is stored in the records listed above.

8.2.4 How to Extract Fields

The records in an NTF Level 4 plain text dataset are stored as ASCII strings in a file. For most records it is therefore a simple matter of reading these records according to their definition in the standard to extract the relevant field and converting from ASCII into an appropriate internal format. However, for some fields, their definition is actually defined in the header records of the dataset itself. For example, ATTDESC records contain the definition of the storage of the attribute values in a dataset, while the number of digits in a coordinate is specified in the section header. The coordinate fields also have to be transformed using the origin and scaling information from the section header.

8.3 Data Sorting and Merging

The polygon overlay and vector-to-raster conversion operations require the spatial coordinates of the boundaries of the features and the attribute values on the left and right side of these boundaries. That is, they require the GEOMETRY1 records with their corresponding left and right attribute values. In addition, as outlined in section 8.3.2, both operations require that the spatial information be sorted in the y-direction. The level of indirection implies that several join operations are required to relate the data correctly.

Since attribute values are not stored with the GEOMETRY1 records, it means creating a list of attribute values for each FACE_ID from the AREA and ATTREC records for the attributes of interest. Then, for each EDGEREC, the left and right FACE_IDs have to be used to extract the left and right attribute values from this list. The GEOM_ID in each EDGEREC can then be used to match these left and right attribute values to the appropriate GEOMETRY1. These GEOMETRY1 records and left/right attribute values can then be sorted in the *y*-direction ready for the operations. Sorting the records by *ymax* value ensures that the attribute values will appear at a worker process in the same sequence as the GEOMETRY1 records.

Dealing with very large datasets, as specified in Chapter 5, implies that there will be no complete set of data in memory and therefore records will be held in files on disk. This therefore requires significant random I/O in locating the attribute values in the list and also much repetition, since the same FACEREC will be referenced by a number of EDGERECs. If this was to be performed in parallel then the random I/O would be spread across a number of processes, all effectively searching through the records a number of times. This is not very efficient and would affect any performance gain from parallelisation.

The parallel vector data input algorithm described later in this chapter uses a more efficient method of sorting the data to minimise the random I/O generated when matching and combining records. It also produces data sorted in the *y*-direction. The method goes through a number of sorting and record merging stages which are described below.

1.   Stage 1 - Data Extraction and Initial Sorting
        The processing for each type of record is described separately.

        a)   The ATTDESC records for those attributes of interest to the
                calling GIS operation are extracted and used to determine how
                the values of these attributes are stored in the ATTREC records.
                The values of these attributes are extracted from the ATTREC
                records and stored in a structure, known as Attribute Values,
                consisting of the value itself and the ATT_ID of the ATTREC
                from which the value was extracted. When more than one
                attribute value is to be found, the structure also contains a field
                indicating to which attribute the value corresponds. These
                structures are sorted by ATT_ID.

        b)   For every FACE_ID and every ATT_ID in an AREA record, a
                structure consisting of a FACE_ID/ATT_ID pair is created,
                known as FaceAttributes. They are sorted by ATT_ID.

        c)   Each EDGEREC is split into two structures, with one consisting
                of the left FACE_ID and the GEOM_ID and the other containing
                the right FACE_ID and the GEOM_ID. These structures are
                sorted by GEOM_ID and are known as GeomFaces.

d)   Each GEOMETRY1 record which lies wholly or partly within the area of interest is extracted. These records are split into segments which are monotonic in *y* and sorted by the uppermost *y*-coordinate. Monotonic in *y* means that segments attain their maximum and minimum *y*-coordinates at their end-points. This reduces the number of intersection tests in polygon overlay. In vector-to-raster conversion, it ensures that geometry records do not re-enter a raster cell, thus reducing the amount of work involved in the conversion. To be able to match these records with their corresponding left/right attribute values, a GeomYMax structure is created which contains the GEOM_ID and the lowermost and uppermost *y*-coordinate. The uppermost and lowermost *y*-coordinates are necessary to deal with those cases where a GEOMETRY1 record has been split in two or more pieces, for example where one leaves and re-enters the area of interest. This ensures the left/right attribute values are matched up with all the pieces of a split GEOMETRY1 record. The GeomYMax structures are sorted by GEOM_ID.

2.   Stage 2 - Record Merging and Sorting 1

a)   The FaceAttributes and the AttributeValues, both sorted by ATT_ID in Stage 1, are merged to produce new structures, each consisting of a FACE_ID and its associated attribute values. This new structure is known as a FaceValue. When there is more than one attribute of interest, the field indicating which attribute each value refers to are also included. When the merging is complete, the resultant structures are sorted by FACE_ID.

b)   The GeomYMax and GeomFaces, both of which are sorted by GEOM_ID, are matched and combined to form GeomFaceYMax structures. These are then sorted by FACE_ID.

3.   Stage 3 Record Merging and Sorting 2

a)   The FaceValues and the GeomFaceYMaxs, both of which are sorted by FACE_ID, can now be merged to create a GEOM_ID with its left or right attribute value and uppermost and lowermost *y*-coordinates. These structures are known as GeomValueYMax and are sorted by uppermost *y*-coordinate.

4.   Stage 4 Final Record Merging

a)   The sorted GeomValueYMaxs can now be matched with their corresponding sorted GEOMETRY1 records to produce finally

the spatial coordinates of the geometry records with their
left/right attribute values.

This method removes almost all random I/O. Since all EDGERECs which
reference the same FACE_ID are grouped together, it also removes the repetition
overhead. Using this method parallel processes can also sort and merge the records
without interfering with each other's access to the files on the disk holding the
records, thereby further improving performance.

8.4 Design Overview

One of the prime considerations in the design of this algorithm was the ability for it
to deal with datasets larger than the total memory available on the host machine.
To do this the algorithm makes use of parallel I/O techniques to produce blocks of
data, sorted within the block, which are then merge-sorted to complete the sort.

The algorithm could be modified to deal with linear and point features and to deal
with other vector-topological data formats where attributes are similarly stored
separately from the spatial information.

8.4.1 Phases of Vector Input

The algorithm uses the method described in section 8.5.3 to extract and merge-sort
data efficiently from one or more NTF Level 4 datasets. It proceeds through three
phases: Sort, Join and Geom-Attribute Distribution. The Sort phase corresponds to
Stage 1 described in section 8.4.3, the Join phase to Stages 2 and 3, and the Geom-
Attribute Distribution phase to Stage 4.

Figure 8.2 illustrates the data flow between the three phases, each of which is
described further below.

   Sort Using the ATTDESC records for the attributes of interest, the relevant
        GEOMETRY1, AREA, ATTREC and EDGEREC records are extracted
        from the input datasets, processed and sorted. The output from this phase
        comprises various temporary files containing the processed and sorted
        contents of these records. In addition, information on data distribution is
        produced.

   Join The temporary files derived from the AREA, ATTREC and EDGEREC
        records are merge-sorted to derive the left and right attribute values for
        the geometry records. There is one Sort and Join phase per input dataset.

The output from Join is a temporary file containing the left/right attribute values for the geometry records.

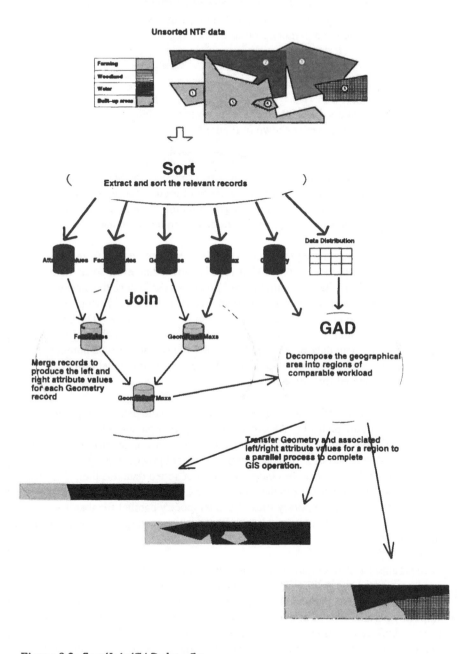

Figure 8.2: Sort/Join/GAD data flow

Geom-Attribute Distribution (GAD) The data distribution information generated by Sort is used to decompose spatially the geometry records generated by Sort, and their associated left/right attribute values generated by Join, into regions of comparable workload. Each region is distributed to a parallel process executing a parallel GIS operation.

8.4.2 Process Structure

The algorithm requires different processes to perform different roles in each of the Sort, Join and GAD phases as shown in Figure 8.3.

During the Sort and Join phases these processes are split into two different groups, the Source and Pool Groups. There is only one process in the Source group which coordinates the actions of the processes in the Pool groups during the Sort and Join phases and gathers the information on data distribution. The Pool processes perform the extraction and merge-sorting of data from the input datasets.

During the Sort phase, Pool processes extract, sort and output data from an input dataset concurrently. Jackson (1995) has clearly shown that, even when the input dataset is a serial file, having multiple Pool processes can reduce the elapsed time to sort a dataset. However, with greater than a certain number of Pool processes, the elapsed time no longer decreases and instead starts to degrade. This is the point at which data can no longer be extracted quickly enough to keep all Pool processes occupied. If the input dataset is instead a parallel file spread across a number of disks, each served by a separate processor, then using more parallel processes decreases the elapsed time further.

The vector input algorithm is designed to be able to handle datasets larger than the total memory of the platform. As a result the temporary parallel files output by the Sort phase are not completely sorted. Instead, they each contain *atoms* of sorted data. Each atom contains sorted data output by a Pool process during the Sort phase. When a Pool process has output an atom of sorted data it is then able to read and sort more records from the input dataset. These records can then be output as atoms in the temporary files. Each temporary parallel file therefore contains a number of atoms output by the Pool processes where the data held within each atom is sorted.

Since parallel files are output by the Sort phase, the Join phase can use parallel processes to read and process data more quickly in a similar way to the Sort phase. The Join phase uses a parallel merge-tree sort to merge the contents of the temporary parallel files to derive the left and right attribute values for the geometry records. Different sections of the merge-tree are located on different processes, so spreading the load of reading and processing the data. A temporary parallel file containing the left and attribute values for the geometry records is the final output of the Join phase; it also consists of atoms where the data within each atom is sorted. The structure of such parallel files and the utilities necessary to support them are described in Chapter 4.

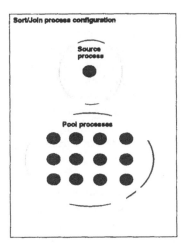 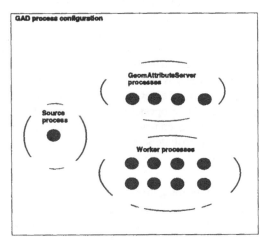

Figure 8.3: Process configuration

During the GAD phase the Source continues to coordinate the actions of the Pool processes; however, here the Pool processes split into two groups, the GeomAttributeServer group and the Worker group. The Worker processes perform the parallel GIS operation on the data for a geographic region sent to them by the GeomAttributeServer processes. This aspect of the Worker processes is explored in more detail in Chapters 11 and 13. The Source process uses the data distribution information generated during the Sort phase to determine regions of comparable workload. It sends the details of a region to the GeomAttributeServer processes along with the identifier of an idle Worker process. The GeomAttributeServer processes use this information to read the geometry and corresponding left/right attribute values from the temporary parallel files for the specified region. These data are then sent to the idle Worker process.

The GeomAttributeServer processes are used to read the data from temporary parallel files for a variety of reasons. As stated in section 8.3, both operations require spatially sorted geometry with left/right attribute values. The temporary file output by the Join phase is only partially sorted since only within each atom are the left/right attribute values sorted spatially. Similarly, the temporary file containing the geometry information generated by the Sort phase is only partially sorted. Only within each atom are the geometry structures sorted spatially. If Worker processes were to read the geometry and left/right attribute values from these parallel files, then each process would have to access every atom, reading each record until it came to the first that was outside the lower bound of its region. This would be mean many processes reading the same records and so interfering with each other.

A more efficient alternative is to further partially sort the files and distribute the data to the Workers to finish the sorting. One or more GeomAttributeServer processes each deal with access to the atoms from those particular parts of the parallel temporary files located on their local disk. Each GeomAttributeServer merges the data from its atoms and distributes the data for a region to a specified Worker. Each Worker receives all the data for the region from each GeomAttributeServer and is then able to complete the merging before carrying out the parallel GIS operation. In fact, the worker can start processing as soon as merged data becomes available provided the priority is given to receiving data from the GeomAttributeServers to prevent the flow of data to other workers from being delayed.

It should be noted that even if all the Worker processes were to read the geometry and left/right attribute values from two completely sorted serial files, this results in random I/O access to these files. The Worker processes would be trying to read from different parts of the file at the same time. This is inefficient and would result in Worker processes sitting idle waiting to read data. If the files were completely sorted parallel files then the random I/O could be reduced since processes would be wishing to access different parts of the file. However, to completely sort the files prior to the Workers reading them adds to the run-time of the algorithm and workers would initially have to determine where in the file the relevant data for their region begins which would also add an overhead. Having the GeomAttributeServer processes means the sorting and processing can be overlapped and the Workers no longer have to read and locate data in the temporary files.

A minimum of three processes are required during the GAD phase, a Source, GeomAttributeServer and Worker process. The Source process determines the number of GeomAttributeServer processes based on how the geometry and left/right attribute values generated from the Join phase are stored in a parallel file spread across a number of disks. Essentially there is one GeomAttributeServer for each processor which is connected to a disk which contains part of the parallel files. This removes any interference between GeomAttributeServer processes.

8.4.3 Use of Library Software

The algorithm makes use of library functions provided by the PUL Key Technology Programmes (KTPs) at the Edinburgh Parallel Computing Centre

(EPCC). This section briefly describes the PUL-PF utility which is relevant to the algorithm.

PUL-Parallel File This utility provides support for parallel file access. PUL-PF considers a parallel file to be composed of a sequence of user-defined *atoms*, which may be fixed or variable length. The atoms are distributed between the disks in a number of possible ways. An *interleaved* file is one in which atoms are distributed so that atom i appears on disk i mod n, where n is the number of disks. A *partitioned* file is one that has been divided into as many contiguous chunks as there are disks, and one of these is placed on each disk. Typically there will be one server process for each physical disk. PUL-PF is particularly useful during the Sort and Join phases when dealing with datasets larger than the core memory of the target platform. It allows I/O by a number of processes to sorted file segments spread across disks to be managed in a coherent and stable fashion.

8.5 Processes

8.5.1 The Source Process

The Source process coordinates the Sort, Join and GAD phases of the algorithm. It initially sends a message containing the information necessary for the Pool processes to extract the required attribute and spatial information from the input dataset(s) for the GIS operation. This information includes details such as the attributes of the geographical feature to be extracted and the bounding box of the area of interest. The Source process extracts this information from the header records of the NTF Level 4 input datasets and from the calling GIS operation. This includes the ATTDESC records which describe the format of the attribute values.

During the Sort phase, each Pool process generates a PartitionList as a by-product. This describes the distribution within the geographical area of interest of the data it extracts and sorts.

The PartitionList contains Partitions, each of which represents a full-width horizontal strip of the input vector dataset. The y-coordinate ranges of these Partitions are of equal size.

Each Partition in a PartitionList contains the total number of Geoms (geometry records) that cross its horizontal strip, the total number of vertices including the first vertex in each Geom over the strip boundary, the total number of Geoms wholly inside the horizontal strip, and the total number of vertices associated with the enclosed Geoms. Each Partition also contains the y-coordinate of its lower boundary.

Figure 8.4 illustrates a Partition in a PartitionList. This Partition contains 7 geometry records which start, end or cross the partition. To define fully the path of these geometry records through this partition, a total of 18 vertices are required. This total includes those vertices which are required more than once since they occur at the junctions of geometry records. It should be noted that some of the vertices in this total do not reside within the partition, for example, consider geom 1, where both the vertices required to define its path lie outside the Partition. It should also be noted that in some cases not all the vertices in a geom are required to define the path of the geom through the Partition. For example consider geom 4, which has five vertices. Only four of these vertices are included in the total vertices in Partition 4. The vertex in Partition 5 which is shared with geom 5 is not required and so is not included in the total.

At the end of each Sort phase the Source collects and merges all these PartitionLists into one to describe the distribution of all the data in the geographical area. There is one merged PartitionList for each input dataset.

After the PartitionLists are collected and merged the Source waits until it receives a message from each Pool process indicating that it has completed both the Sort and Join phases for all input datasets. The Source process must receive these messages before allowing any output file from Join to be read by processes during the GAD phase, (so preventing reading of a file before it has been completely written).

Once the Sort and Join phases are complete for all input datasets, the GAD phase is entered. Here the Source process sends messages to split the Pool process group in two. These messages instruct some processes to become GeomAttributeServer processes and others to become Worker processes. This is based on the number of processors with local disks over which the parallel geometry and left/right attribute value files generated by the Sort/Join phase are spread. For optimal performance, one GeomAttributeServer per processor is necessary since with this configuration there is no competition for disk or CPU resources.

The PartitionList generated by the Sort phase is used to decompose and distribute the geometry and left/right attribute values generated by the Sort and Join phases to Worker processes which perform the parallel GIS operation. The distribution information in the PartitionList allows packets of comparable workload to be generated to assist in load-balancing the GIS operation.

Once all the geometry and left/right attribute values have been distributed and processed, the algorithm is finished.

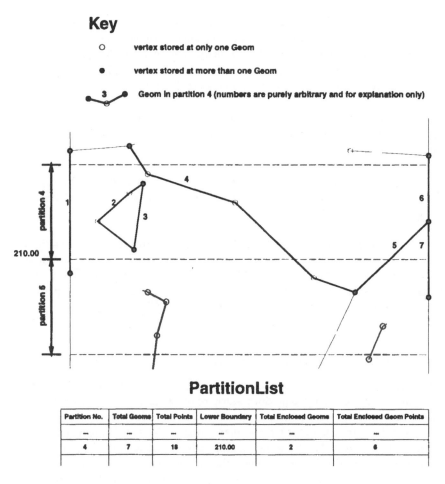

Key

○　　　vertex stored at only one Geom

●　　　vertex stored at more than one Geom

3　　　Geom in partition 4 (numbers are purely arbitrary and for explanation only)

PartitionList

Partition No.	Total Geoms	Total Points	Lower Boundary	Total Enclosed Geoms	Total Enclosed Geom Points
---	---	---	---	---	---
4	7	18	210.00	2	6

Figure 8.4: A partition in a PartitionList

8.5.2 The Pool Processes

All Pool processes initially receive from the Source process a message containing the information necessary to extract the required spatial and attribute information and to perform the desired GIS operation.

For each input dataset, first the Sort phase is initiated. All Pool processes must complete the Sort phase before the Join phase can start. This is to prevent an output file from Sort being read by Join before it has been completely written.

When Sort finishes each process sends its PartitionList to the Source before it enters the Join phase. When there is more than one input dataset then when a Pool process completes the Join phase it can proceed to execute the Sort phase for the next dataset. That is, once each Pool process completes a Join phase it may proceed to Sort dataset n+1 whilst dataset n is still undergoing Join elsewhere.

This procedure of Sort and Join is repeated once per dataset. When all datasets have been processed, a message is sent to the Source by each Pool process indicating completion.

Finally, the GAD phase is entered. Here each Pool process receives a message from the Source process instructing it to become either a GeomAttributeServer process or a Worker process.

A GeomAttributeServer process reads the parallel geometry and left/right attribute value files created by the Sort and Join phases. It distributes the contents of these files to Worker processes based on instructions from the Source process. A Worker process receives the geometry and left/right attribute values for a particular region and performs the remaining components of the GIS operation.

8.5.3 The GeomAttributeServer Process

The GeomAttributeServer processes extract from the Geom and GeomValueYMax files all the data necessary to produce a strip of output data and send them to the specified Worker process as a Geom parcel and GeomValue parcel. Where an input geom crosses one of the strip borders it is split. To avoid rounding error, new vertices are not created at the border but the geom is instead split at the first vertex lying outside the strip. To resolve the ambiguity when a vertex lies on a strip border, the following convention applies: a vertex lying on a Northern border is considered outside the strip (except for the Northern border of the Northern-most strip, which is considered inside that strip); a vertex lying on a Southern border is considered inside the strip.

Each GeomAttributeServer process constructs a local sort merge-tree for each Geom and GeomValueYMax file. The leaves on each tree are attached to atoms in these files with each atom assigned to only one GeomAttributeServer process. The atoms are assigned evenly over all the GeomAttributeServer processes. The local trees are used to merge-sort, by uppermost y-coordinate in descending order, the Geoms and GeomValueYMaxs from the atoms attached to the leaves. The files are merge-sorted in a similar way to the temporary files in the Join phase, except the root of each local tree is not attached to another process to perform a full sort. Instead, each GeomAttributeServer process extracts from the root of its local tree the data which lie within the y-coordinate range specified by the Source process and sends them to the Worker process also specified by the Source process.

Should a partition in the PartitionList(s) consist of too much work for a single worker, the Source process can send a message to the GeomAttributeServer processes instructing them to produce a MiniPartitionList. This describes the data distribution within the over-loaded partition by means of mini-partitions. These mini-partitions are just like the partitions in a full PartitionList but the y-coordinate range of the mini-partitions is a fraction of the full partition. The Source process merges these MiniPartitionLists into one. The Source uses the MiniPartitionList to determine a more appropriate y-coordinate range for a Worker process. In this way the overloaded partition is spread across a number of Worker processes.

8.5.4 The Worker Process

Each Worker process receives one Geom parcel per dataset and one GeomValue parcel per dataset from each GeomAttributeServer process for the current strip. The final sort/join of the GeomValue and Geom records, to produce a stream of geometry records with left/right attributes which are sorted by descending uppermost y-coordinate, is performed on the Worker. If more than one input dataset exists, the data are labeled with dataset number.

Each Geom parcel is only ordered within itself, with no global ordering, so the contents of all the parcels are merge-sorted using a local merge-tree similar to that used on the GeomAttributeServer processes. The ordering of the GeomValue records is not important. Pointers to the GeomValue records are stored in a hash-table, using the GEOM_ID field as the key, for the final join operation. The left/right attribute values for a particular GEOM_ID may be required a number of times, since the original GEOMETRY1 record was split at maxima and minima in y to reduce the overhead of intersection tests. Storing these GEOM_ID and left/right attribute values in a hash-table allows quicker direct access.

The basic task-farm model of GAD involves Worker processes signaling to the Source that they are ready for more work (see Figure 8.10). When the Source allocates a strip to a Worker it sends a strip-descriptor to the Worker. On receipt of this strip-descriptor, the Worker can set up a tree for merge-sorting the geometry records and a hash-table for storing the GeomValues. If more than one input dataset exists, the data from all datasets (with dataset number encoded into GEOM_ID) are merged in one tree and entered into one hash-table.

GeomAttributeServers then send data to the Worker. On arrival of these data, the Worker process can fill the hash-table with GeomValue data and process the merge-tree to produce sorted geometry records in the merge-tree output buffer. If more than one input dataset exists, these are merged into a single stream. Using the hash-table and merge-tree, the geometry records with their associated left/right attribute values can be extracted by the calling GIS operation as and when required.

When the hash-table and merge-tree have been emptied, typically when the strip has been processed by the calling GIS operation, they can be destroyed to free up memory for the next strip.

This model repeats until a 'flush' strip-descriptor is received from the Source, at which point the Worker closes down and sends a 'closing-down' message to the Source.

8.6 Phases

In this section the Sort, Join and Geom-Attribute Distribution (GAD) phases are described in more detail.

8.6.1 Sort Phase

Section 8.3 described the sorting of records into intermediate files. This section contains a detailed description of the role of the Pool processes in this phase. The purpose of this phase is to extract all the data necessary to create the left and right attribute values for the geometry records of the input data sets. An overview of the input and output of each process is shown in Figure 8.5.

Each participating process reads and processes data derived from the AREA, ATTREC, EDGEREC and GEOMETRY1 records in the input dataset. The processed data are placed in an internal parcel according to its record origins. A parcel is a collection of data derived from records of the same type which is stored in the internal memory of a process.

When data are added to the Geom parcel, the PartitionList (see section 8.5.1) held by that process is updated.

Once the combined size of the parcels reaches a limit, based on the amount of memory available to a process, the largest of these parcels is sorted and written to the appropriate temporary file as a parallel file atom (see section 8.5.3). The sort keys for each file are indicated in Figure 8.5. Figure 8.6 indicates how the sorted parcels are written as atoms stored in a parallel file. Choosing the largest parcel to be output maximises the size of the atoms in the parallel intermediate files and ensures that the maximum amount of memory is released for processing new records.

When the largest parcel is a Geom parcel, then after it has been written to the Geom file, a GeomYMax parcel is derived from the contents of the Geom parcel, sorted by GEOM_ID and written as a PUL-PF atom to the GeomYMax file. A

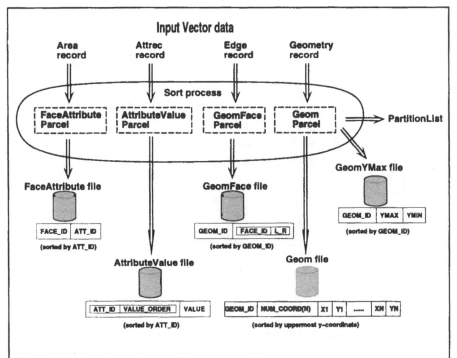

Figure 8.5: Sort phase input and output overview.
Note in the GeomFace file, the L_R (left or right) value in stored in a single bit in
the FACE_ID. This is indicated by the box surrounding both the FACE_ID and
L_R fields in the diagram. Similarly in the AttributeValue file, the
VALUE_ORDER value is stored in a single bit of the ATT_ID.

GeomYMax parcel contains a GeomYMax record for every Geom record in the Geom parcel. Section 8.3.1 explains how the YMax values are used to ensure that the attributes arrive at a worker at the right time to be combined with corresponding Geom.

After the largest parcel has been written, the memory used by the parcel is released and the process continues reading and processing records until once again the combined size of the parcels reaches a limit and the largest parcel is sorted and output as a parallel file atom to the appropriate file.

This cycle of reading, processing and output continues until no more records can be derived from the input vector dataset. At this point any remaining contents in the parcels are sorted and written to the appropriate file and a PartitionList is sent to

the Source process by each participating process. These are merged together by the Source process and the Sort phase is then complete for that dataset.

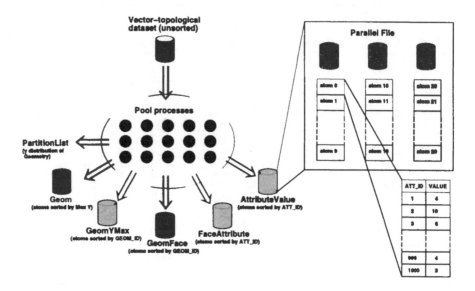

Figure 8.6: Atoms produced by participating processes in Sort phase

8.6.2 Join Phase

Once all Pool processes have completed the Sort phase for an input dataset, these processes enter the Join phase. Here, all the temporary files, other than the Geom file, which were created by the Sort phase are globally sorted via merge-trees and combined to produce the GeomValueYMax file (see Figure 8.7). The GeomValueYMax file is a parallel file which consists of atoms containing GeomValueYMax records. Within each atom, the records are sorted in descending order by uppermost y-coordinate.

To join the various temporary files to produce the GeomValueYMax file, merge-tree sorts are used. Firstly, the GeomYMax and GeomFace are merge-sorted and combined using GEOM_ID as the join key to produce the GeomFaceYMax file. This is a parallel file containing atoms within which the records are sorted by FACE_ID. Next, the FaceAttribute and the AttributeValue files are merge-sorted and combined using ATT_ID as the join key to produce the FaceValue file. Again, this is a parallel file containing atoms within which the records are sorted by FACE_ID. Finally, the GeomFaceYMax and the FaceValue files are merge-sorted and combined to produce the desired GeomValueYMax file.

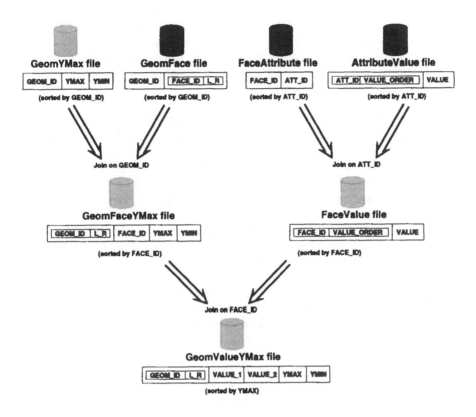

Figure 8.7: Overview of the Join sub-phase

All the Pool processes are used to merge-sort and combine each pair of files. Each pair of files is executed in sequence with the merge-tree for each file being distributed across the processes. Each process executes sections from both merge-trees. This spreads the load of reading the temporary files and processing the data, particularly when there are large numbers of atoms.

The merge tree is a binary tree where each leaf of the tree corresponds to an atom of sorted data, while the internal nodes correspond to comparisons between pairs of values and record the results of the last comparisons. The "winner" of a comparison is propagated up the tree with the effect the overall "winner" emerges from the root of the tree. Once the "winner" has been processed, the next record at the leaf from which "winner" came is retrieved and used to reevaluate the comparison at the parent internal node. The number of comparisons required for each records is therefore minimised and the overall merging time is $O(k\log_2 n)$ where k is the number of records and n is the number of atoms. (See Knuth, 1973, for more details on replacement sorting, comparison trees and multiway merging.)

Pool process 0 owns the root of the merge-trees and some of the tree near the root. Sub-tree beyond this point are distributed over the other pool processes. The leaves of the tree are distributed evenly across the participating processes, where the number of leaves is equal to n, the number of atoms, and the number of nodes (including leaves) in the tree is $2n - 1$.

The tree is initially filled by a single preorder traversal of the tree (see Figure 8.8). At each leaf node on a Pool process, a record is extracted from an input buffer. This input buffer can be attached to either another Pool process or an atom. At each non-leaf node visited in the tree the key values of the children are compared and the lowest one propagated to the parent. When a value reaches the root its record is placed in an output buffer which may be attached to either another Pool process or an area of local memory. This is the first record in the sort. To find the next record in the sort a new record is extracted from the same leaf and the empty path is filled by traversing up the tree, propagating the winner of the two children at each level in the tree. This repeats until each input buffer is exhausted.

At the Pool process 0, the records with matching join key are combined and stored in internal memory. When there is insufficient memory left to store any more records, these records are sorted by the appropriate sort key and output as an atom to the appropriate temporary parallel file.

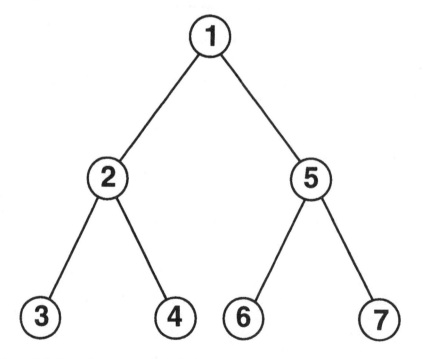

Figure 8.8: Preorder traversal of a binary tree

8.6.3 Geom-Attribute Distribution Phase (GAD)

As explained earlier, on completion of the Sort/Join phase for each input dataset the Source and all the Pool processes enter the Geom-Attribute Distribution Phase, otherwise known as GAD. Here, the Source process instructs each Pool process to become either a GeomAttributeServer process or a Worker process.

The role of the GAD phase is to data-decompose the output data space over a group of Worker processes and then distribute the data to the Workers as required. The space is decomposed into strips which are rectangular and span the full width of the output data space. This is achieved by using the PartitionList generated by the Sort phase, so that the geographical area is decomposed into strips of a size suitable for the Workers to process. Each strip is assigned to a Worker process which is sent the geometry and associated left/right attribute values necessary to perform calling the GIS operation on the strip. Many strips can then be operated upon in parallel by different worker processes.

GAD has three input items per dataset: the Geom file, the GeomValueYMax file and the PartitionList (see Figure 8.9). GeomAttributeServer processes are assigned atoms of the Geom and GeomValueYMax files. The Source process is assigned the PartitionList(s).

A GeomValueYMax file is created by the Join phase. It contains a number of atoms which each consist of a variable number of GeomValueYMaxs (see Figure 8.9). Each GeomValueYMax record comprises the GEOM_ID of a geometry record and the uppermost and lowermost y-coordinate of that record. It also contains one or more attribute values on the left or right side of the geometry record as indicated by the left/right flag. Within each atom, GeomValueYMaxs are sorted in descending order by uppermost y-coordinate. The number of attribute values is dependent on the calling GIS operation.

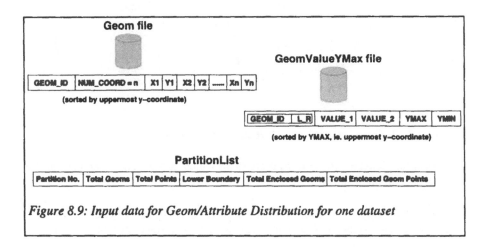

Figure 8.9: Input data for Geom/Attribute Distribution for one dataset

An overview of the phase is shown in Figure 8.10. In GAD, the Source examines the PartitionList(s) (one per input dataset) to determine a y-coordinate range of sufficient size for a ready Worker process to process as a single strip. The Source sends a strip-descriptor message (containing, amongst other things, the y-coordinate range and the ID of the Worker process) to all the GeomAttributeServer processes.

8.7 Design Notes and Limitations

8.7.1 Data Distribution

It should be noted that the distribution of geometry and attribute information for strips from the GeomAttributeServer processes is essentially sequential in that only one strip is sent to one Worker at a time. As a consequence Workers must have the memory resources to accept *all* the data sent to them - data cannot be read from disk as required. This unfortunately does not take advantage of the memory-saving features of scan-line and plane-sweep algorithms (described in Chapters 11 and 13) where, when given a sorted stream of data, a worker need only read and store a fraction of the input data for the strip at a time. It follows from this that strip sizes will be smaller than if the Workers read data in parallel (e.g. directly from random-

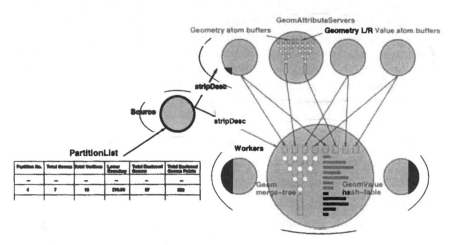

Figure 8.10: Overview of Geom-Attribute Distribution (GAD)

access file). Smaller strips means more strips which increases the overheads of, for example, communications.

However, the major advantage of the essentially sequential distribution is the removal of the overhead of random access to disk. A further advantage is that all those geometry records which start above a strip do not have to be read to check whether or not they enter the strip. Instead, by maintaining a split-ends list in each GeomAttributeServer, all those geometry records which enter the strip from above can be sent out with all the other strip information.

8.7.2 Merge-sorting Data from Multiple Datasets

The polygon overlay operation requires a single stream of sorted data. This therefore requires the merge-sorting of the data from the two datasets. In GAD, this merge-sorting occurs solely on the Workers where the data from all the GeomAttributeServers are merge-sorted anyway. This requires the Workers to have a merge-tree which contains two leaf nodes per GeomAttributeServer, that is one for each dataset. An alternative approach is for the merging to occur initially on the GeomAttributeServers, this would mean the merge-trees on the Workers would only require one leaf per GeomAttributeServer. To do this means merging the output from trees located on the GeomAttributeServers and putting this into a new buffer which is attached to the Worker. This would require a redesign of the tree merging mechanism due to the increased space requirements of such a method.

8.7.3 Full-width Horizontal Strips

At present the design assumes full-width horizontal strips can be processed. Should there be too much work in a Mini-Partition then the module will be unable to proceed.

To get around this the data distribution described in the PartitionList could have an additional x component. This could be incorporated in the Sort phase where the extra overhead of producing this information is parallelised. However, it could be argued that this extra information is rarely needed. In this case it could be incorporated in the building of the MiniPartitionList where problems with overloaded partitions become apparent in GAD Distribution.

8.8 References

Harding, T.J., Hopkins, S., Sloan, T.M., Dowers, S., Gittings, B.M. & Healey, R.G., 1993a, *Polygon Overlay Concepts*. Edinburgh Parallel Computing Centre and University of Edinburgh Department of Geography EPCC-CC93-06, Edinburgh.

Harding, T.J., Mineter, M.J., Sloan, T.M., Wilson, A.J., Dowers, S. & Gittings, B.M., 1993, *Raster to Vector and Vector to Raster Conversion Concepts*. Edinburgh Parallel Computing Centre and University of Edinburgh Department of Geography EPCC-CC93-07, Edinburgh.

Jackson, R.O., 1995, *Parallel merge-sort for GIS vector data*. Edinburgh Parallel Computing Centre Summer Scholarship Programme, SS-95-10, Edinburgh.

Piwowar, J.M., LeDrew, E.F. & Dudycha, D.J., 1990, Integration of spatial data in vector and raster formats in a geographical information system environment. *International Journal of Geographical Information Systems*, 4 (4), 429-444.

9

Creation of Vector-Topological Data Structures

M.J. Mineter

9.1 Introduction

GIS operations such as raster-to-vector conversion (Chapter 12) and polygon overlay (Chapter 13) generate vector-topological data. This chapter explores the algorithms, data structures and software architecture which are required to support such operations on MIMD parallel computers. Referring to the Level 4 NTF data model described in Chapter 6, it can be seen that creating vector-topology includes, amongst other tasks, the building of lists of vertices along the edges between nodes, of lists of edges comprising the boundaries of faces, and of links between geometry, topology and attribute records. Working within the context stated in Chapter 5:

1. The data comprising the inputs to the GIS operations are distributed by sub-area to a number of Workers.

2. The Workers create the vector-topology, by invoking operation-specific algorithms (for example, the Intersection phase of polygon overlay, Chapter 13) which in turn invoke the generic algorithms for building topology in parallel. The latter are the subject of this chapter.

The scope of the algorithms is as follows:

1. The sizes of datasets may be large in comparison with the memory of the parallel architecture.

2. Non-overlapping areal data only are considered. Linear features are not included in the discussion.

3. The initial dataset(s) can be ordered from top-to-bottom of the dataset. For raster data this is straightforward; for vector data this was discussed in Chapter 8.

The benefits of parallelism are those of speed: many Workers create the vector-topological data concurrently. The challenge is to 'stitch' together the data held by these processes to create the final vector-topological records, without negating those benefits (Harding *et al.*, 1993). The methodology proposed in this chapter is thus entitled 'TSO' - topology building, stitching and output (Mineter *et al.*, 1993 and 1994). A further challenge is that the complexity of the parallelism should be encapsulated within TSO, with the stitching and output of records hidden from the developer of a GIS operation. The developer should thus be able to use a simple interface to TSO and to focus upon building topology rather than upon the parallelism.

The major issues to be addressed in creating vector-topological data in a MIMD environment are now introduced by exploring the extent to which a task farm (introduced in Chapter 4) needs to be amended to support the necessary requirements. In a task farm the data would be distributed amongst the Workers, such that each receives one sub-area of the initial data in turn, and creates the vector-topological data for that sub-area. Such an approach is consistent with the declared scope of this chapter, to explore algorithms which support processing where the complete dataset can be large in comparison with the available memory. A task farm is characterised not only by the existence of source (or data distribution), sink and Worker processes, but also by the Workers being independent of each other, so that each communicates only with the source and sink processes. However, the task farm will be seen to be too simple a software architecture for topology creation.

a) **Workers must interact with each other.** The resultant dataset includes, in terms of the NTF level 4 data model, Geoms within each of which is held the vertices along an edge, and chains, each of which lists the edges forming a boundary of a face. Each Worker can generate only those parts of the objects which fall within the sub-area allocated to the Worker, so that many of these objects may extend across the data held by several Workers. Co-operation between Workers is therefore required. This can be considered in two phases. 'Stitching' is necessary not only to recognise which parts of objects within which Workers comprise each object, but also to support the discovery of the relationships between these objects. Additional communication is then required to collate the parts of an object which are held in different processes in order that a complete record can be written out. In general, related objects will be distributed across different subsets of processes, so that a complex pattern of communication might result.

b) **The allocation of record identifiers entails additional communication.** Relationships between vector-topological objects, for example between a geom and the corresponding edge, or between an edge and its faces, are expressed in terms of identifiers. If many Worker processes are generating topology in parallel, then each will need to insert identifiers into the records which it generates. To achieve their purpose identifiers for each type of record must be unique. The simplest method of achieving this is to assign one additional process the responsibility of allocating identifiers. The consequence is that Workers must communicate not only with each other but also with this Identifier Generator process (IDG).

c) **Multiple threads of processing exist in each Worker.** The inter-process communication is accomplished by the exchange of messages. In a task farm one thread of processing exists. It is invoked by the task farm code on receipt of a message from the Data Distributor. It processes the data held within the message and requests the next subset of data from the Data Distributor. In contrast, within TSO, a Worker engaged in building vector-topology needs to respond to messages from the Data Distributor, from the IDG and from other Workers engaged in stitching. Each type of message requires a different reaction from the recipient of the message, so that the processing within a Worker is multi-threaded.

The following section explores the functionality required for building vector-topology. Subsequent sections develop these ideas with emphasis upon one possible design.

9.2   Overview of Topology Building, Stitching and Output (TSO)

TSO (topology building, stitching and output) is the name given to a module which supports the creation of vector-topology within a parallel GIS operation. The previous section discussed the context of TSO and several of the major issues raised by its purposes. It outlined the processes which would be present in such a GIS operation. These are shown in Figure 9.1 and comprise one or more Data Distribution processes, multiple Workers to derive the vector-topology for sub-areas of the initial dataset, and an IDG (Identifier Generator) process to generate identifiers, which are used within the resultant records. This figure also shows the exchange of messages between the Data Distributor process and Workers, and between the IDG and Workers. Within this section each major component is described in turn. Figure 9.1 is the first of a series of hierarchical design diagrams used in this chapter. More detailed diagrams are given in Figures 9.2 and 9.3. (These diagrams use some symbols similar to those of the Yourdon methodology (Yourdon, 1989), but do not seek to implement that methodology.)

9.2.1   Data Decomposition

Support for the processing of large datasets entails distributing the input data in subsets, where more subsets exist than there are processors, and each subset is sent to one Worker. In a manner similar to that of a task farm, the distribution of a subset is initiated by each Worker, by the sending of a 'data request' message to the

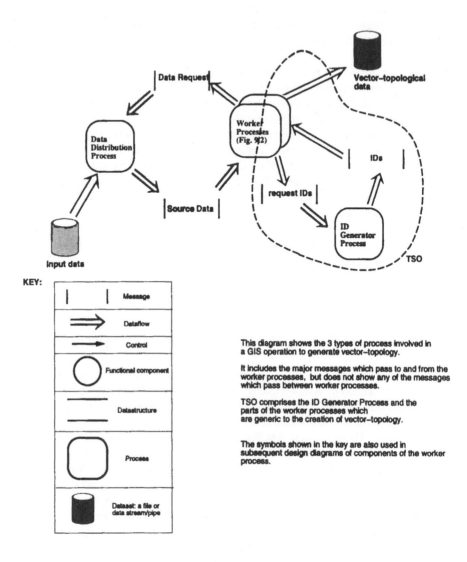

Figure 9.1: Processes required for vector-topology creation

data distributor processes. This message is sent after initialisation, and subsequently on completion of the processing associated with the previous subset.

The strategy for decomposition of data will need to take the following into account:

1. Efficiency of an implementation will depend in part upon minimising the total length of time for which the processes are inactive whilst their requests for data are enqueued in the distribution process. At any time a number of Workers may be available to receive new data. Distributing small datasets will minimise the time for which each request is enqueued, but will lead to a larger number of requests being issued during a complete run.

2. Efficiency will also be increased by minimising the processing and communication entailed in stitching and collation of objects. A large subset of data has a greater chance of some vector-topological objects falling entirely within that subset. Such objects do not need to be stitched.

3. Stitching and collation of objects requires that the parts of objects held in neighbouring subsets of data, usually on different processors, can be matched with each other across the borders of the subsets.

4. Each Worker may have a different amount of memory available for processing a new subset of data. The Worker must thus exercise some control over the amount of data which it receives in a subset.

The various issues show that optimal strategies of decomposition will be subject to the specific characteristics of the GIS operation and the computer architecture upon which it is being run. (In such cases the use of run-time parameters can allow code to be both portable and efficient.)

The third requirement suggests that a rectangular decomposition is most suitable. This simplifies the processing related to the data along subset borders, is relatively trivial for cases where raster datasets are to be distributed, and is appropriate where more than one input dataset is in use. (For example, two vector datasets are used by polygon overlay. These can be separately distributed once the corresponding subset extents are determined.) Within each Worker, and associated with each rectangular sub-area, termed a 'strip', must be held information on its alignment with its neighbours and identification of those neighbours. These data are either distributed with the strip data or communicated between Workers.

The shape of the strips is now discussed, a matter which is explored in more detail in the context of raster data management in section 10.3.1. If a dataset is split into strips which are long and narrow, then it would be expected that relatively few objects would be fully determined within one strip. A decomposition into squarer strips, of the same area but of less than full-width and also less than full-length may reduce stitching, but additional processing is necessary to achieve the data distribution. This comprises additional sorting, not only from top to bottom but also across the dataset. This is relatively straightforward with raster data in a simple array, for example, which can readily have each scan-line split into several pieces corresponding to the strip widths, but it would be an additional phase of sorting for vector input (Chapter 8).

9.2.2 The Identifier Generator Process (IDG)

Each vector-topological record is required to have a unique identifier which is used to express unambiguous links with related records. Consequently the identifiers for each record type must be unique within that set of records. Imposing an additional requirement that identifiers be contiguous is not only more aesthetic, but would facilitate the transfer of the vector-topological data to a GIS package. The simplest method of allocating identifiers under these restrictions is to use a dedicated process, here named the 'IDG'.

Alternative strategies avoiding the need for the IDG could be contrived, especially if the requirement for contiguity is lifted, by defining rules for generating IDs as a function of the process number. For example, on an architecture running with 256 Workers, these processes would be numbered from 0 to 255. Process 233 could allocate identifiers all of which have 233 as their least significant three digits. However, in the case where contiguity is required a subsequent and cumbersome processing phase would be needed to overwrite these locally generated IDs.

The proposed IDG process necessitates message-passing, with Workers sending requests for IDs, and the IDG responding with the IDs. It is only after the IDs are received that records can be written out. Although there is a trivial amount of processing within the IDG to compose a reply message, the fact that all ID requests are serviced by a single process is a potential bottleneck. This risk can be reduced by the appropriate use of non-blocking messages (so that a Worker is not idle pending receipt of its IDs), and by minimising the number of messages sent by each Worker, for example by Workers requesting a range of IDs for each type of object in one request, rather than by requesting an ID for one object at a time.

9.2.3 The Worker

Figure 9.2 shows the major components which are required to exist within the Worker. These are described below, distinguishing between those components which are within TSO, and those which are specific to a GIS operation.

9.2.3.1 *Data receipt*

This component will be specific to a particular data distributor process; it is outside the TSO code and linked with TSO within the Worker executable image. Its role is to send messages requesting data to the Data Distributor process, and to receive

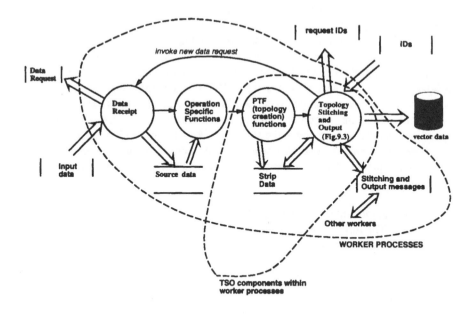

Figure 9.2: Design diagram: components of the Worker process

data from that process. The function which sends the message to request a new strip of data will be invoked by TSO code after initialisation and subsequently when the Worker is otherwise unoccupied, after completion of the stitching immediately associated with the previous strip. The function which is invoked on receipt of requested data will organise the data in memory for processing by the operation-specific functions.

9.2.3.2 *Operation-specific functions for strip building*

The building of the vector data for a strip is operation-specific. In raster-to-vector conversion, for example, the 'operation specific functions' represented in Figure 9.2 are those which will analyse the received sub-raster block and derive the corresponding strip vector-topology by calling the 'partial topology functions' (PTF) discussed in the next subsection. The effect of these functions is that the operation-specific code processes one strip in a manner closely analogous to that of a sequential program processing the complete dataset.

9.2.3.3   'Partial Topology' Functions (PTF)

To build topology GIS operations must perform basic functions such as generating lists of vertices, joining two such lists, defining a node, etc. The term 'partial topology' is given to these functions because topology is built in strips, and each strip will contain parts of many objects. By providing these functions as a part of TSO, the amount of code which must be developed specifically for each operation will be minimised. A sequential version of PTF could be implemented to allow prototyping of a new operation on one workstation, with straightforward porting to a parallel architecture. Figure 9.2 illustrates that these functions are called by the operation-specific code to create the strip data. An additional TSO function, *ptf_complete* is called by that code when the entire vector-topology within a strip has been completed. This function initiates stitching of the strip with its neighbours.

 PTF comprises the interface between the GIS operation and the vector-topological data model. The full details of the implementation of the latter should not need to be known to the operation.

9.2.3.4   Stitching and Output

The component labelled 'Topology Stitching and Output' in Figure 9.2 constructs the complete topology from partial objects which are held within processed strips, and writes out the completed records. This component is shown in more detail in Figure 9.3.

Communication between processes is necessary to:

1. Match partial objects across strip borders, and recognise topological relationships.

2. Obtain identifiers for completed objects from the IDG.

3. Pass the IDs to any completed or incomplete objects which require them. (For example each Edge identifier must be held in two Chains, associated with the polygons on either side of the Edge.)

4. Collate data for output as vector-topological records.

Stitching (Figure 9.3) can be considered to be 'area based', in that the discovery of topological relationships requires information from a number of connected objects. Consequently it is best initiated after all partial topology has been fully derived for a strip. The output component (Figure 9.3) is 'object-based', in that it collates in turn the data for objects which were recognised to be fully described during stitching. Data specifying these objects are passed from the stitching component to

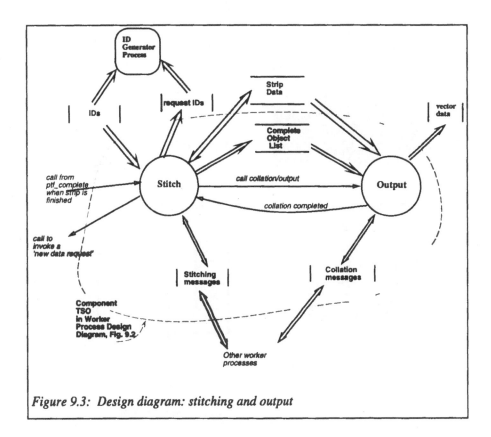

Figure 9.3: Design diagram: stitching and output

the output component via the 'complete objects list' shown in the same figure. Examples for which collation is required include the lists of Edges in a Chain and of vertices in a Geom, and the two IDs of Nodes at either end of an Edge. The writing out can happen either by gathering all parts of an object into one Worker before the record is written, or else through cooperation between processes so that consecutive parts of records can be sent in order to a file or other output stream.

Once a Worker has completed all stitching, collation and output possible at that stage of processing, it will request additional data from the Distributor processes so that the Worker can generate partial topology for an additional strip. This is further discussed in section 9.4.2. The initiation of this request is represented in Figures 9.2 and 9.3 by the control arrow labelled 'invoke new data request'.

The stitching algorithms are discussed in section 9.4.4, but at this stage in the discussion Figure 9.4 is given to illustrate facets of the problem. In particular it shows that communication about data held in strips is necessary to distinguish between a number of possibilities. Firstly, a part of a chain within a strip could, after stitching, be discovered to comprise either an external boundary of a polygon

BEFORE STITCHING

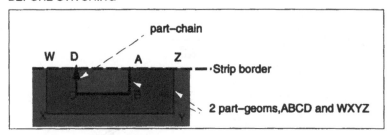

AFTER STITCHING: results depend upon the neighbouring strip(s)

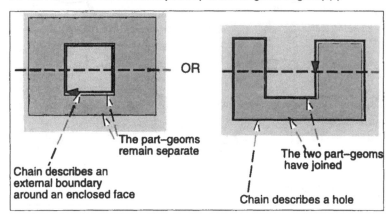

Figure 9.4: Example of stitching: communication between processes is required to build topology

or else a hole within a containing polygon. Secondly, data held within one strip cannot recognise those part-Geoms which at a later stage will need to be connected.

9.2.3.5 Strip data structures

The strip data structures represented in figures 9.2 and 9.3 are created by the 'partial topology' functions, modified when IDs are allocated to objects, accessed to initiate stitching, and destroyed when their data have been used by the output components of TSO to create the resultant records. Each Worker is likely to process many strips during a run. In general some objects in each strip will be completed within that strip, and so can be written out, and other objects will be incomplete pending the processing of strips in other Workers, and subsequent stitching. The data structures within a Worker must accommodate partial topology from all strips processed by the Worker, and information related to the strip borders to allow stitching and collation to occur. These are discussed further in section 9.3.

9.2.3.6 Multi-threading

The major threads within the Worker are as follows:

1. On receipt of the data from the source process, a strip of vector-topological data is derived. Messages are then sent to other Workers to initiate the stitching together of objects which are shared between them, and to the IDG to obtain identifiers of objects which are fully within a strip.

2. During stitching, messages comprising data about shared objects are exchanged and processed. Once complete objects are recognised to be held within the strips of the cooperating Workers, a message is sent to request identifiers from the IDG.

3. On receipt from the IDG of the identifiers for the completed objects, the identifiers are distributed to those Workers which require them to establish the topological links with other records. In order that all the data for completed objects can be collated, requests are passed to all processes holding parts of those objects.

4. On receipt of requests for data from other Workers, the requested part of an object is sent to the collating process.

5. On receipt of IDs from other Worker processes, the IDs are inserted in the appropriate records. If the records are completed then they are written out.

6. On receipt of parts of objects from other Workers, those parts are collated. Once complete, the corresponding record is generated and written out. Once all such objects are collated a new request for data is sent to the Data Distributor process.

The processing within any particular Worker is subject to the timing of message sending, and hence processing, in other Workers and in the IDG. It is an unpredictable pattern, often termed 'event-handling': the receipt of a message by a Worker constitutes an event. Each type of message (for example, from the IDG returning identifiers, or from the source sending new data) needs to be recognised, so that it can be handled by invoking a particular thread of processing. This is a style of software design frequently met in communications, operating systems and also in user-interfaces. The different Workers will be running the same executable image, but at any one time are likely to be invoking different functions in response to different received messages.

The structure of TSO is further discussed in section 9.6.

9.3 Topology Building

Section 9.2.3.3 introduced the possibility of defining a set of 'partial topology' functions (PTF) which could be invoked by a GIS operation to create a strip and then to initiate the stitching of objects and output of records.

The following subsections outline:

1. The interdependencies of the NTF objects.

2. Extensions to the NTF data model. Although the NTF level 4 data model was developed as a transfer format, it can be modified and exploited in the creation of vector-topology also.

3. Some necessary characteristics of the data structures used to hold the strip data.

4. Further details of the proposed PTF functions.

Some terms used in the following sections are now defined:

• The term 'closed' is used to describe the object when its full extent is known.

• The term 'completed' is used when all data required for the record of an object is held. Thus a Geom is closed when all its vertices are found, and it is completed when its ID is assigned.

• The term 'part-Chain' refers to the part of the data of a Chain which is derived from one strip. In general a Chain will comprise several part-Chains which require collation prior to writing out. Similarly, the term part-Geom refers to a part of a Geom held in the data of one strip.

• The NTF record type Geom is used both to locate a Node and to list a number of vertices which define the location of an Edge. To avoid ambiguity the former is termed a 'node Geom' in the discussion below.

• The complete dataset is viewed as being held in one or more holes within a 'Universal Polygon' (Figure 6.4).

9.3.1 The Interdependencies of Vector Objects

The relationships between vector-topological objects influence the time when each object can be output, and influence the data structures used to build the vector-topology. They are as follows:

- *Node and node Geom* records are in one-to-one correspondence. The node ID must be placed in the relevant Edge records.

- *Edge and (vertex) Geom* records are in one-to-one correspondence, but

  Geom records contain only vertices and therefore complete as they close.

  An *Edge* record can only complete when its left and right Polygon IDs are available.

- *Face, Area* and *Attrec* records are in one-to-one correspondence, so the same ID is used for each, and in further discussion the triplet is referred to as a Polygon record. As well as the external Chain, a polygon may have any number of hole Chains. The Polygon ID must be written into all Edges bounding the polygon.

- Each *Chain* record comprises a list of Edge IDs, where the Edges define a boundary of a Polygon. The structures associated with each part-Chain can be used to propagate the ID of the Polygon with which the Chain is associated. Chains thus close when they form a continuous loop, complete when all Edge IDs are known, and are written out when the Polygon ID associated with the Chain has been propagated to any associated Edges.

In summary:

1. Polygon records require Chain IDs.

2. Chain records require Edge IDs.

3. Edge records require Polygon, Node and Geom IDs.

9.3.2 Transitions

Figure 9.5 shows a simple topological dataset split into two strips. The places along the strip border where polygon edges intersect with the strip border are labelled in the figure. As the border is scanned from left to right for example, the transition from face (or polygon) F3 to face F4 occurs at label T2.

ORIGINAL DATASET

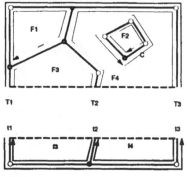

VECTORISED DATASET

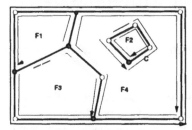

VECTORISED DATA IN 2 STRIPS

F2 is a hole in F4:
  Chain C is associated with the Face 4 record.

T1   T2   T3

t1   t2   t3

KEY:

O	Vertex
●	Node
——	Edge
—▶	Chain
— — •	Strip border

The transitions in the upper strip are

T1: from Universal Polygon to Face 3.
T2: from F3 to F4, noting that chain C exists as
    a hole in the part of F4.
T3: from F4 to Universal Polygon.

Corresponding transitions in the lower strip are shown in
lower case.

At each transition parts of an edge and 2 chains cross the
strip border. Also attribute data and hole data are held
for the part-polygon on the right-hand side of the
transition.

*Figure 9.5: Vector-topology for a dataset comprising 2 strips: the use of
'transitions'*

At each transition the relevant objects comprise:

• edge

• left-hand part chain

• right-hand part chain

- attribute of the right-hand region

- a collection of holes within the right-hand region.

By 'right' is meant the side of the transition when viewed from infinity to the east (for vertical borders) or south (for horizontal borders). Pointers to the data structures corresponding to these objects can be grouped together within a transition data structure.

The following data must be held to allow objects to be followed from strip to strip during stitching or output:

- An index of transition structures along the strip border. This allows random-access to a specified structure.

- Information to allow matching of transitions in adjoining strips.

- Each part-chain and part-geom structure needs to include pointers to any transitions at which that object exits the strip.

During collation of a geom, a message such as 'Send process number 9 data about the 5th geom crossing the south border of strip 4' may be received by a particular Worker. This Worker should use the index of transitions for the strip's south border to obtain the transition structure, and use the transition-to-geom pointer to access the structure for the part-geom. The Worker can then send process 9 its vertices of that geom. By use of the geom-to-transition pointers the transition structure for the 'other' end of the geom can be located. Strip alignment data allow an appropriate message, e.g. 'Send process 9 data about the 3rd geom crossing the west border of strip 5', to be sent to the process holding the next part of the geom.

9.3.3 Partial Topology Data Structures

Strip data comprises both complete and partial objects. Partial objects are those which are held in a number of strips, such as a Chain which crosses strip borders. In the resultant vector-topological dataset, IDs are used to express relationships between records. Similar relationships between incomplete objects also need to be held. This is done by means of pointers between the data structures of the part-objects. For example, associated with each end of a part-Edge is a pointer to either a transition or a Node structure. In the latter case, once the Node ID is known and the Node record is written out then the ID can be used to overwrite its pointers in the edge record. In order that this replacement can occur, the Node structure must include pointers to the attached part-edge structures, a relationship additional to those specified in NTF Level 4.

During the building of a strip and during the stitching of strips partial objects will be created and subsequently modified. The evolution of polygon objects is particularly noteworthy because of the existence of chain structures. The major consequence of this is that the connectivity between regions of part-polygons does not need to be maintained, so avoiding extensive processing as the shapes of polygons evolve. (The discussion in section 12.3 concerning raster-to-vector conversion illustrates this.) Instead, the chain data structures allow the polygon shape to evolve by the building of new part chains, and the joining of two existing part-chains. Once a chain encloses a polygon completely then the polygon ID can be requested. The chain structures can then be used to propagate the polygon ID to all the edges of the chain.

9.3.4 Processing of Holes

A hole comprises one Polygon, or a cluster of connected Polygons, completely contained within another Polygon. For example, the unclosed part-polygons, f4 and F4 in Figure 9.5 will eventually close as stitching proceeds. Any holes in that unclosed Polygon, in this case formed by chain C around F2, must be used in stitching in order that:

1. The enclosing Polygon object (for F4) can refer to any contained holes. This is done by inclusion of the IDs of Chains bounding holes within the Face record containing the holes.

2. The Chains bounding the holes can be used to propagate the Polygon ID to the Edges in those Chains.

The more detailed discussion of stitching returns to this topic, since processing must be efficient even in cases where a large number of holes exist in one polygon.

A further issue associated with holes illustrates how 'simple' functionality in a sequential environment can cause some complexity in a parallel design. With one exception, nodes are created where two edges join, a circumstance which is recognised within a strip, not during stitching. The exception is when, after stitching, an isolated polygon is found to exist (e.g. the first possibility shown in Figure 9.4). In this case stitching must insert a node, requiring additional communication between processes so that the associated edge, geom, chain and face records can be created correctly.

9.3.5 Building of Strip Topology - the PTF Library

The PTF functions have been introduced as a collection of functions which would be used by code specific to a GIS operation to build strip topology (section

9.2.3.3). In section 9.2.1 it was stated that the decomposition of the source data into strips would require top-to-bottom sorting of data. A consequence is that each strip received by a Worker will also be ordered, so that topology can be built from top to bottom of a strip. During derivation of the vector-topology of a strip there is in effect a moving border as indicated by Figure 9.6. Above the border are the data known to TSO via previous calls to PTF. Below the border are data as yet unprocessed, in a form known to the operation-specific components of the Worker.

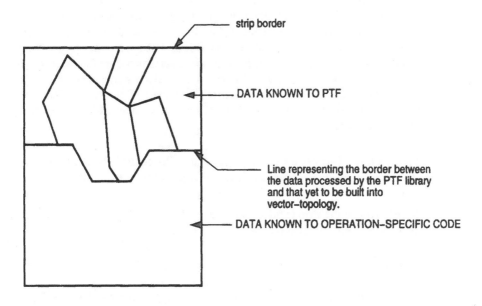

Figure 9.6: Vector-topology is derived from top to bottom of the strip, resulting in a border between initial and vectorised data

The transition structures (section 9.3.2) are used to identify the data along this moving border. The PTF functions use pointers to one or more transitions as arguments. A consequence of this is that the manipulation of vertex lists in a geom and of edge lists in a chain can be handled by the PTF functions and does not need to be controlled by the operation-specific code. The operation-specific components within the Worker should implement the moving border in such a way that transitions can be inserted as new objects are encountered during the scan of the strip, and removed when edges join. Use of a doubly-linked list is proposed in section 12.3 for raster-to-vector conversion.

The operation-specific components in the Worker use PTF functions as shown in Figures 9.7 and 9.8 to:

- Create nodes, edges, Geoms, chains and transitions.

- Extend lists of vertices and of edges within geom and chain structures respectively.

- Join two chains. This function tests whether the joined chain forms a closed loop. The arguments of this function specify the part-chains being joined by referring to the transitions where those part-chains are found, and by specifying the right or left-hand part-chain at each transition. The transition data structures include pointers to a part-chain; if both specified part-chains are the same then a closed loop is formed. In this case, either a hole or else a polygon boundary has been formed, and appropriate actions are taken by the PTF functions. The effect is that the operation-specific code does not create polygon structures or hole structures explicitly.

- Connect edges. This has the consequence that the associated chains also join, and the transitions on the 'moving border' merge and are eliminated.

- Inform TSO that the strip is fully processed so that stitching and output (tasks 3, 4, 5 in the list below) can commence.

As a consequence the PTF functions:

1 Build data structures to hold both partial and 'in-strip' topology. ('In-strip' topology refers to objects which are fully completed within the strip.)

2 Recognise closure and completion of in-strip objects.

3 Request IDs for the closed objects.

4 Receive the IDs and write out completed in-strip objects.

5 Invoke stitching and output functions to cause the completion and writing out of objects shared between a number of strips.

The use of the PTF set of functions is further discussed in chapters 12 and 13 on Raster-to-Vector Conversion and Vector Polygon Overlay.

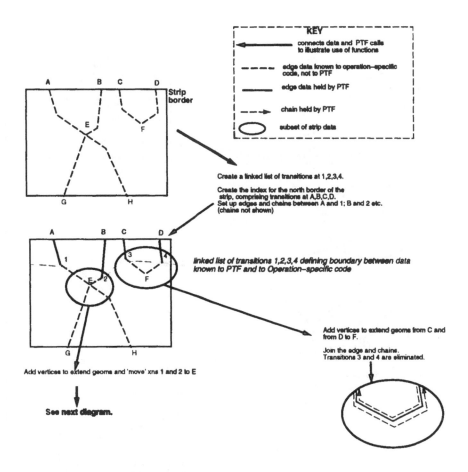

Figure 9.7: Use of PTF functions to create vector-topology - 1

9.4 Stitching

Previous sections described how the distribution of data necessitates that strips of data be 'stitched' together to create vector-topology. This section explores the functionality required for stitching and describes facets of one possible design. Several factors influence this design:

1. A geom or chain can cross any number of strips in the dataset, any number of times. Consequently there is scope for creating a design which will incur such overheads that a parallel GIS operation would not only take vastly longer to develop than a sequential implementation, but also to execute. An example of

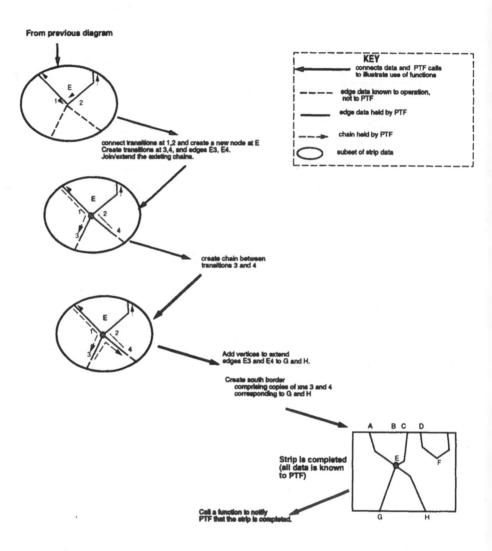

Figure 9.8: Use of PTF functions to create vector-topology - 2

this might be if, on an object-by-object basis, the processes holding strips with parts of each chain or geom were repeatedly to test for closure by passing messages between them. More likely would be the introduction of a pattern of processing whereby interdependence of processes led to a deadlock or else a bottleneck.

2. Objects cannot be treated in isolation: the relationships between objects need to be derived.

3. Data distribution (section 9.2.1) is intended to utilise the memory of the individual Workers as fully as possible, allowing for architectures in which dataset sizes are large in comparison to available memory. Consequently the communication of large amounts of data (e.g. complete strips of data) between processes not only is inefficient, but also risks causing memory to be exhausted on one or more processors.

Recalling from section 9.2.3.4 that stitching is 'area-based' in that it seeks not only to identify the parts of one object, but also to recognise topological relationships, the concept of a 'domain' is introduced. A domain is defined to be a rectangular area of the dataset comprising one or more strips. Its data comprise the minimum information necessary to allow stitching and then collation of objects to be initiated. Initially a domain comprises information about objects which are incomplete within one newly generated strip. As stitching progresses, domain data are passed between two Workers to eliminate a shared domain border, so creating a new, larger domain. This domain is held by one of these processes to be used in further stitching.

Stitching requires knowledge of objects which are incomplete within a domain, and knowledge of the transitions associated with such objects. Domain data structures do not duplicate data held in strips. Stitching does not redistribute the strip data, which remains in the process where it was derived, in accordance with the third comment in the list above. During stitching, only the domain border data are sent between processes.

Several rules are proposed to control the complexity both of the processing within a Worker and of the interactions between the Workers:

1 A strip can only be within one domain at a time, so that domains do not overlap.

2 Each process can only stitch across one domain border at a time.

3 Different pairs of processes can be stitching different domains at the same time.

4 No more than one domain can be held by a Worker.

The resultant effect of these rules is to control the number of messages and the quantity of data which must be exchanged during stitching. It is also to impose constraints upon Worker activity, thereby reducing the scope for deadlock and for conflicting, multiple processing of parts of the same object.

9.4.1 Domain Data

The functionality which domain data structures need to support is similar in several ways to the case of strip borders, section 9.2.3.5. The similarity is particularly evident in the use of transitions.

The domain structures need to allow:

- Matching of objects in adjoining domains. This can be accomplished by means of an ordered list of transitions.

- Recognition of object closure during stitching. During stitching of two domains, each chain which crosses the stitched border must be followed to test for closure of the chain. Hence, bi-directional links are needed between the transitions at which each chain exits the domain. Similar data are required for an edge which enters a domain, but with the additional possibility that the edge terminates at a node rather than leaving the domain.

- Creation of a new domain by merging two adjoining domains.

- Collation of objects which are found to be closed during stitching of two domains. This requires cross-reference from the domain border transition to data in a strip. This is because an object is recognised as being closed during stitching of two domains, but collation of the object requires access to the details which are held in the strip data structures of two or more processes. (For example, 'part of the Geom at the 15th transition on the northern domain border is held in the 6th strip received by process number 10, at transition 5 on its northern border'.)

- Pre-allocation of space for objects requiring collation. When a Geom or Chain closes in stitching, its vertices or edge IDs (respectively) are held in a number of strips, and require to be collated into one list. Different lists will have different lengths. Knowing the number of the vertices or edges in the object allows scheduling of collation in such a way that use of memory remains controlled.

- Efficient sending of domain data in messages. For example, if the domain data for a border are held in one contiguous array of flat structures (i.e. with no pointers to other data structures), they can be sent in one message with no local buffering of the message being necessary. If pointers were to be used, then either a number of messages would need to be sent for each border, or else data from the different structures would need to be copied and buffered prior to sending. The characteristics of the parallel processor will determine the overheads of the different possibilities.

- Processing of holes within unclosed polygons. These holes must be associated with regions of polygons which cross domain borders. The data will include the total number of holes within the region associated with the transition, and information specifying the location of these holes within other processes.

9.4.2 Stitching - a Design Outline

Stitching is initiated by TSO within a Worker when no more important event requires processing (priorities are discussed in section 9.6), and when a new domain has been generated, either from processing of a new strip or from previous stitching. The initiation entails identifying a partner process. Three approaches can be considered for this. Firstly, domain evolution could follow a sequence which is imposed upon all processes by Data Distribution processes, so that one Worker will wait until its next partner is available. Secondly, a 'stitching control' process could monitor activity across all Workers and dictate the pattern of stitching as processes holding neighbouring domains become available for further stitching. This is slightly more complicated to implement, but because a domain may have more than one border across which the next stitching could occur, the flexibility in determining the next partner might reduce waiting times. The third approach is similar to the second, but seeks to achieve the coupling of processes without a 'stitching control' process by additional message-passing between Workers. In this case care must be taken to avoid conditions whereby conflicting decisions are made in different processes.

Once the partner is identified then domain border data are exchanged and the stitching algorithms are invoked. The two Workers involved in stitching together two domains assume different roles, and so for the purposes of describing stitching are termed the Domain Stitching Process and the Output Stitching Process (DSP and OSP):

- The OSP causes to be written out those objects which completed during stitching. It recognises those objects, requests IDs from the IDG (section 9.2.2), receives the IDs and associates them with appropriate objects. At this stage the collation and writing out of objects can be initiated (section 9.5). After freeing all memory used by domain objects it has then concluded its present role in stitching and output as an OSP, and will send a request to the Data Distributor to obtain another strip for processing. This request for a new strip is sent when all completed objects have been written out, and no messages are waiting to be read. This helps to maximise the amount of memory available for the new strip. It also reduces the likelihood of the reading of the data being interrupted. (If one Worker is so delayed then the sending of data to other processes might also be delayed.)

- The DSP constructs the new domain comprising the objects which are still unclosed. It then stitches again, as either the DSP or the OSP, with a different partner.

Figure 9.9 gives an example of the pattern of processing and domain evolution

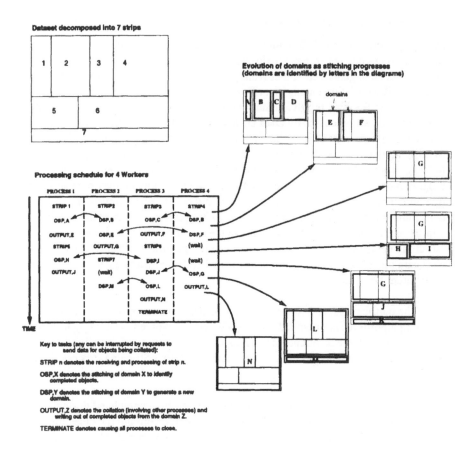

Figure 9.9: An example of domain evolution and scheduling

which could result.

9.4.3 The Stitching of Holes

During the derivation and stitching of topological data, holes in a polygon can be found before the extent of that polygon is known (for example, in Figure 9.5 F2 comprises a hole in the unclosed F4). The boundary of a hole is described by a chain. Many such holes might be found in one unclosed polygon, for example if a dataset represented a cluster of islands. In order that topology can be expressed in the records of the edges and polygon, there is the need to build a list of holes associated with each polygon or part-polygon. During strip processing this list is held associated with strip data; during stitching it is associated with domain data. The remainder of this sub-section outlines two approaches designed to reduce the consequences of cases where a large number of holes exist. The first is intended to allow stitching to proceed without the passing of long lists of holes, and the second is to allow records associated with holes to be written out in some cases, before the extent of the enclosing polygon is found.

9.4.3.1 Hole 'Collections'

To eliminate the need for passing lists of holes between processes, a structure comprising a collection of holes within an unclosed polygon can be used. Each of these structures is associated with a region of a strip or else a domain border. The data comprise pointers to a part-chain of a hole, or data to identify a collection known to another process from earlier stitching. The effect of this is that a large number of holes can be held in a distributed tree of collections, that the domain data required at stitching remains concise, but that the tree must be scanned once the ID of the enclosing polygon is known.

9.4.3.2 Early Allocation of IDs

Section 9.2 stated that a desirable goal was for TSO to be designed to allow the memory of a host architecture to be smaller than that required to hold a complete dataset. This goal implies that on completion of the stitching resulting from the vectorisation of one strip, a Worker would require to be able to write out sufficient completed objects that space in memory could be freed for a subsequent, though perhaps smaller, strip to be processed. The operation would fail if the memory of all Workers were to become full, and additional strips had yet to be processed. Additional complexities, such as writing intermediate disk files, could be considered but it is preferable to facilitate the early closure and writing out of vector objects.

Usually an object is granted an ID when it closes. This is to prevent distinct IDs being granted for two parts of the same object. The ID of the record, and all IDs held within that record are needed before it can be written out. The NTF Level 4 data model is an ally in that vertices are held distinct from their edges and so once the nodes delimiting a geom are found, the vertices can be written out, thereby freeing memory. (A different model might use one record for both the NTF geom and the NTF edge data, which would delay the writing of the vertices until the IDs

of the adjoining polygons were known.) The edge records cannot be written out
until the IDs of the adjoining polygons are known; the chain structures are used to
propagate the polygon IDs to the edges along the hole borders, so that neither edge
nor chain records can be written until the polygon ID is allocated. In the case of a
dataset representing a large number of islands, each island would be described in
the resultant vector-topological dataset as a hole in the polygon representing the
sea. The chains and edges would be held in memory until the 'sea polygon' closed,
which might be only after all data have been processed. This section proposes that
TSO would benefit from the concept of 'super-IDs'. In the circumstance described
above, an ID would be allocated to the polygon representing the sea prior to its
closure.

Care would be necessary in the use of 'super-IDs' to ensure that these were never
allocated separately to different parts of the same polygon. Consider a dataset for
South America. As Cape Horn is approached from the north, the Atlantic and
Pacific oceans apparently comprise different polygons: a super-ID could be
allocated to only one of these oceans, for argument's sake, to the Pacific. Once the
polygons are seen to connect, then the 'super-ID' of the Pacific ocean can be
propagated along the chain representing the Atlantic coast, and also along chains
representing any islands off that coast. Recalling that the dataset exists as holes in
the Universal Polygon, the ID of the Universal Polygon will be defined as a
constant, and thereafter treated like a super-ID.

In summary the following are needed:

- The IDG must, in addition to allocating IDs to closed objects, allocate super-
  IDs subject to these conditions:

    - Requests for super-IDs specify the attributes of the polygon for which
      the super-ID is requested.

    - Only one super-ID per attribute is allowed to avoid the possibility of
      two part-Polygons with different super-IDs merging.

- To constrain the use of super-IDs to occasions when they would be significant,
  a super-ID is requested by a DSP process if the number of Edges in holes or
  along one Chain passes a threshold.

- On receipt of a super-ID by a DSP, the super-ID is propagated along all Chains
  adjoining the relevant unclosed polygon, allowing the Edges of each Chain to
  complete (by insertion of the super-ID as one of the required Polygon IDs into
  the edge records) and be written out. Completed Chain records will also be
  written out, but their IDs must be held for insertion in the Face record of the
  enclosing polygon.

9.4.4 Algorithms for Stitching

Stitching uses a number of facts about topological objects and domains:

- If a Geom enters a domain then either it exits that domain elsewhere or it terminates at a Node within the domain.

- If a Chain enters a domain, it also exits that domain. (Chains eventually close to form continuous loops.)

- Whether a Chain describes a hole or whether it describes an external boundary of a polygon can be determined only when the Chain is closed (e.g. Figure 9.4).

- During stitching across a particular border, Chains and Geoms cannot close if they exit the domains being stitched at any other border.

- All Chains which exist within the area encompassed by a closed Chain must themselves be closed. (I.e. there cannot be an unclosed polygon within a closed polygon.)

The algorithms used by the OSP and DSP processes are now described further, firstly dealing with those relevant to both the OSP and the DSP, then taking each in turn. The algorithms use the fact that the domain border comprises a list of transitions, ordered from left to right along the border, and account for the fact that some chains will cross the border several times, with transitions along the domain border existing for each crossing.

Two algorithms are used by both the DSP and OSP processes:

1. **Identification of those regions along a domain border which are within the area bounded by a closed chain** (Figure 9.10). When a chain is found to close during stitching, a list of its transitions on the domain border is generated. These transitions can be ordered from left to right along the border. Regions between these ordered transitions must be alternately inside and outside the area bounded by the chain.

2. **Recognition of holes or polygon chains.** When part-Chains within a domain form a closed chain it is necessary to distinguish whether that chain forms a hole in a larger polygon, or whether it describes the boundary of a polygon. Recall from section 9.3.2 that at a transition the left-hand chain defines the boundary of the region to the left of the transition. If the closed chain is on the left of its left-most transition then it cannot enclose its region (or it would not be the left-most transition) and so the chain describes a hole; if it is on the right of that transition then it must enclose that right-hand region and it describes a polygon.

Holes and polygons may exist within larger polygons and so the algorithm must
be applied recursively to regions of the domain border which are within a
closed chain.

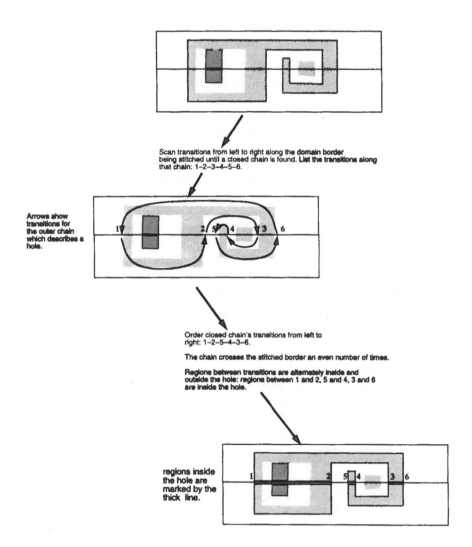

*Figure 9.10: Identification of those regions along a domain border which are
within the area bounded by a closed chain*

The above algorithms are used differently in the OSP process (which seeks to find all newly closed objects) and the DSP (which seeks to identify unclosed objects including holes in unclosed polygons). This is illustrated in Figure 9.11.

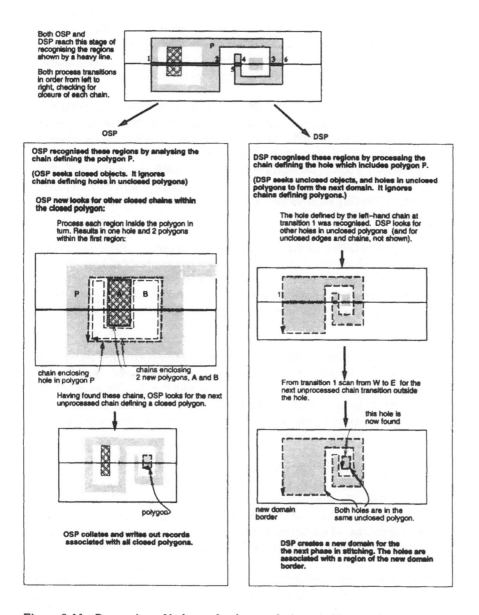

Figure 9.11: Processing of holes and polygons during stitching

9.5   Object Collation and Output

The design proposed in this chapter requires that all processors be able to output data. This is because at any one time many processors are deriving strip data, and others are stitching derived strips. Data records are completed in both of these phases. Output might be achieved by means of an I/O Library (Chapter 4), or by sending completed records as messages to one or more sink processes. The method used must take account of the fact that some records are of variable length, e.g. the number of vertices in a Geom, or the number of Edges in a Chain. This section now considers the collation of records which occurs prior to the writing out of data.

Following stitching, the distributed objects which are recognised by the OSP process (section 9.4.2) as being completed within a new domain must be collated. Collation entails the OSP initiating an 'output request' (OPR) message which is passed on from strip to strip, in effect 'fuse-burning' along the object. Each process holding a part of the object will receive the OPR and send its part of the object to the OSP, and then pass on the OPR to the process with the next part of that object. The OSP receives the OPR replies and builds the complete object ready for writing out.

Processing of an OPR is an asynchronous activity in that an OPR is initiated by an OSP process with no regard to the state of the Workers which hold part of the relevant object. Each Worker checks frequently (section 9.6) for receipt of an OPR. After replying to the OPR, memory used by that Worker for parts of the object being collated can be freed, and the previous activity of that Worker resumed.

Prior to initiating collation, the OSP process must ensure that sufficient space is available to hold the complete object in memory. If not then two options arise. Either the OSP can collate smaller objects, expecting that this will result in more memory becoming available for the larger objects, or the data for one record can be written out in fragments from one process then the next, entailing additional communication between processes.

OPR messages are of the following types:

- Geom OPRs. These initiate collation of the vertices along a Geom between two nodes, and cause the Node ID at the end of the Geom to be sent to the process holding the start of the Geom, which will then be responsible for building the corresponding Edge. The Edge ID is included in the OPR message.

- Chain OPRs, which cause the Edge IDs to be collated. These OPRs are also used to pass a polygon ID to the Edge records. (Two such OPRs are required to be received before an Edge can be written out, one for each polygon bounded by the Edge.)

- Hole OPRs. Section 9.4.3 proposed that collections of holes could be held as a distributed tree for each polygon. The top level of that tree is held in the domain data. When a polygon closes, the polygon ID is propagated to all processes sharing that collection. In addition to the polygon ID, the IDs of all the hole Chains in the collection are distributed through the tree. (To support this, the total number of holes in the polygon, and details to specify the next-level collections of holes are held in the domain structures.) No reply to the process initiating a Hole OPR is needed. On receipt of a Hole OPR a Worker process will split the IDs between the chains which it can itself collate and those in lower levels of the hole-tree. It sends Chain OPRs to collate the former, and Hole OPRs to the latter.

Amongst the issues which could be considered in a more detailed description of the collation are:

- The amalgamation of Geom and Chain OPRs. A Chain closes when its Edges close. Consequently Geom and Chain OPRs can be sent as one message.

- The avoidance of duplicate 'fuse-burning'. Some Edges will close before the corresponding Chains close. In this case the propagation of the Chain OPRs can be optimised by allowing OPRs to jump from one Edge to the next, rather than following the path of the corresponding Geom. To allow this, a Geom OPR must, in effect, cause pointers to be set up between the processes and structures holding the ends of that Edge. Care must be taken to avoid race conditions between the establishment of these links and the propagation of a Chain OPR initiated at a later stage in stitching.

- When a Geom closes during stitching, the stitched domain border by definition intersects that Geom. Either the Geom OPR can be sent to one end of the Geom and then propagated along the Geom, or else two OPRs can be sent along the two stitched part-Geoms from the stitched border towards the nodes. On some architectures the second approach would have two advantages. The first is that several OPRs might more effectively be buffered together after stitching (as a number of closed Geoms will exist in the strips on either side of the stitched border), so reducing the number of messages. The second advantage is that two OPRs might achieve the collation more quickly than one.

9.6 Structure of the TSO Framework

This section explores aspects of a software architecture capable of supporting the processing outlined in sections 9.3, 9.4 and 9.5. Within TSO, Workers have three foreground activities:

1. Receipt of data and creation of vector-topology for a strip

2. Stitching

3. Collation of distributed data and writing out of records.

In addition, there is a background activity. At any stage Workers may receive requests for data for objects being collated by other processes.

To support these activities each Worker exchanges messages with other Workers, the ID Generator, data distribution processes and, subject to implementation, processes responsible for file I/O. As discussed in section 9.2.3.6, TSO provides an event-handling framework, in which each Worker invokes functions in response to these messages. The different foreground activities are initiated either by the completion of a previous activity, or by the receipt of messages.

Although the existence of domains has the effect that not all messages are expected by a Worker at any time, the pattern of processing is such that messages of various types from several processes may be enqueued at the same destination at any time. For example, during receipt of data for a strip there may be messages from the Data Distribution processes providing more source data, messages from the IDG with identifiers for objects already recognised within the strip being processed, perhaps messages from those processes which are ready to stitch with the strip being derived, and perhaps OPR messages from processes that are collating objects partly held in previously vectorised strips. The order in which messages should be processed can be determined by allocating a priority to the different types of message. If more than one type of message has been received by a process, then the highest priority messages are processed first. The following ordering is proposed on the assumption that freeing memory and avoiding delays in other processes are the most important criteria. (Implementations could allow these priorities to be reordered according to the phase of processing, and according to architecture characteristics.)

1)   Requests for data for collation and output (OPRs). This is likely to be the highest priority for the following reasons:

- Disk I/O might be a constraint on performance. Writing data to disk as soon as objects are completed is likely to spread the output load.

- Once data for parts of objects have been sent for collation, then those memory structures can be reused.

- The OPR in general will be passed on to other processes. Delaying it in one process thus preclude the freeing of memory in other processes.

- The collating Worker in general will have 'ring-fenced' memory for the complete object being collated. Delays in collating one object will, if memory is short, delay the issuing of other OPRs, which are queued pending memory being available.

- The process responsible for issuing the OPR cannot receive another strip until all currently completed records are written out.

- These messages can be initiated by a number of processes, and are thus the most likely candidates for causing message queues to grow.

- Processing of these messages is simple and quick.

2) Receipt of identifiers from the IDG process.

3) Data Distribution messages. (These will not compete with stitching in practice; a new strip is only requested once the previous domain has been passed on to another process.)

4) Stitching messages.

5) Hole-OPR messages. Recipients of these messages may themselves then have to initiate collation of Chain data. This may be more easily organised if it is not interwoven with other processing. However, in order that memory used by these Chains and the associated Edges can be freed, processing these messages could be made a high priority between activities (i.e. between stitching and strip derivation).

A Worker process includes application-specific processing of source data, supported by the PTF functions (section 9.2.3.3) and invoked by TSO upon receipt of messages from Data Distributors. It is because TSO is both called by, and makes calls to, application-specific code that TSO is described by the term 'framework', rather than 'library'.

Implementations might also include an additional set of messages, related to process control, generation of trace output, and the gathering of statistics. Management of this nature would be the highest priority of all, and could, for example, allow a reallocation of priorities according to circumstances within TSO.

MPI (1994) introduced in section 3.9.3.2, provides a range of functionality which is capable of supporting the TSO framework, including:

- Non-blocking messaging. For example, a process can send a message and then continue with other processing without synchronising with the receiver. This reduces both delays and the danger of deadlocks. Tests for completion of the send can be issued at a later stage.

- Groups of processes can be defined. Different groups may comprise the same or different sets of processes. Messages can be sent between members of a group, and can be selected by group for reading. This allows TSO to support different priorities of messages. In TSO some groups will comprise all Workers (for stitching and OPRs), another all Workers and the IDG, yet another all Workers and the Data Distribution processes.

- A process can issue one call to MPI to wait for either the receipt of any message or the completion of any send. This is necessary when no messages are enqueued, and a Worker has no other processing to perform.

- Messages can be tagged. A tag identifies the message type. In TSO, messages sent between members of a group might be of various types and lengths. For example, messages sending data to a collating process will differ according to the object being collated. Both the tag and the length of the message can be obtained from MPI prior to reading that message, so that an appropriate memory structure can be allocated for the data.

9.7 Conclusions

This chapter has explored some of the issues encountered in designing software to create vector-topological data within parallel GIS operations in which data are distributed by sub-areas. Areal data only were considered, expressed in terms of the NTF Level 4 data model. Complexity arises from the distribution of the dataset and the fact that vector-topological data comprise multiple, related record types. The major components of topology building, stitching and collation were identified and one possible design was outlined.

The chapter concludes by considering the goals of parallel libraries and the extent to which TSO might accomplish these in practice:

1. **Encapsulation of complexity**. TSO is intended to enable parallel architectures to be exploited for GIS operations by developers who have little expertise in parallel processing. The PTF functions, proposed to create geometric and topologic objects within a strip, would allow applications to generate these objects in simple steps. The TSO framework hides the issues associated with message-passing and multi-threading. The details of the vector-topology data structures, the creation of polygons, the management of holes in polygons, and the need for stitching and collation of data would all be hidden from the developer of an application. Sequential versions of the PTF functions and TSO framework would allow prototyping of an operation upon one workstation, and then exploitation of a parallel architecture without further code development.

2. **Portability between parallel architectures**. This could be achieved by building an implementation upon MPI (or another messaging standard).

3. **Speed and scalability**. The need for stitching and for collating distributed objects becomes more demanding as the data decomposition becomes more fragmented. The use of domains was proposed to help to control these overheads.

4. **Reusability**. TSO has potential application in a number of GIS operations, as is shown by Chapters 12 and 13 which investigate raster-to-vector conversion and vector polygon overlay operations respectively.

9.8 References

Harding, T.J., Mineter, M.J., Sloan, T.M., Wilson, A.J.S., Dowers, S. & Gittings, B.M., 1993, *Raster to Vector Conversion and Vector to Raster Conversion Concepts*, EPCC-PAP-GIS-R2V2R-CONC, EPCC PUB. NO. EPC-CC93-07, EPCC and The University of Edinburgh, Department of Geography, Edinburgh.

Message Passing Interface Forum, 1994, *MPI: A Message-Passing Interface Standard*, University of Tennessee, Knoxville, TN.

Mineter, M.J., Harding, T.J., Dowers, S., Healey, R.G., Chapple, S.R. & Trewin, S.M., 1993, *Raster to Vector Conversion: High level Design*. EPCC-PAP-GIS-R2V-HLD EPCC Pub No. CC94-09, EPCC and The University of Edinburgh, Department of Geography, Edinburgh.

Mineter, M.J., Harding, T.J.,Welsh, E., Dowers, S., Healey, R.G., Trewin, S.M. & Chapple, S.R., 1994, *Topology Creation, Stitching and Output Component of Raster-to-Vector and Polygon Overlay Operations: Design Document*. EPCC-PAP-GIS-TSO-DDD-EPCC and The University of Edinburgh, Department of Geography, Edinburgh.

Yourdon, E., 1989, *Modern Structured Analysis*, Prentice Hall Inc.

10

Partitioning Raster Data

M.J. Mineter

10.1 Introduction

Partitioning raster data for parallel algorithms requires methods for reading data
from disk, distribution of data amongst processors, gathering resultant data from
processors, and writing data to disk. A number of alternative approaches exist.
Each imposes a different pattern of coordination and communication between
processes, suited to some algorithms and unsuited to others. This chapter
investigates these options. Following the theme established in Chapter 7, the
emphasis is upon techniques for dividing a raster into rectangular areas, termed
'sub-raster blocks'.

Section 10.2 explores the characteristics of different types of raster algorithms and
their implications for parallelisation, from the viewpoint of designing software
comprising a source process to distribute the data, worker processes, each of which
manipulates one or more SRBs, and a sink process to gather and write out the
resultant data. The intention is to illustrate some of the key issues, rather than to
categorise raster algorithms comprehensively, or give detailed designs. Different
standard methods of decomposition, well-known from applications using grid data
(Chapter 4), are explored in section 10.3. Against the background of the previous
sections, a hybrid method of SRB decomposition is then proposed in section 10.4
to meet the requirements of the conversion from raster to vector-topological data
(Chapter 12).

10.2 Implications of the Characteristics of Algorithms

10.2.1 Scope of Data Required by Algorithms

Some algorithms process one pixel at a time, with no regard to any other data;
others process each pixel in association with a defined extent of neighbouring
pixels; yet others require that all pixels within a region be identified and processed
in association with each other. (A region is an area of a raster comprising

connecting pixels, either containing the same value or else segmented according to another criterion.) Several examples related to remotely-sensed data are given in Chapter 18. These three types of algorithm, which correspond closely to those identified by Berry (1987), are discussed below.

Single pixel algorithms are defined to be those in which each pixel is processed in isolation. They are simplest to parallelise, as they require data for each pixel to be held by one processor only. Data decomposition can be arranged to achieve maximum efficiency between source and workers and between workers and sink processes, as no other communication need occur.

Defined-extent algorithms are those in which each pixel must be processed in the context of a pre-determined number of surrounding pixels. For example, the recognition of boundaries between regions entails comparison between adjacent pixels, and moving-window filtering (Chapter 14) processes each pixel with regard to a surrounding window, usually of sides of between 3 and 15 pixels long. The major implication is that the processing of pixels near SRB borders requires information from neighbouring SRBs. Each process thus holds its own block in a larger array configured to include a 'halo' of data assigned to surrounding blocks (Figure 10.1). In order that the contents of all pixels are uniquely determined each halo is read but not updated during the processing of a particular SRB. In iterative algorithms, the haloes must be updated after each iteration. The haloes either are distributed with each SRB by a source process, or, especially in the case of an iterative algorithm (e.g. moving-window filtering), are communicated in messages between the Workers.

Region-based algorithms present a dilemma. If decomposition into SRBs is used then many regions will span SRB borders, and additional messaging, and perhaps additional algorithms, will be required to account for the region fragmentation (as in raster-to vector conversion, Chapter 12). The alternative approach avoids these complexities by decomposing the raster by region, but recognising the pixels within a region entails an additional, preliminary phase of processing and gives rise to other problems which are now discussed. Consider a design for this second approach, in which the source process identifies and distributes regions. This would constitute an initial phase of processing which could act as a bottleneck. To achieve scalability, the time taken by Workers to process regions would need to be large in comparison with the time taken to recognise and distribute them. In cases where all regions were known to be small, and their number was large in comparison with the number of processes, then this approach, built upon a Task Farm (Chapter 4), might be effective. In a more typical case these constraints are not met, so that, for example, a small number of regions might extend across the complete raster reducing the likelihood of either efficient data distribution or load balancing. A further issue to be overcome concerns cases where it is necessary to generate a resultant raster dataset from the region-based decomposition. Each scan-line will comprise pixels from a number of regions, and require assembling

a) Regular decomposition into 16 SRBs for a 'single pixel' algorithm

Data held by one processor comprises one SRB

b) Regular decomposition into 16 SRBs for a 'defined extent' algorithm

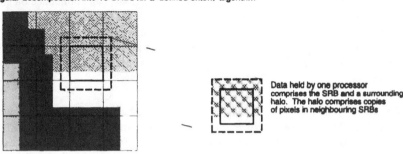

Data held by one processor comprises the SRB and a surrounding halo. The halo comprises copies of pixels in neighbouring SRBs

c) A dubious decomposition for a 'region–based' algorithm

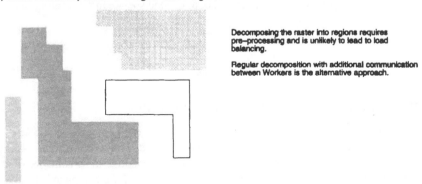

Decomposing the raster into regions requires pre–processing and is unlikely to lead to load balancing.

Regular decomposition with additional communication between Workers is the alternative approach.

Figure 10.1: Possible strategies of decomposition for single pixel, defined-extent and region-based algorithms

prior to writing out. Subsequent scan-lines would require pixels from different regions, and different processes. Ensuring that the necessary control data were made available to the Workers, as well as the collation itself, would entail a significant overhead. It is for these reasons that this chapter focuses upon decomposition into blocks rather than regions.

10.2.2 Iterative or One-pass Algorithms

The choice of the manner of data distribution is influenced by whether the algorithm to be parallelised is iterative, as in moving-window filtering, or is one-pass, as in a conversion, for example to vector-topology. The nature of iterative algorithms is that, in the absence of virtual memory, all data must be held in memory following one iteration before the next iteration can commence. In contrast, most one-pass algorithms, such as conversions, can commence on some processors whilst initial data are still being received by others. This is particularly trivial in 'single pixel' or 'defined extent' algorithms. Decomposition can be organised so that the writing out of the resultant database can be commenced before all the initial data have been read, so giving rise to 'streaming' of data through the algorithm.

To achieve load balancing, some differences in emphasis result. In the case of one-pass algorithms the focus will tend to be upon activating all processes quickly, and avoiding queues for SRBs from the Source. A task farm can be an effective way in which a large number of small SRBs may be distributed, especially for single-pixel algorithms. In the case of iterative algorithms, a coarser decomposition will generally be advantageous, especially in the case of defined-extent algorithms when using fewer, larger SRBs reduces the amount of time spent communicating 'halo' data.

10.2.3 Sensitivity of Processing to Raster Contents

The simplest way to decompose a raster is to allocate each processor one SRB, where each SRB comprises the same number of pixels. However, in some cases the amount of processing required by each SRB will be sensitive to its contents. For example, in converting raster to vector-topological data if the complete SRB lies within one region, then little processing is required. If many regions exist in the SRB then the converse is true. In these cases, regular decomposition will rarely lead to load balanced performance.

A number of approaches are used in such cases, as discussed in section 10.3. Either they divide the raster into smaller SRBs, so that each processor receives more than one SRB, or the raster is pre-processed to allow one SRB only to be sent to each process, with the number of pixels in each SRB being different, so that the amount of work is balanced.

10.2.4 Raster Data Formats

Raster data can be held on disk as a simple array, or as a compressed dataset. The compression can be line-based (e.g. run-length encoding) or area-based (e.g. quadtree). The format influences the ease of reading and distribution of data, and of collating and writing out of data, as well as potentially influencing the parallelisation of particular algorithms.

10.2.4.1 Simple Array

Holding a raster on disk as a simple array allows straightforward reading, distribution, collation and writing of data. The techniques are identical to those used in other grid-based applications. If parallel I/O is available, Chapter 4, then one possibility is for each Worker to read its own data from disk. This is because the simplicity of the data format allows random access to pixels. The more general approach is to use a source process to read data and then to distribute SRBs in messages. In writing out data which are decomposed across a number of processes, similar issues apply and the same two approaches are evident. The first is for the data held by each process to be written using random access. The second is to collate data in the memory of a sink process and then write data out.

The use of source/sink processes to distribute/collate blocks is likely to be preferable, as disk I/O tends to be slower than the passing of messages. Random access to disks is particularly slow as the buffering of file contents will be less effective than for sequential access. However, if the number of Workers is sufficiently large that the sink and source comprise a bottleneck, then alternatives may be necessary. For example, if parallel I/O is available then the work of the source and sink processes could be shared by a number of processes. These could read (or, for the sink, write) data in chunks corresponding to several SRBs, and be responsible for the distribution and collation of data for a subset of Workers, so avoiding the problems of a large number of disk accesses.

10.2.4.2 Scan-line Compression

In the discussion which follows, the compression is assumed to conform to the format of section 6.6.2, i.e. each scan-line is compressed into one variable-length record, and the scan-lines are held on disk in their correct order. Recalling the discussion of the previous section, in this case random access to scan-lines is only possible if a separate scan-line index is generated, giving the length and start location in the disk-file of each scan-line; random access to individual pixels is not feasible. In writing data, due to the variable length of the compressed scan-lines the location of each scan-line cannot be anticipated and so random writing of compressed scan-lines is not feasible.

In the event that decomposition is such that consecutive parts of a scan-line are held in different processes (as in Figure 10.1), scan-lines themselves need to be assembled before they are written out. In this case the following options arise:

- One process assumes the role of 'Sink' and receives data from all others, assembling the complete dataset. This could be a bottleneck, as each partial scan-line in each SRB would comprise one message.

- Each process assembles a subset of scan-lines, and writes these out in turn. This requires a small amount of additional synchronisation, so that after the first subset has been written, a message is sent to the process holding the second subset, and so on. In order that scan-lines be collated, each piece of a scan-line within an SRB must be sent to the relevant process in a message with sufficient data that the complete scan-line can be assembled. On architectures such as the Cray T3D where each process can write to the memory of other processes, then a synchronise-write data-synchronise sequence would be most effective, using a simple array to collate the data. Once collected, each scan-line can be compressed prior to writing out.

In decomposing a run-length encoded raster the length of the record of each scan-line will give some indication of the number of different spans which are present on each scan-line, and so aids attempts at decomposition with regard to data complexity, an approach used in section 10.4.

The discussion of compressed scan-lines has thus far focused upon the problems in distributing and in collating data. There are also some aspects of this format which are helpful in parallelising GIS algorithms: whether they compensate for the additional effort in managing the compressed data would require testing. The data in an SRB may be held either as a simple array, or in a compressed form. In some cases, such as in the raster-to-vector conversion of Chapter 12, maintaining the run-length encoding will speed processing. This is because the extents of spans can be obtained directly, without inspecting pixels.

Were the data to be run-length encoded with wrap-around from one scan-line into the next, then the possibility of building an index into scan-lines would not arise, and the data would need to be read in chunks and interpreted span-by-span. This illustrates the general pattern, that the more compressed is a dataset then the more pre-processing is required before data can be distributed or written.

10.2.4.3 Quadtrees

Four of the possible options for reading and distributing a quadtree-encoded dataset are now considered:

1. to expand the raster into a simple array in a preliminary phase.

    This reduces the function of the quadtree to one of compression, although to avoid a bottleneck multiple processes might need to be involved, so the following options would then be considered for accomplishing the expansion.

2. to distribute the leaf-nodes of the quadtree so that each process receives an approximately equal number.

3. to use an SRB decomposition which maps directly onto a mid-level of the quadtree. If multiple input datasets are used, then this approach seems most viable, in order that corresponding data can be held on the same processor, otherwise excessive sorting and communication would result.

4. to assign to each Worker a branch of the quadtree comprising a similar number of nodes.

The second option will be effective in some cases, especially for single-pixel algorithms. However, for defined-extent and region-based algorithms, it is necessary to distribute sufficient information that for each leaf-node the process holding its neighbours can be identified. The fragmentation of the dataset can give rise to significant communication overheads, as the work of Mukai (1990), discussed further below, found.

The third option seems to offer most scope, but the question of how to select the appropriate level of decomposition arises (Figure 10.2). For example, supposing that 16 processors were available, the immediate inclination is to select a mid-level at which 16 nodes exist (corresponding to regular decomposition in the section 10.3), so that each processor receives one sub-tree, corresponding to an SRB holding 1/16th of the data. However, as Mukai noted, neither the quadtree nor, therefore, the loading of processors is likely to be balanced. This approach would rule out use of a number of processes which was not a power of 2. Consequently a lower mid-level is likely to be preferable, so that each processor receives a number of small quadtrees rooted at that mid-level. Use of an arbitrary number of processes would result in some processes receiving one more sub-tree than others. Some leaf-nodes of the quadtree will occur at higher levels than that used for decomposition; these would need to be divided in order that each processor held a node at the same level. This would be similar in effect to the scattered spatial decomposition discussed in the next section. The same difficulty exists: a more fragmented decomposition (e.g. a lower level in the quadtree) improves chances of load balancing, but increases the overhead of parallelism for defined-extent and region-based algorithms especially.

The fourth option corresponds to irregular decomposition, also discussed in the next section. Information about the structure of the tree, especially about the sizes of the sub-trees below each node, is needed. The decomposition will typically be more intensive in processing and communication than would be expected from the third option. Information allowing communication with processes holding neighbouring leaf-nodes would be similar to that of option 2 above, more complex than in the third option.

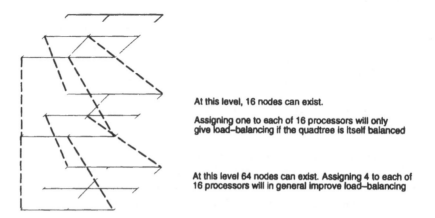

At this level, 16 nodes can exist.

Assigning one to each of 16 processors will only give load–balancing if the quadtree is itself balanced

At this level 64 nodes can exist. Assigning 4 to each of 16 processors will in general improve load–balancing

Figure 10.2: Selection of the node-level for quadtree decomposition

Mukai (1990) processed binary images, and recognised the advantages of using a linear quadtree, in which each leaf-node is identified by the coordinates of one corner and the size of its corresponding square. Decomposition by leaf-node was used so that 'single pixel' algorithms such as the computation of the area and moment of the black regions of the image were trivial, entailing each process generating the contributions of its own leaf-nodes, and communicating these to a Sink process for summing. Efficiency is dominated by the speed of disk I/O and leaf-node distribution. In terms of the decomposition methods used in the next subsection (regular, irregular, scattered), this decomposition, the second option mentioned above, can be seen to comprise a hybrid "irregular, scattered" decomposition in which the degree of scattering is controllable by a source process. The suitability of the method for a defined-extent algorithm is thus a matter of concern, which was investigated by Mukai (1990) using algorithms to compute perimeters.

The first approach Mukai described followed Bhaskar *et al.* (1988) and sent all leaf nodes to all processes. The costs of communicating the entire dataset to all processes, and the memory demands which result, led to a second approach being tested. This entailed using a network of INMOS T414-20 transputers, each with 256 Kbytes of memory on a network with bandwidth 20Mbits/sec, and distributing a subset of leaf-nodes to each processor. The problem to be overcome is that of identifying the locations, and hence the values, of neighbouring leaf-nodes. Mukai used a parallel search, initiated by a source process for each leaf-node in turn. Efficiency of the search relies upon multiple searches being completed at the same time, which in turn requires that neighbouring leaf-nodes be deliberately scattered across different processors, losing the spatial contiguity. A speed-up of 11 was achieved with 32 processors and a quadtree comprising 8192 nodes. With 16 processors a speedup of 8.5 was attained. It would be interesting to compare the behaviour of the perimeter computation using the third option noted at the start of this section. Firstly, the scattering of leaf-nodes would be reduced, so reducing the load of communication, and secondly communication could be between workers, rather than initiated by a source process, so eliminating a potential bottleneck.

The generation of a quadtree encoded raster from a regularly decomposed raster array is also a problem which the 3rd option might address. By redistributing the data so that SRBs corresponded to mid-levels in the quadtree, each Worker could then generate its own sub-tree.

10.2.5 Multiple Datasets

A variety of algorithms require that multiple datasets be analysed, for example to generate an overlay. The issues here are the same as for one dataset, but the likelihood of a bottleneck is enhanced as more data have to be read and distributed. Using one source process for each dataset is thus an obvious possible approach, ensuring that corresponding SRBs are sent to the same Worker. This in turn requires either that the Source processes perform preliminary analyses of data, before jointly determining the pattern of decomposition, or, more simply, that a pattern of SRBs be determined using regular or else scattered spatial decomposition.

10.3 Methods of Decomposition into SRBs

Different modes of decomposition into SRBs are discussed in this section. Section 10.3.1 discusses the most frequently used method, suited to well-balanced problems, which for raster data are those in which processing time for an SRB depends on the number of pixels and not on their contents. The remaining three methods are suited to unbalanced problems. Irregular decomposition attempts to maintain the advantages of having one SRB per process; scattered spatial decomposition and task farm are alternative approaches to achieving load balancing by distributing many SRBs to each process.

10.3.1 Regular Decomposition into One Block per Process

Regular decomposition is favoured in cases where processing time is not very sensitive to the contents of pixels, so that each process can receive one SRB, and each SRB can contain the same amount of data without seriously compromising load-balancing.

It is interesting to compare two simple modes of decomposition in the case of one-pass algorithms. If a source process divides each scan-line into equal-width sections, one per process, and distributes these sections in turn then all processes will be able to commence processing relatively quickly, and they will each receive (eventually) one full-length column of data. If instead, one or more full-width rows are sent to each process in turn, so that each process receives a strip of data comprising several full-width scan-lines, then the number of messages is lower but, especially at initialisation, processes are in effect queuing for data. The extent to which this queue would in fact impact upon performance depends on the speed of disk access, of data distribution, and, after initialisation, of strip processing.

A further compromise concerns the competing goals of minimising the total block perimeter and minimising the number of messages required to communicate data across block borders. On many architectures, and in particular when networked workstations are being used, performance will benefit from decomposition requiring the minimum number of messages to effect an exchange of data at block boundaries. This is because the initiation of an extra message is more costly than increasing the size of a message. However, region-based algorithms, for example, may need to take some account of SRB shape. The closer a block is to being square, the greater may be the percentage of processing which can complete without the need for inter-process communication amongst workers. In such cases the benefits of minimising perimeter length might be more apparent.

Figure 10.3 compares two approaches. The library PUL-RD was used in processing a raster 2401 pixels square (see Chapter 14). It produced a compromise between square-ish blocks and minimisation of the number of messages being exchanged. Lee & Hamdi (1995) investigated a method of decomposition termed 'heuristic partitioning' in which the goal is to minimise block border length, despite the need for additional messages. This decomposition is accomplished by recursively dividing the dataset into two sub-images, and assigning each sub-image half of the available processes until each process is assigned its own sub-image. In each division the direction of the new border is chosen to minimise the new border length. Where a sub-image is assigned to n processes, the division is into areas of the ratio $m:(n-m)$, where $m=(integer)n/2$.

Decomposition used by PUL–RD:

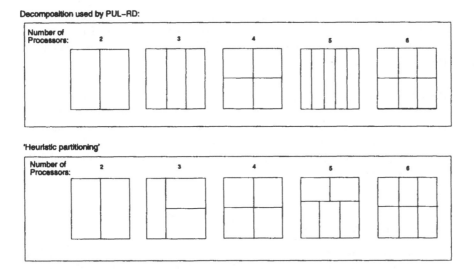

'Heuristic partitioning'

Figure 10.3: Two methods of raster decomposition: PUL-RD seeks a compromise between number of messages and block shape. 'Heuristic' partitioning seeks to minimise block border lengths.

10.3.2 Irregular Decomposition

Irregular decomposition is suited to cases where the processing time associated with a pixel is dependent on the contents of the pixel, or on the complexity of data, for example the number of spans or regions in an area of the raster. Raster-to-vector conversion (Chapter 12) is an example of such an operation. In irregular decomposition, one or more source processes distribute SRBs, one per process, with the size of each SRB being chosen with regard to the processing it will require. Hence different processes will receive different sizes of SRBs in an attempt to load each processor to the same extent. This approach is less suited to those iterative algorithms in which the load balancing will be modified during processing: making adjustments to that load balancing would have a high overhead.

Decomposition to achieve the same number of spans in each SRB would require that the complete image be analysed, in a preliminary pass through the whole dataset. This analysis, facilitated by run-length encoded data, would determine the number of spans in different parts of each scan-line, and from this determine suitable SRB boundaries. The amount of processing could be reduced by making the assumption that neighbouring scan-lines were similar, so that not all need to be examined. Again, run-length encoding as described in Chapter 6 would give a clue as to the possible accuracy of this assumption, in that the number of spans in a scan-line can readily be assessed. Once the distribution of the data is derived in some manner, then the issue of dividing the raster into SRBs needs to be addressed.

The 'heuristic partitioning' method, in the previous section, could perhaps be developed to divide data according to complexity rather than area. The effect would be similar to the method of Dynamic Alternating Recursive Bisection in Chapter 17.

It is clear that the disadvantages of this method are that the preliminary analysis is necessary and non-trivial, and that in iterative algorithms the distribution of work must not change if load-balancing is to be maintained. The advantage is that once the data decomposition is accomplished, then the overheads in defined-extent, iterative algorithms would be lower than in the alternatives, because the total SRB perimeter length would be smaller.

10.3.3 Scattered Spatial Decomposition

Scattered spatial decomposition entails decomposition into a large number of SRBs of the same size. Each process has the same number of SRBs, scattered throughout the data-space in the hope that the SRBs needing more intensive processing are distributed evenly. The additional fragmentation of the dataset can impact on performance by imposing demands for additional messaging, especially in defined extent or region-based algorithms, but in iterative algorithms where the distribution of intensively processed data might change this approach can provide a simple, fairly effective strategy.

10.3.4 Task Farm

A task farm also decomposes the dataset into a number of SRBs which is large in comparison to the number of processes. In this case, each Worker process holds one SRB at a time, and requests its next SRB on completion of processing of the previous SRB. Load-balancing is achieved because the different processes receive different numbers of SRBs during a run. An advantage is that little analysis is required by the source process. The standard task farm does not include communication amongst Workers, and so is suited to single-pixel algorithms, where only the controlling process needs to be aware of the spatial relationships between SRBs. Iterative algorithms would require that data be written to disk and then redistributed at each iteration.

10.4 SRB Distribution for Raster-to-vector Conversion

This section describes an approach to the distribution of raster data which is suited to the requirements of the raster-to-vector conversion operation in Chapter 12, namely to:

- Distribute a run-length encoded raster dataset which could be large in comparison to the memory of one (or all) processes.

- Support an operation in which processing is strongly data-dependent

- Support a one-pass algorithm

- Provide enough information that Workers can carry out a defined-extent algorithm, recognising borders of regions.

The following aspects of a standard task farm source process functionality are appropriate for this data decomposition:

- One process acts as a source distributing data for worker processes. This requires that the source 'Sub-raster Block Distributor' (SRBD) process is relatively fast and simple: it will, especially at initialisation, be a sequential bottleneck. (If it remained a bottleneck after initialisation, then the design would have to be extended to allow multiple processes to use parallel I/O, each serving a subset of Workers, and communicating with each other to distribute information about the extent and worker process identification of each SRB.)

- The SRBD can read-ahead in accessing the raster dataset, preparing data in anticipation of requests from worker processes.

- The SRBD decomposes the data into more SRBs than there are processors. It distributes one to each process in turn at initialisation; each subsequent SRB is distributed in response to a 'request-for-SRB' message from a worker.

- A worker sends such a request when its last received SRB has been processed. (In practice, coding the worker process such that it anticipates the completion of processing of the last SRB would reduce the idle-time pending receipt of the next SRB.)

The above functionality requires extensions to modify standard task farm functionality in a compromise with the goals of irregular decomposition (section 10.3.2 above). The compromise is due to the benefit to be gained in allowing the sizes of SRBs to be determined according to:

- The goal of minimising the overhead arising from polygons which cross SRB borders. This can be achieved by using larger SRBs and to an extent by avoiding elongated shapes. (Such shapes are less likely to contain full polygons.)

- The number of requests already received by the SRBD. The goal is to reduce the time for which Workers are waiting for data. This can be achieved by using smaller SRBs.

- The complexity (i.e. density of spans) in different areas of the raster.

- The availability of memory in a Worker. In memory-limited architectures, it is advantageous for the worker processes to specify, in each request for an SRB, the maximum effective-size of SRB which they can accommodate. The 'effective-size' of an SRB is determined by the number of spans within it. (In raster-to-vector conversion this requires a relationship between the number of spans and the memory required to hold the vector data to be derived from experience.)

These requirements are in obvious conflict with each other. It is proposed that parameters determined from experience can resolve this conflict, allowing weightings to be applied to the different factors in determining the sizes of SRBs (Welsh *et al.*, 1993).

The SRB Distributor process:

- Receives messages requesting data.

- Establishes the vertical borders between SRBs according to the number of idle processors and scan-line complexity.

- Distributes each scan-line in turn, splitting it according to the SRB vertical borders. On some architectures, minimising the number of messages by buffering several rows of an SRB into one message will improve the speed of data distribution.

- Terminates SRBs either before any parameter-imposed or worker-imposed limits on SRB size are broken, or between consecutive scan-lines which have very different numbers (or distributions) of spans. In this last case, terminating SRBs at this change in complexity helps to reduce communication during the processing of SRBs. In the case of run-length encoded data in the format described in section 6.8, although the number of spans along each scan-line is directly available from the dataset, or can be estimated from the length of the scan-line record, the distribution of the spans along a scan-line is not available without span-by-span analysis. This analysis is a task which the SRBD could do as a background activity, so that it improves but does not delay SRB decomposition. All concurrently distributed SRBs are terminated at the same scan-line.

- Ensures that SRBs overlap by one pixel, otherwise region borders which fall on SRB borders would not be recognised.

- Data concerning the borders of the SRBs are distributed to the workers in order that communication of data across SRB borders can be performed when required.

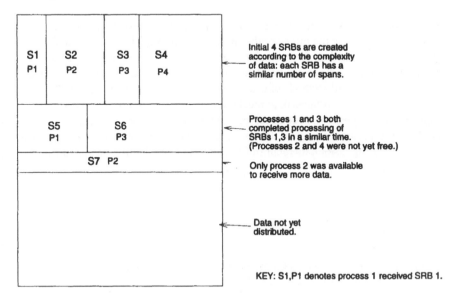

Initial 4 SRBs are created according to the complexity of data: each SRB has a similar number of spans.

Processes 1 and 3 both completed processing of SRBs 1,3 in a similar time. (Processes 2 and 4 were not yet free.)

Only process 2 was available to receive more data.

Data not yet distributed.

KEY: S1,P1 denotes process 1 received SRB 1.

Figure 10.4: Example of SRB decomposition using 4 processors. The horizontal decomposition is arranged according to the data complexity and the number of available processes. The length of SRBs is determined by available memory and the number of processes awaiting further data.

The data structures within the SRBD hold:

- The block of the raster currently being distributed.

- SRB decomposition control information.

- A queue of SRB parts awaiting transmission in messages to each worker currently receiving an SRB.

- A list of process identifiers for worker processes awaiting an SRB, with associated worker-defined limits on SRB sizes.

An example decomposition is shown schematically in Figure 10.4. (Figure 9.9 used the same example to illustrate the discussion of TSO.)

10.5 Summary

This chapter has discussed a range of possibilities for processing raster data in parallel, emphasising the often conflicting goals of minimising the overheads of communication and data decomposition, avoiding bottlenecks, achieving load-balancing, and providing near-neighbour data for region-based and defined-extent algorithms. Standard methods of decomposition were described, and finally a hybrid approach was proposed, to meet the requirements of raster-to-vector conversion (Chapter 14). The decomposition into SRBs leads to the need for stitching the vector-topological data from each; this is the topic of Chapter 9. The issues of creating raster data recur in Chapter 11, in considering vector-to-raster conversion.

10.6 References

Berry, J.K., 1987, Fundamental operations on computer-assisted map analysis. *International Journal of GIS*, **1** (2),119-316.

Lee, C.K. & Hamdi, M., 1995, Parallel image processing operations on a network of workstations. *Parallel Computing*, **21**, 137-160.

Bhaskar, S.K., Rosenfield, A. & Wu, A.Y., 1988, Parallel processing of regions represented by linear quadtrees. *Computer Vision, Graphics, and Image Processing*, **42**, 371-380.

Mukai, R., 1990, Parallel processing of quadtree images. *Proceedings of the 3rd Transputer/OCCAM International Conference*, Tokyo, 257-278.

Welsh, E., Mineter, M.J. & Harding, T.J., 1993, *Sub-raster-block Distribution Component of Raster-to-Vector Conversion: Detailed Design.* Document ID: EPCC-PAP-GIS-R2V-DD-SRB, EPCC Publication No. EPCC-CC94-10, EPCC and University of Edinburgh Department of Geography, Edinburgh.

Part Three

Parallelising Fundamental GIS
Operations

11

Vector-to-Raster Conversion

T.M. Sloan

11.1 Introduction

This chapter discusses the design of a parallel algorithm for a vector-to-raster conversion operation which utilises the parallel vector data input algorithm described in Chapter 8. This chapter therefore provides an example of the exploitation of a modular approach to parallel GIS software.

Section 11.2 describes the various sequential algorithms which can be used in the operation, with section 11.3 discussing possible parallel algorithms based on these and section 11.4 assessing the feasibility of these approaches. Based on the recommendations in section 11.4, sections 11.5 to 11.8 describe a parallel design for the operation.

11.2 Vector-to-Raster Conversion

The operation converts sets of non-overlapping polygons (see Chapter 6), each with an associated attribute, into a grid of arbitrary resolution where each grid cell, also known as a pixel, contains the relevant attribute data. The grid is also called a raster.

The conversion consists of two basic steps:

- the transfer of polygon boundaries to the raster grid;

- the filling of the pixels within the transferred polygon boundaries with polygon attribute values.

11.2.1 Existing Sequential Algorithms

Three classes of algorithm may be identified for rasterising the polygon boundaries:

- Frame Buffer Algorithms

- Image Strip Algorithms

- Scan-Line Algorithms

In a sense, this list represents a continuum of algorithms, with the Frame Buffer and Scan-Line algorithms at opposite ends. From the viewpoint of sequential programming, moving from the top of the list to the bottom, the running memory requirement for the image under construction decreases, but this must be traded off against an increase in processing (sorting) time.

If one chooses to parallelise one of these algorithms, there are further considerations. Prominent among these is whether one expects to use data decomposition on either polygons or the raster image or both. For example, the geographical area of the raster image may be split into a number of smaller geographical areas or the input polygons may be split into groups of, say, 100 polygons. The chosen data decomposition and the chosen algorithm should result in as few data dependencies between the sub-sets as possible in order to minimise communication. Therefore this chapter discussed a decomposition in sub-areas: raster output is more manageable than would be the case with polygon-based decomposition (Chapter 10) and the vector-topological data can be sorted as shown in Chapter 8.

References for the algorithms below are Foley & Van Dam (1984), Rogers (1985), Ackland & Weste (1981), Bouknight (1970), Jordan & Barret (1973).

11.2.2 Frame Buffer Algorithms

This class of algorithm is usually used in applications where the whole raster image can be held in memory. It is generally considered as an effective algorithm, but due to its high memory usage it is an unlikely candidate for handling very large datasets. It runs through the polygons, drawing edges according to any standard line-drawing algorithm (such as Bresenham's in Rogers (1985)). Foley & Van Dam (1984) contains a useful description of this type of algorithm.

As described in Ackland & Weste (1981), the polygon is then filled, using for example:

- **seed fill algorithms:** The fill is propagated from a seed point known to be interior to the polygon. Seed fill can be subdivided into:

    - **flood fill:** The fill is propagated two-dimensionally.

- **scan-line seed fill:** The seed fill proceeds a scan-line at a time (but note that the scan-line immediately preceding and immediately following must be checked for unfilled areas).

- **parity fill algorithms:** These algorithms are based on the idea that a given point is interior to a given polygon if and only if the number of edges of that polygon between the point and the raster boundary is odd. They include:

  - **edge fill:** This involves complementing all pixels to the right of each edge. Each pixel can be visited many times so I/O is high and it is only suitable for two-value attributes or monochrome images.

  - **edge flag fill:** This algorithm works a scan-line at a time, setting a flag to represent "inside" or "outside" as it moves across the scan-line and filling when the flag is set to "inside". It is more efficient than edge fill, can be adapted for multiple-value attributes or colour images and can produce run-length encoded output.

11.2.3 Image Strip Algorithms

This class of algorithm, described in Jordan & Barret (1973), exploits the edge coherence inherent in raster images of polygons. This occurs because the process of rasterising a straight line is a simple iterative loop, which is easy to break off and recommence so long as the following descriptor is stored:

- Δy, the remaining number of scan-lines that the line intersects,

- x, the intercept of the line with the current scan-line and

- Δx, the increment in x at each new scan-line (corresponding to the signed reciprocal of the gradient).

For example, with the polygon in Figure 11.1, for edge e1 at scan-line 2.5, then $x = x1$, $\Delta y = 2$ because the edge terminates at scan-line 0.5 and $\Delta x = (0.8 - 1.5) / (4.0 - 0.5)$.

The frame buffer is divided into image strips (groups of scan-lines). Before rasterising a given strip, an *active edge list* (AEL) is set up. It consists of a linked list of all the edges which intersect the strip (or rather a list of the descriptors above). These are then plotted and filled as in the frame buffer algorithm. Any unfinished lines are transferred to the AEL for the next strip.

The edges are initially bucket sorted (usually as an indexed linked list) into an *ordered edge list* (OEL) according to the image strip in which they will first appear

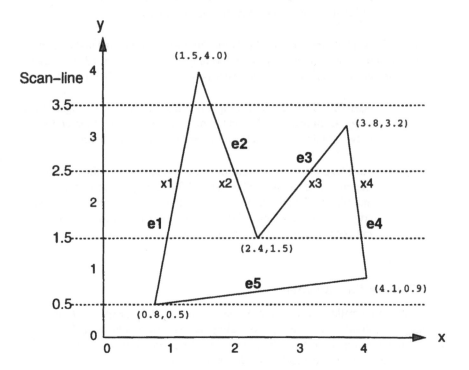

Figure 11.1: Example polygon for image strip and scan-line algorithms

using the maximum y coordinate of the edge. This simplifies the creation of the AEL for each new image strip since all that is required is to copy a section of the OEL. The use of the AEL and OEL in image strip and scan-line algorithms is the same and is illustrated in an example in Section 11.2.4.

Table 11.1: The x-sorted active edge list at scan-line 2.5 for polygon in Figure 11.1

edge	e1	e2	e3	e4
x	x1	x2	x3	x4
Δx	$\frac{0.8 - 1.5}{4.0 - 0.5}$	$\frac{2.4 - 1.5}{4.0 - 1.5}$	$\frac{2.4 - 3.8}{3.2 - 1.5}$	$\frac{4.1 - 3.8}{3.2 - 0.9}$
Δy	2	1	1	1

11.2.4 Scan-Line Algorithms

When the image strip consists of a single scan-line, the image strip algorithm becomes the scan-line algorithm. The AEL starts off null. At each scan-line, any new edges are added from the OEL and any finished edges are deleted. We have to be careful when a scan-line intersects the polygon at a vertex or we are left with an odd number of intersections. The solution is to count two intersections when the vertex is at a local y-minimum or y-maximum of the polygon and one otherwise.

The obvious fill algorithm to use is scan-line seed fill, edge flag fill or scan-line fill. For the last, the AEL has to be re-sorted on x coordinate for each new scan-line, although coherence ensures that the required re-sorting is usually small. The scan-line is filled between the resulting pairs of intersections, naturally producing run-length encoded output.

As an example, consider the scan-line conversion of the polygon in Figure 11.1. The OEL for the image consists of the five edges bucket-sorted as in Figure 11.2. At scan-line 2.5 there are four active edges which are sorted in order of x-intercept as in Figure 11.1. The scan-lines are numbered 0.5, 1.5, etc. because they are usually considered to run through the centre of the pixel line.

11.3 Parallelisation Strategies

This section discusses the merits of performing the operation in parallel using methods based on the sequential algorithms described in section 11.2. Each algorithm requires vector-topological input data consisting of the spatial location and attributes of the geographical features to be converted. As explained in Chapters 6 and 8, within vector-topological datasets, spatial and attribute information may be stored separately. As a result, the input data must be pre-processed to extract the information necessary for the algorithms. That is, in terms of BSi NTF Level 4, the original AREA, EDGEREC, ATTREC, ATTDESC, and GEOMETRY1 records of the input dataset must be processed to obtain the left/right attribute values for the GEOMETRY1 records. The GEOMETRY1 records are sorted by uppermost y-coordinate and split at maxima and minima in $\underline{y}$. This assists the conversion, since it ensures that once a GEOMETRY1 record leaves a pixel it never re-enters. A GeomYMax is created for each part of a split GEOMETRY1 record. These are sorted by GEOM_ID. This pre-processing is explained in section 8.4 and summarised in 11.3.1.

Sections 11.3.2 to 11.3.4 describe parallel algorithms based on Frame Buffer, Image Strip and Scan Line algorithms respectively. These usefully illustrate how a modular approach can be used in the design of parallel algorithms, since each

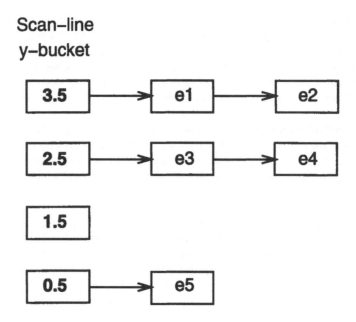

Figure 11.2: Ordered edge list for the polygon in Figure 11.1

utilises to some extent the parallel vector data input algorithm described in Chapter 8.

The strategies assume that the vector input dataset conforms to BSI NTF Level 4 and that run-length encoded raster data is the product of the conversions. Both these formats are described in Chapter 6.

11.3.1 Data Preprocessing

Preprocessing is accomplished by the Sort/Join/GAD functionality of Chapter 8 and results in each worker receiving vector-topological data for a subraster block. This approach is directly suited to the Image Strip and Scan-line algorithms.

The Frame Buffer algorithm requires slightly different processing as it does not explicitly require sorted spatial data and therefore less preprocessing is required. The Sort/Join stage can be modified to generage a Geom file sorted by GEOM-ID. The attribute processing would omit the GeomYMax file and instead the GeomFace file sorted by FACE_ID would be merged with the FaceValue file. The output of this merge stage would be sorted by GEOM_ID to be combined with the Geom file as part of the Geom Attribute Server or on the worker processes.

11.3.2 Frame Buffer

11.3.2.1 Data Distribution

Here, the area of the raster output is split into sub-rasters. The rasterisation of a sub-raster is dealt with by a single process, termed a sub-raster process. To aid load-balancing, the PartitionList created in the Sort phase of the Vector Input algorithm is used to create sub-rasters of comparable conversion workload.

A source process distributes records from the Geom and GeomValue files to sub-raster processes. It is not known which sub-raster process they should be sent to prior to their reading by the source process.

For those Geom records which cross more than one sub-raster block there are two options:

- They can be split and the parts sent to the relevant sub-raster processes.

- They can be sent in their entirety to all the relevant sub-raster processes, where each process converts only those parts which reside within its block.

An alternative, using parallel I/O, is for each sub-raster process to read the Geom file, retaining those that lie within its sub-raster. Those that do not, or only partially cross it can be sent to the relevant sub-raster process.

11.3.2.2 Data Conversion

Within each sub-raster process, Geom and GeomValue records are converted to a run-length encoded raster representation by a frame buffer type algorithm. Records are converted immediately and then discarded. Other approaches, such as image strip or scan-line, are not able to start converting as soon as data arrive at a sub-raster process but must wait until all data for a scan-line has been received. Since the input data used by the frame buffer algorithms is unsorted, all the input data must be read and converted before any line of the raster can be said to be complete and so be output.

When no more memory is available to a sub-raster process, the sub-raster block is written to a temporary file on disk. This allows the sub-raster process to continue receiving and converting any remaining data.

11.3.2.3 Data Output

Once all the GEOMETRY1 records with left/right attribute values are read and converted, each sub-raster block consists of at least one version resident in the memory available to a process and if the data is too large to fit in the process memory, one or more versions are also resident on disk. Before the raster image of the vector dataset can finally be output, not only must all the input data be read and converted but any sub-raster blocks consisting of more than one version must be merged into a single version. Each individual process performs this by writing its version stored on internal memory to disk. It then reads the multiple versions of its sub-raster, merging these versions on a line by line basis. These lines are output when complete, so freeing memory for merging further lines.

To remove this merging and the associated I/O overhead, the Source process could use the PartitionList to attempt to create sub-rasters of a size suitable to fit in the memory of a sub-raster process. However, since the input data is unsorted, it must all be read before a sub-raster can be said to be complete and therefore output. For particularly large datasets where more sub-rasters than processes are created, this means some input data is read more than once. This would likely negate any benefit from removing the merging of sub-rasters.

Since it is preferable for the raster lines to be output in the correct order, some form of synchronisation must take place between the sub-raster processes to ensure this. An alternative is for complete raster lines to be passed onto a sink process. This sink process stores lines in its available memory and in a temporary file until they can be output in the correct order. The source process can perform this role, since at this stage it has completed reading the input data and sending it onto the sub-raster processes.

11.3.2.4 Drawbacks

Primarily these are related to the accuracy issues of how pixels intersected by more than one attribute value are assigned. A Frame Buffer algorithm leaves a pixel assigned to the last attribute value that intersected it. Other issues of concern are the high memory usage of Frame Buffer algorithms since they must maintain a complete buffer of pixels representing the sub-raster until all the data is read. This is because raster lines cannot be said to be complete until all the input data has been read.

The accuracy issues can be addressed by only converting pixels when all the relevant data is available. Unfortunately this increases the memory usage further since input data on the sub-raster processes cannot be converted on arrival and immediately discarded. On the other hand, it is possible to determine when a raster scan-line is complete and so output, freeing memory.

With these modifications, the algorithm effectively becomes a variant of the Image Strip or Scan Line algorithms. In such circumstances the parallel vector data input algorithm, as described in Chapter 8, can be used in its entirety since it spatially sorts the input data before sending all the relevant data for a sub-raster to a process. Using this also means that suitable sized sub-rasters for a process are created, so removing the need for merging multiple versions of sub-rasters. Further, this is achieved without having to read the input data more than once in those circumstances where there are more sub-rasters than processes.

11.3.3 Image Strip

The parallel vector data input algorithm, described in Chapter 8, quite naturally sorts the Geom and GeomValue records into image strips (groups of scan-lines). In the GAD phase of vector data input, the Source process uses the PartitionList to assign sub-rasters of a suitable size for a Worker process to convert. These sub-rasters can be considered as image strips. These are sized to ensure that the memory of a Worker process is not swamped and no temporary disk space is required. All the records necessary to convert a sub-raster are sent by GeomAttributeServer processes to the appropriate Worker process.

At this point in the parallel vector data input algorithm, a Worker process would sort this incoming data to complete the spatial sort. However, if the Image Strip algorithm is used this is not necessary. Instead, an active edge list is created and the algorithm proceeds as described in section 11.2.2 with a Frame Buffer algorithm converting the data to its raster representation.

The disadvantages of this method are again related to the accuracy and memory concerns of Frame Buffer algorithms.

On completion of the conversion, as with the parallel Frame Buffer algorithm, the complete raster lines could be passed onto the Sink process, which could in fact be the Source process, for output in the correct order.

11.3.4 Scan-line

The scan-line algorithm requires Geom and GeomValue records completely sorted by uppermost y-coordinate. The parallel vector data input algorithm, described in Chapter 8, produces such data for the sub-rasters specified by the Source process.

Due to the possible sinuous nature of a GEOMETRY1 record the left/right attributes values may be required a number of times not only in different scan-lines but also within a scan-line. The ordering of these values is therefore not important and so these are placed in a hash table to reduce memory requirements and to enable quick access.

On a worker process, the conversion proceeds sequentially down the sub-raster a scan-line at a time. The conversion for the next scan-line in the sequence can start as soon as the last GEOMETRY1 record which crosses it emerges from the merge-tree. Conversion can therefore start before all the GEOMETRY1 records are sorted.

Once a scan-line is converted, those GEOMETRY1 records not necessary for conversion of the next scan-line can be discarded, so freeing memory. Scan-lines can be output, as for the Frame Buffer and Image Strip algorithms, via a sink process which could be the source process or instead via synchronisation between processes to ensure correct ordering.

11.4 Parallel Algorithm Selection Criteria

In choosing a possible parallel algorithm, the criteria suggested by Piwowar *et al.* (1990) for assessing sequential vector to raster algorithms together with the ease of efficient parallelisation and the ability to handle large datasets can be considered. The criteria are described below.

11.4.1 Quality of Polygon Fill

Seed fill has the disadvantage of requiring a seed point for each polygon. Polygons which become *pinched off* (a land area with a narrow isthmus may, for example, be rasterised as two separate areas) need extra seed points. Seed fill is also susceptible to catastrophic failure if boundary pixels are missing. Image strips algorithms with seed fills tend to produce fill errors because pixels on the boundaries of the strip may be difficult to allocate to polygons.

Overall, the most reliable fill algorithms are scan-line fills.

11.4.2 Accuracy

Accuracy can be assessed by comparing a layer in one format (vector or raster) with the same layer after conversion to and from the alternative format. Perimeter lengths tend to grow with all rasterisation algorithms, but in general accuracy reflects the assignment of *cut cells* and *small polygons*.

Cut cells:

A choice has to be made as to how pixels intersected by more than one polygon are attributed.

Small polygons:

A choice as to whether to represent polygons smaller than one pixel also has to be made.

Scan-line algorithms can make decisions about the allocation of a particular pixel with knowledge of all the edges intersecting that scan-line, whereas frame buffer algorithms leave a pixel assigned to whichever polygon intersected that pixel last.

Using the scan-line algorithm, the choice can be made a number of different ways, some of which are listed below.

- The attribute value associated with the polygon identifier with the largest area in the pixel is assigned.

- The attribute value with the largest area is assigned to the pixel. For example, an attribute can be associated with two areas present in the pixel, each smaller than a third area but together having a greater area.

- *Attribute weighting* is often incorporated into a vector-to-raster conversion operation. Polygon attribute values are assigned certain weights and these are used to determine the polygon attribute to be assigned to a pixel which is crossed by more than one polygon. To determine the attribute value to be assigned to a pixel crossed by more than one polygon, the area of each attribute value in the pixel is multiplied by its weight and the attribute value with the highest aggregate weight is assigned to the pixel.

- *Attribute priority* is where priorities are assigned to polygon attribute values. When a pixel is crossed by more than one polygon, the polygon attribute value with the highest priority is assigned to the pixel. Where two attribute values with the same priority cross a pixel, the attribute value with the greatest area is assigned to the pixel. This method maintains topological correctness at the expense of areal accuracy.

Following from this further considerations are:

- Areal accuracy;

- Perimeter accuracy;

- Displacement accuracy.

11.4.3 Efficiency

There are three parameters to consider:

- Memory;

- Processing time;

- Amount and nature of I/O, e.g. random access.

The memory requirement of the algorithms can be traded off against the processing time spent sorting edges. According to Piwowar *et al.* (1990), the sequential scan-line algorithm takes between two and three times as long as the other two algorithms, which are of roughly equal speed, and all the sequential algorithms have roughly equal disk I/O requirements.

11.4.4 Data Volume Limitations

GIS packages are being used to manipulate larger and larger datasets. Consequently, the parallel algorithm for the operation cannot assume that either the entire raster or vector-topological dataset can be held in the main memory of the operational platform.

Furthermore, it is desirable that as the volume of the data increases, the performance should degrade predictably, and if possible smoothly, rather than reaching a threshold beyond which drastic degradation occurs.

The Frame Buffer algorithm retains a pixel buffer until all the input has been read and converted. The performance of this algorithm degrades when this buffer is not completely held in memory. Parts of the pixel buffer must be read and written to disk when input data are not located in the part currently held in memory. All the parallel algorithms described in section 11.3 use the PartitionList generated by the parallel vector data input algorithm, to load-balance the conversion across processes. For the Frame Buffer algorithm with larger datasets, this can be used to ensure that the pixel buffer always fits in memory. However, this means that smaller areas are converted by Worker processes compared to other algorithms, so potentially slowing the performance of the algorithm in parallel compared to the other algorithms. The Image Strip algorithm also uses a pixel buffer and so also suffers from this.

11.4.5 Recommendations

The scan-line algorithm has reduced memory requirements compared to the Frame Buffer and Image Strip algorithms since a pixel buffer is not required and raster lines can, if necessary be output immediately on conversion. Also, only the data required to convert the current scan-line need be held in memory. This reduced memory requirement means larger sub-rasters can be converted by the Worker processes.

Whilst the sequential scan-line algorithm is slower than the others due to its dependence on sorted data, it produces better quality raster imagery. It is able to address the accuracy issues of cut cells and small polygons since all the information on the GEOMETRY1 records which cross a pixel are available at conversion.

It is clear that the scan-line algorithm is the most likely candidate for parallelisation due to its better quality output, reduced memory requirements and ease of handling large datasets. Section 11.5 describes a parallel algorithm for vector-to-raster conversion which utilises the parallel vector data input algorithm of Chapter 8 with the scan-line algorithm.

11.5 Design Overview

This section discusses the design of a parallel algorithm for the vector-to-raster conversion operation using the scan-line algorithm and the parallel vector data input algorithm.

The algorithm utilises two main modules: the parallel Vector Input module and the Rasterisation module. The Vector Input module reads, pre-processes and distributes the input vector dataset for conversion. It uses the parallel vector data input algorithm described in Chapter 8. The Rasterisation module interacts with the Vector Input module to perform the conversion operation using the scan-line algorithm.

Two groups of parallel processes are required - the Source and the Pool groups. The Source group has only one process as a member, which is from now on referred to as the Source or Source process. The Pool group has at least two processes as members since the GAD phase of the Vector Input module has at least one GeomAttributeServer process and one Worker process. A process cannot be a member of both the Pool and Source groups.

The Vector Input Module coordinates the actions of all the processes throughout the operation. As described in Chapter 8, the algorithm for the module consists of three phases; Sort, Join and Geom-Attribute Distribution (GAD). On the Source process, the module coordinates the Sort, Join and GAD phases.

On the Pool processes, it firstly coordinates the Sort and Join phases. For the GAD phase, it splits the Pool processes into two groups, the GeomAttributeServer and Worker groups, and behaviour differs between the new groups. On the GeomAttributeServer processes, the Vector Input module continues to coordinate their actions throughout the GAD phase. However, on the Worker processes the Rasterisation module takes over.

In summary, for vector-to-raster conversion, the Vector Input module transparently performs the complete operation on all processes except for the Workers during the GAD phase, where the Rasterisation module takes over.

11.6 External Interfaces

11.6.1 Vector Input Data

As explained in Chapter 8, the algorithms for GIS operations such as Vector-to-Raster conversion and Polygon Overlay require the spatial coordinates of the polygon boundaries with the corresponding attribute values of the polygons on their left and right sides. These polygon boundaries must also be sorted by uppermost y-coordinate.

The Vector Input module extracts this information from an input vector-topological dataset in BSI NTF Level 4 format and sorts it into scan-line ranges for conversion by the Rasterisation module. The Rasterisation module stores the spatial coordinates in Geom data structures and the left and right attribute values in GeomValue data structures. The contents of these structures are shown in Figure 8.9.

11.6.2 Raster Output Data

The raster representation of the vector data is output in the raster format which is fully defined and described in Chapter 6.

11.7 Process Structure

This design assumes the operation is written in the SPMD style, i.e. there is a single executable image for all processes. This is illustrated in the pseudo-code program below which is executed by the Source process and all the Pool processes.

```
/*******************************************************************
 *          main()
 ******************************************************************* */
 int main()
{
    switch(Process group)
    {
      case Source :
        get parameters from gis_VectorToRaster() function;
        VectorToRasterSource(parameters);
        break;
      case Pool :
        VectorToRasterPool();
        break;
    }
}
```

The Source process calls the VectorToRasterSource function whilst the Pool processes call the VectorToRasterPool function.

11.7.1 The Source Process and the VectorToRasterSource function

The role of the function is essentially to coordinate the Pool group processes through the various phases which make up the vector-to-raster conversion software. It does so by calling the VectorInputSource function from the Vector Input module.

11.7.2 The Pool Processes and the VectorToRasterPool Function

This function is called by all the Pool processes. It calls the VectorInputPool function of the Vector Input module to coordinate its actions during the Sort and Join phases. In the GAD phase, the Pool processes are split into two groups, the GeomAttributeServer and the Worker groups. Those processes which become Workers exit the VectorInputPool function and call the RasteriseWorker function of the Rasterisation module. This function performs vector-to-raster conversion using the scan-line algorithm on the data sent to a Worker process by the GeomAttributeServer processes.

The GeomAttributeServer processes remain under the control of the VectorInputPool function until all the input data has been converted. Below is a pseudo-code version of the VectorToRasterPool function which helps to illustrate how the VectorInputPool function returns a value indicating that a Pool process has become a Worker or a GeomAttributeServer process. If the process is a Worker, it calls the RasteriseWorker function. If the process is a GeomAttributeServer process then the function can exit, since the GeomAttributeServer processes do not exit the VectorInputPool function until all the data has been converted.

```
/* ****************************************************************
 *   VectorToRasterPool()
 *
 * Effect: Called on all Pool processes to perform Vector-to-Raster
 * conversion of a dataset.
 *
 * Side-Effect: Allocates memory.
 *
 * I/O: Creates output dataset and various intermediate files which
 * are removed before exit.
 *
 * Various messages are sent to the Source.
 *
 * Returns: Success or failure
 *
 * ***************************************************************** */
void VectorToRasterPool(void)
{
   /* **************************************************************** *
    * The Sort and Join phases of Vector Input Module are executed.  At
    * the end of those phases, those processes which are to execute
    * the Rasterisation module exit the Vector Input Module as
    * members of the Worker group of processes.  The remaining processes
    * perform the GAD phase of the Vector Input Module      *
   ***************************************************************** */
   poolGroupType = VectorInputPool();
   switch(poolGroupType)
   {
      case Worker :
      /* *************************************************************** *
       * Perform as a Rasterisation Module worker
       * *************************************************************** */
       RasteriseWorker();
       break;
      default :
       break;
   }
}
```

On each Worker process, within the RasteriseWorker function of the Rasterisation Module, the following functions from the VectorInput Module are called to request strips from the Source process and to receive Geom and GeomValue data from the GeomAttributeServers. These function calls have the GAD prefix since it is with the GAD part of the Vector Input module that the Rasterisation module interacts. This usefully illustrates the modular nature of the design since the RasteriseWorker function does not need to handle any of the intricacies of data input.

GADWorkerInit initialises the structures for strips and Geom and GeomValue data at the start of the Operation. This includes creating the connections to the GeomAttributeServers for data transfer. For Geoms one buffer is required per GeomAttributeServer per dataset and these buffers will form the input to the leaf-nodes of the Geom merge-tree. For GeomValues a single buffer per dataset collates data from all the servers.

GADWorkerClose tidies up and destroys the structures at the end of the Operation.

GADWorkerReceiveStripDesc performs a blocking receive of a strip-descriptor message from the Source. The strip-descriptor message contains the y-range of the strip, the ID of the strip (allocated North to South starting at 0) and the ID of the Worker which received the previous strip (Northern neighbour). (It also contains the ID of the Worker which has been allocated the strip, although this is only used by the GeomAttributeServers).

GADWorkerEndOfOperation tests a strip-descriptor to see if it contains a "*close-down*" or "*flush*" message from the Source instead of strip information.

GADWorkerStripInit "*opens*" a new strip, by creating an empty Geom merge-tree and an empty GeomValue hash-table.

GADWorkerStripClose tidies up after a strip has been processed.

GADWorkerRequestNextStrip sends a non-blocking request for a new strip containing a specified amount of work. A flag argument causes a "*closing-down*" message to be sent instead.

GADWorkerFillValueTable fills the GeomValue hash-table from the GeomValue input buffer(s).

GADWorkerFillGeomOutBuf attempts to re-fill the Geom merge-tree output buffer by performing a merge-sort. The function returns when either the output buffer is full or the merge-tree is exhausted.

GADWorkerGetGeomDesc returns a pointer to a single geom-descriptor. The geom-descriptor is created by taking a geom from the output buffer of the Geom merge-tree and looking up the attribute value in the GeomValue hash-table. The function returns NULL if the output buffer is empty.

GADWorkerDestroyGeomDesc frees the space used by a GeomDesc.

GADWorkerGetApplicArgs returns a pointer to a structure containing user-defined application-specific information sent to the Worker by the Source.

GADWorkerGeomOutBuf, **GADWorkerGeomValueInBuf**, **GADWorkerStripDescBuf** these functions return pointers to the respective internal buffers.

GADWorkerGeomValueInBuf returns a pointer to the GeomValue input buffer.

11.8 Modules

11.8.1 Vector Input Module

The algorithm for the parallel Vector Input Module is described in detail in Chapter 8.

11.8.2 The Rasterisation Module

The Rasterisation module executes on the Worker processes. It creates a file of raster data which is the output from the operation. The module interacts with the GAD phase of the Vector Input module to perform the conversion. Here, the Workers are allocated strips of input data in a modified form of the classical taskfarm. A strip of data corresponds to a number of contiguous raster scan-lines. The details of the GAD phase are explained fully in Chapter 8. Essentially, each Worker sends a message to the Source process, which is under the control of the Vector Input module, indicating it is available for work. The Source allocates an available worker to a strip for conversion and instructs the GeomAttributeServer processes to send the data for that strip to the Worker.

The Source process allocates strips of comparable workload to each Worker process in the descending y direction and all the data for conversion is sorted in descending y direction. That is, the strip allocation and the data are already essentially in scan-line order. This introduces a degree of synchronisation since no worker can receive the data for its strip until the worker responsible for the previous strip has received all its data. This sequential startup of the conversion means the algorithm potentially has limited scalability. However, since this staggers the conversion, this proves useful in helping to ensure the scan-lines are output to the raster file in order.

11.8.2.1 Receiving the input data

Each Worker process receives a parcel containing Geoms and GeomValues from each GeomAttributeServer for the same strip. The ordering of the Geoms by uppermost y-coordinate is crucial to the performance of the scan-line algorithm. However, each parcel is only ordered within itself, therefore all the parcels are merge-sorted using a merge-tree as explained in Chapter 8. The ordering of the GeomValues is not important and so a pointer is stored to each in a hash-table using the GEOM_ID.

Figure 8.10 illustrates the sending of data parcels and the use of the merge-tree and the hash-table on the Worker processes.

Conversion can proceed as soon as all the data for the first scan-line in the strip has emerged from the sort merge-trees. That is, the data does not need to be completely sorted before the conversion starts.

11.8.2.2 Performing the Conversion

Starting from the top of the strip, those Geoms that begin on the current raster line contribute a segment to an ordered list for consideration. A segment refers to the line between two adjacent points in a Geom. The ordering of the Geoms by descending uppermost y-coordinate means no searching is required. Segments on this list that no longer intersect the current raster line are deleted. By stepping along this list, attributes can be assigned to pixels using one of a number of different methods. These methods are outlined in section 11.4.2.

A second ordered list is maintained in order to hold a list of segments that range over the set of pixels currently being considered in the x-dimension.

11.8.2.3 Raster Line Output

Data are held in many workers as subraster blocks. The problem of writing a coherent raster dataset is discussed in Chapter 10. The following approaches are worthy of consideration:-

- Exchange of messages between processes to coordinate output. When the Source process allocates a strip to a Worker w, it gives that Worker the process identifier, v, of the Worker responsible for the strip above it. Process w sends a message to v, to make this relationship known to v. When v has completed its output, it sends a non-blocking message to w. w can then write its data. This coordination completely removes the overhead of reading the file, ensuring that scan-lines are written in the correct order..

- A parallel I/O utility which buffers up the scan-lines and ensures they are output in the correct order.

Of these suggestions, the third is the most attractive, since it removes all extraneous I/O and does not involve the extra overhead of another utility.

11.9 Conclusions

This chapter has considered various classes of algorithms for conversion from vector format to raster format, both from the perspectives of the strengths and weaknesses of the algorithms and in terms of the implications for parallel implementation. The scan-line algorithm appears to have the most appropriate characteristics for a parallel implementation. One approach to an implementation has been described concentrating on the process structure and the relationship between the workers implementing the operation and the raster output and the vector input modules.

11.10 References

Ackland, B.D. & Weste, N.H., 1981, The edge flag algorithm - a fill method for raster scan displays, *IEEE Transactions on Computers*, **30** (1), 41-47.

Bouknight, W.J., 1970, A procedure for generation of three dimensional half-toned computer graphics representations, *Communications of the ACM*, **13** (9).

Foley, J.D. & Van Dam, A., 1984, *Fundamentals of Interactive Computer Graphics*, Addison-Wesley, Reading, MA.

Jordan, B.W & Barret, R.C., 1973, A scan conversion algorithm with reduced storage requirements, *Communications of the ACM*, **16** (11), 676-682.

Piwowar, J.M., LeDrew, E.F. & Dudycha, D.J., 1990, Integration of spatial data in vector and raster formats in a geographical information system environment. In *International Journal of Geographical Information Systems*, **4** (4), 429-444.

Rogers, D.F., 1985, *Procedural Elements for Computer Graphics*, McGraw-Hill, New York.

12

Raster-to-Vector Conversion

M.J. Mineter

12.1 Introduction

The conversion of raster datasets to a vector-topological representation is significant in that it enables analysis of these datasets to occur within vector-based GIS. This is particularly important in the processing of classified remotely-sensed images, where a combination of trends has led to recognition of a role for parallel processing, as Chapter 18 describes. These trends include increasingly sophisticated and time-critical analyses (Fritz, 1996), and the large data volumes being produced by new generations of sensors.

As described in Chapter 5, this operation was investigated in order to identify issues inherent in the parallel processing of raster- and vector-topological data. This chapter builds upon the modular approach of Chapter 7, the raster decomposition methods of Chapter 10, and the TSO framework for vector topology creation described in Chapter 9 to discuss a design for the parallel conversion of raster data to vector-topological data in the NTF Level 4 data format (Chapter 6). Raster data is assumed to consist of contiguous, non-overlapping regions, where a region is a set of horizontally or vertically (but not diagonally) connected pixels of the same value. (Where pixels are only diagonally connected, topology cannot be unambiguously constructed.) The goal is to describe algorithms for areal data only. As stated in section 5.1, these should be capable of processing datasets which are arbitrarily large, and which therefore require data-streaming, so that the conversion of the top of the dataset can be accomplished whilst the bottom of the dataset is still being read in.

As Nichols (1981) noted, if many small regions exist in a raster then after conversion the number of vector-topological records can be unmanageable. Consequently a preliminary analysis is required to produce a classified image with an appropriate size of region, for example by filtering small polygons (Chapter 14).

The component tasks of the conversion, shown in Figure 12.1, are (Harding *et al.*, 1993):

Boundary extraction: comprises the creation of lists of vertices corresponding to the boundaries between regions.

Boundary simplification: involves the avoidance or removal of redundant collinear vertices from vertical or horizontal segments of a Geom. These would significantly add to the volume of data, but not to the information held in the final dataset.

Boundary smoothing: involves attempts to remove artefacts of raster representation, for example 'staircase' effects for sloping polygon borders.

Topology construction: creates the vector-topological objects required to describe the polygons and their relationships, i.e. Nodes, Edges, Chains, etc. as described in Chapters 6 and 9.

Generalisation: entails removing detail from the vector-topological dataset which is considered inappropriate for a particular application of the data. Unlike boundary smoothing, generalisation can lead to changes in topology. This is beyond the scope of this chapter.

In section 12.2 established sequential approaches to the conversion are reviewed and options for a parallel algorithm are discussed. A design is proposed in section

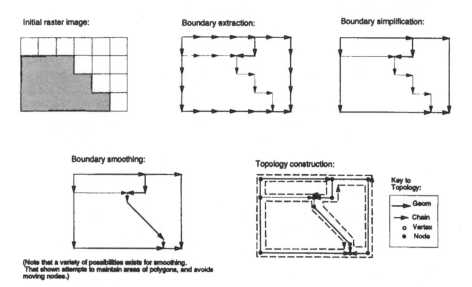

Figure 12.1: The component tasks of raster-to-vector conversion

12.3 emphasising boundary extraction and topology construction. In section 12.4 boundary smoothing is discussed.

12.2 Methods of Raster-to-Vector Conversion

12.2.1 Sequential Approaches

Following Piwowar *et al.* (1990), three classes of sequential algorithm for boundary extraction and topology construction can be identified. These are summarised in the light of the goals stated in the introduction:

1. **Polygon cycling algorithms** step around the boundary of each region, recording the coordinates of each pixel. This style of algorithm is inefficient unless all of the raster data can be held in memory, as the order in which the data is accessed cannot be anticipated. Performance is compromised because each boundary is followed twice - once for each polygon on either side of an edge. Examples of this approach include the following (Connealy, 1992).

2. **Edge stepping algorithms** begin by assigning polygon identifiers to the regions in the complete dataset. They then extract a sorted list of all line segments on region boundaries, together with associated polygon identifiers, and finally construct topology from the segments. The allocation of polygon identifiers requires access to the complete raster dataset. In effect the connection of the line segments comprises a second pass across the dataset. Nichols (1981) exploited virtual memory to hold the intermediate data inherent in this method.

3. **Boundary linking algorithms** compare the values of pixels on adjacent scan-lines, beginning at the top of the dataset. As subsequent scan-lines are read in, so the description of both the boundaries and the topology related to all regions is generated. Only two scan-lines need to be stored at a time. Capson (1984) used this approach for binary images, and demonstrated that boundary simplification may be efficiently incorporated into the procedure. As new scan-lines are received and processed, so some regions which were distinct can merge to form one polygon (e.g. a 'U' shaped polygon), and others (e.g. 'n' or 'o' shaped) split around another region (which may become a hole). Hence Capson's implementation of the algorithm explicitly supports relationships between regions in the currently held scan-lines. A single pass across the data is sufficient, so that the approach is suited to processing large datasets and to data streaming, whereby some objects can be completed and written out before some scan-lines have been read in. The comparison between scan-lines is well suited to raster data which have been compressed using run-length encoding.

Piwowar *et al.* (1990) tested the three approaches using criteria of correctness and performance and favoured Boundary Linking.

12.2.2 Parallel Approaches

Decomposition into 'sub-raster blocks' to meet the requirements of large dataset size, of one-pass processing, and of load-balancing for raster-to-vector conversion was presented in Chapter 10. The processing of an SRB can then be accomplished by a modified sequential algorithm, followed by a phase of stitching (Figure 12.2). Such an approach is also consistent with the fact that, in general, the most effective use of parallel processing occurs when the processors are working independently. This section explores the scope for parallelising the three types of algorithm mentioned above.

Methods based on polygon cycling. If each processor was to attempt to follow in turn the border of each of the regions within an SRB, then several problems would arise as a consequence. Where polygons cross SRB borders strongly inter-dependent processing would result. Two more fundamental issues arise from processing which is focused upon one polygon at a time. The building of full vector-topology requires that relationships between polygons, edges and nodes be established, which, as Chapter 9 noted, demands that a polygon is derived in the

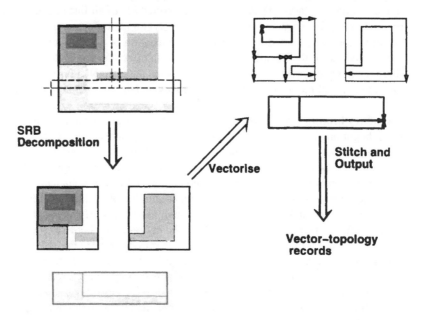

Figure 12.2: Overview of parallel raster-to-vector conversion showing data decomposition into SRBs, vectorisation and stitching of strips

context of its relationships. Furthermore, in cases where the dataset is large, some polygons may not be capable of being held in memory. Using intermediate disk files is a last resort which in general will impair performance.

Methods based on edge stepping. The inefficiencies of the sequential edge stepping algorithm will afflict a parallel implementation, with two passes across the dataset being required. The problems of polygon cycling also would be encountered here in the processing of polygons which cross SRB borders, and in the requirement that the complete dataset can be held in memory, expressed as line segments and attributes, or else that intermediate disk files can be used.

Methods based on boundary linking. The scan-line by scan-line approach of boundary linking facilitates independent processing of an SRB. The creation of topology, rather than simple polygon identification, can readily be supported because each region is processed in the context of complete scan-lines. With this approach, there is also scope for more efficient message passing, because all messages related to objects crossing one SRB border can readily be buffered and sent together in a stitching phase. The 'TSO' module proposed in Chapter 9 can be exploited, so that the code specific to the raster-to-vector conversion becomes relatively simple.

The advantages of Boundary Linking, evident in both consideration of the sequential algorithm and ease of parallelisation, are such that alternatives are discussed no further in this chapter.

12.3 Parallel Boundary Linking

This section further develops an approach based upon boundary linking, presenting a design for the boundary extraction, topology construction and boundary simplification tasks of the raster-to-vector conversion.

The processes comprising the raster-to-vector conversion operation are:

1. The Sub-Raster Block Distribution process (SRBD), described in Chapter 10, which:

- Writes (or else distributes data to allow a Worker to write) the header of the vector dataset. It is suggested that this task is performed by the SRBD because the vector dataset header is derived from that of the raster dataset. The Worker processes do not need to hold these data.

- Receives requests for SRBs from the Worker processes.

- Creates Sub-Raster Blocks from the raster data. The SRBs overlap by one pixel; without this 'halo'. Edges along the borders of SRBs would be undetectable. (This can be seen in Figure 12.1.)

- Distributes the SRBs to the Worker processes in response to requests for data.

- Sends information about SRB alignment to the Worker processes so that objects which cross SRB borders can be matched within TSO.

2. The Worker processes which comprise functions specific to the raster-to-vector conversion, linked to the TSO framework. The specific functions:

    - Send a request for an SRB to the SRB Distributor.

    - Receive an SRB and vectorise it, using the PTF (Partial Topology) functions within TSO.

    - Request a subsequent SRB.

   The creating, stitching and writing out of vector-topological records are accomplished within TSO.

3. The ID Generator process, a part of TSO, which allocates IDs to the vector objects.

The algorithm follows the pattern indicated by Chapter 9. (The above processes are consistent with those shown in Chapter 9, Figures 9.1 and 9.2.) Familiarity with that chapter is assumed, and particularly with the concepts of the PTF library as described in section 9.3 and illustrated in Figures 9.5, 9.6, 9.7 and 9.8. In this description, the first scan-line is considered to be at the north of the dataset, and its first pixel is at its west end. A span is defined to be a consecutive set of cells with the same attribute in one scan-line. Vectorisation proceeds scan-line by scan-line as indicated by Figure 12.3. On receipt of the first scan-line of an SRB, a linked list of transition structures is set up. A transition is created corresponding to the ends of each span. If the first scan-line is not on the border of the complete raster, the linked list is also used to generate strip-border data for stitching at the north border. Each subsequent scan-line is processed by calls to the PTF library to cause the vector-topological objects and the linked list of transitions to evolve. The PTF functions are selected on the basis of comparisons between the attribute values and extents of adjoining spans in the current and previous scan-line, thereby accomplishing both boundary extraction and topology construction (Figure 12.4). Boundary simplification, removing redundant collinear vertices, can also occur within PTF.

AFTER 1ST SCAN–LINE:

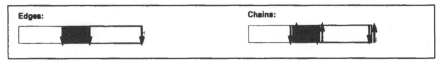

AFTER 2ND SCAN–LINE:

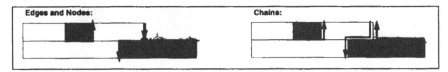

Figure 12.3: Example of scan-line by scan-line vectorisation

In making the comparison between spans and calling the PTF functions the following apply:

1. A Node is created either at a point where three polygons touch, or when a simple hole exists within a polygon. In the latter case a Node is created to both start and terminate the Edge.

2. An Edge exists between cells of different attributes, and it extends in each direction to a Node or to a transition. The transition can be either on the strip border or on the linked list used as a moving border between data known to PTF and that which has yet to be vectorised.

3. A Chain is associated with the part-polygon on either side of an Edge. Chains are directed so that their polygons are to their right. In an implementation of boundary linking for binary images, Capson (1984) maintained the relationships between regions as 'merged' ('u' shaped polygons), 'potential hole' ('n' or 'o' shaped polygons), 'split' ('n' shaped) and 'hole' ('o' shaped). With the exception of the relationship between holes and their containing regions (held in TSO structures), these relationships can be held within the Chain structures: when regions split, additional Chains are defined; when they merge, Chains are concatenated; when regions terminate to close a polygon, the Chain closes in a loop. PTF creates holes or polygons as appropriate when Chains close.

4. The list of vertices in the Geom corresponding to an Edge is created to follow the border between spans with different attributes.

5. Geoms are initially directed from the upper to the lower border of the scan-lines. (The joining of Geoms will generally entail connecting two differently ordered part-Geoms. The PTF functions can achieve this in an optimised fashion, without causing additional complexity for the code which calls those functions.)

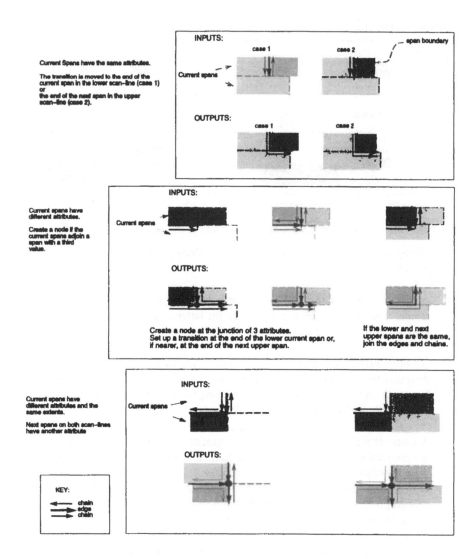

Figure 12.4: The effect of a sub-set of the functions which vectorise span-by-span

On receipt of the second and subsequent scan-lines, a linked list of transitions, ordered from west to east, exists defining an interface to the vector-topology which has already been derived to the north of that scan-line. During processing of the second and subsequent scan-lines the linked list evolves by:

- Amending data associated with a transition, for example by adding vertices to a Geom.

- Removing transitions when two parts of an Edge connect (Figure 12.3).

- Inserting a new transition when a new part-polygon is encountered on the current scan-line. When adjoining spans on the current and previous scan-lines have different attributes then an Edge runs west-to-east between these spans. A temporary transition exists associated with that Edge and the two Chains. As processing of the line progresses, it connects with a transition from the previous scan-line, or finds a place in the double-linked list, or is inserted in the list of transitions at the East border of the strip.

Once the final scan-line of an SRB has been processed then the boundary data of the strip are held in the remaining double-linked list of transitions. These, together with the strip data derived from the SRB, are then used for stitching, collation and writing out of records within TSO.

12.4 Boundary Smoothing

Boundary smoothing comprises the removal of superfluous vertices without disruption of topology. This section seeks to indicate directions for exploration, rather than to offer detailed algorithms. A three-phase approach is possible, with smoothing occurring:

1   During creation of a strip from one SRB, so avoiding redundant collinear vertices and regular staircases during the creation of a list of vertices for a part-Geom. This both reduces the memory required to hold parts of a distributed Geom and reduces the data which must be communicated in messages to collate a Geom following stitching.

2   During collation of the list of vertices. Following stitching, any completed Geoms must be collated for writing out. As and when two adjoining parts of Geoms are received by a collating process, it is possible to remove collinear vertices and regular stair-cases from Geom segments which cross the corresponding strip borders.

3   After collation of a complete Geom to remove vertices from irregular segments of a Geom.

The third phase might be omitted, but if so then the smoothing in the first two phases must be equivalent to each other, so that results from the parallel conversion are independent of the data decomposition. (Varying methods of data decomposition will influence ordering of data output and performance, but should not influence the content of the resultant dataset, e.g. the same vertices should always be derived for the same initial dataset and processing parameters.) If the

M.J. Mineter

first two phases are not equivalent then the third phase is necessary to achieve this
independence, by smoothing the data more strongly than either of the first two
phases.

Use of the Douglas-Peucker algorithm (Douglas & Peucker, 1973) for the third
phase is now discussed. The algorithm begins by representing the vertices of a
complete Geom by a straight line connecting its nodes. The lengths of
perpendicular offsets from this line to each point along the Geom are evaluated,
and the point with the greatest offset is identified. The Geom is then divided into
two segments, both terminating at this point. The processing continues recursively.
For each segment, a new straight line is generated; the evaluation of offsets leads to
further division into two smaller segments. Division of a segment ceases when the
point of maximum offset is within a user-defined tolerance. The intervening
vertices on the segment are then eliminated.

As Vaughan *et al.* (1991) noted, this algorithm has been commended for the quality
of its results, but not for its speed of execution. Consequently Vaughan *et al.*
explored methods of using a shared memory Sequent Symmetry computer to
parallelise the processing of individual Geoms. In early steps of processing the
parallelism was exploited to speed the identification of the point with maximum
offset, and then, once the number of segments was sufficient, different segments
were smoothed by different processors. Using Geoms of 10,000 points, a speed-up
of 7-fold was achieved with 9 processors. Additional processors did not lead to
further speed-up; 15 processors led to a slight degradation in performance, thought
to be due to competition for access to the shared memory. In contrast to this
approach, Mower (1996) used a CM-5, distributing more than 20,000 Geoms so
that each Geom was processed within one processor. This is the best way to
proceed when many Geoms are being smoothed: attempting to share the processing
of each Geom would merely add to the communication overhead thereby degrading
performance. The implementation of Mower demonstrated the need for careful use
of non-blocking communications, discussed for TSO in section 9.6. Despite being
noteworthy, neither of these implementations is directly relevant to the smoothing
that is required by parallel raster-to-vector conversion. In this case it occurs in the
context of other processing, namely the conversion and stitching.

As stated earlier, the smoothing required for the purposes of the conversion must be
sufficiently constrained to prevent topological restructuring, which requires use of a
small tolerance, comparable to pixel dimensions. The smaller the tolerance, the
greater is the processing demanded by the Douglas-Peucker algorithm, as
additional levels of splitting are required. In view of the constraint of avoiding
restructuring topology, a technique weaker but faster than that of Douglas-Peucker
may be feasible. Using any algorithm for smoothing a Geom, the smaller the
number of vertices, the faster smoothing will complete. Hence extending the
smoothing in the first two phases, for example by including some irregularity in
staircases, would improve performance significantly. One further possibility is to

generalise the raster image before conversion so that smoothing in phases 1 and 2 only is sufficient.

Thus far, boundary smoothing has been discussed with no regard to the nature of the boundary, which is assumed to be sharp. One trend in remote-sensing analysis is towards supporting boundaries which in practice are ill-defined, which leads on to the possibility of the extent of smoothing being dependent upon the attributes on either side of the boundary. This investigation is also beyond the scope of this chapter.

12.5 Summary

This chapter discussed the design of the conversion from a raster to a vector-topological dataset. Basing the method upon boundary linking allows exploitation of the raster decomposition and the topology building modules of Chapters 9 and 10. The requirement for simplification of data both before conversion to remove small polygons (Chapter 14) and after conversion to simplify the lists of vertices was stated. In remote-sensing analysis a range of developments including the trend towards 'fuzzy' boundaries makes this operation one which is likely to continue to be refined, with consequent demands for more processing power.

12.6 References

Capson, D.W., 1984, An improved algorithm for the sequential extraction of boundaries from a raster scan. *Computer Vision, Graphics and Image Processing*, **28**, 109-125.

Connealy, T., 1992, A complete and innovative raster to vector conversion algorithm. In: Ordaatje, D.A., Harts, J., Ottens, H.F.L. & Schotten, H.J. (Eds), *Third European Conference on Geographical Information Systems* Vol. 2, Utrecht /Amsterdam, The Netherlands (EGIS Foundation), 1102-1110.

Douglas, D.H. & Peucker, T.K., 1973, Algorithms for the reduction of the number of points required to represent a digitised line or its caricature. *The Canadian Cartographer*, **10** (2), 119-122.

Fritz, L.W., 1996, The era of commercial earth observation satellites. *Photogrammetric Engineering and Remote Sensing*, Vol. 62, No.1, 39-46.

Harding T.J., Mineter, M.J., Sloan, T.M., Wilson, A.J.S., Dowers, S. & Gittings, B.M., 1993, *Raster to Vector Conversion and Vector to Raster Conversion*

Concepts, EPCC-PAP-GIS-R2V2R-CONC, EPCC Pub. No. EPC-CC93-07 EPCC and University of Edinburgh Dept of Geography, Edinburgh.

Message Passing Interface Forum, 1994, *MPI: A Message-Passing Interface Standard*, University of Tennessee, Knoxville, TN.

Mineter M.J., Harding, T.J., Dowers, S., Healey, R.G., Chapple, S.R. & Trewin, S.R., 1993, *Raster to Vector Conversion: High level Design*, EPCC-PAP-GIS-R2V-HLD EPCC Pub No. CC94-09 EPCC and University of Edinburgh Dept. of Geography, Edinburgh.

Mower, J.E., 1996, Developing parallel procedures for line simplification. *International Journal of Geographical Information Systems*, **10** (6), 699-712.

Nichols, D.A., 1981, Conversion of raster coded images to polygonal data structures. In: *PECORA VII Symposium*, Sioux Falls, SD (Falls Church: American Society of Photogrammetry), 508-515.

Piwowar, J.M., LeDrew, E.F. & Dudycha, D. J., 1990, Integration of spatial data in vector and raster formats in a geographical information system environment. *International Journal of Geographical Information Systems*, **4** (4), 429-444.

Vaughan J., Whyatt, J.D. & Brookes, G., 1991, A parallel implementation of the Douglas-Peucker line simplification algorithm. *Software - Practice and Experience*, **22**, 331-336.

13

Vector Polygon Overlay

T.J. Harding, R.G. Healey, S. Hopkins and S. Dowers

13.1 Introduction

Vector polygon overlay is among the most computationally intensive group of GIS operations and frequently proves to be a bottleneck in 'production-line' processing. This chapter aims to examine how parallel methods can be applied within the different stages of the operation, and how these different stages fit into the overall modular design scheme set out in Chapter 7.

Following a brief summary of the nature of overlay operations, the individual stages in the process are outlined and the choice of algorithm for the key stage of geom intersection detection is described. The final parallel approach adopted is prefaced by a review of the limited existing work on parallelisation of aspects of the polygon overlay process. Owing to the significant complexity of the overall operation and its component algorithms, extended example material is included in an appendix, to enable the non-specialist reader to grasp the issues involved more readily.

13.2 Polygon Overlay Background

Overlay is the generic name for a set of closely-related GIS operations. Conceptually, overlay answers queries such as "what areas of land are forested and owned by X?" where one digital map represents land use and the other represents ownership (White, 1978), or "what soil types underlie different land use zones?" In practice, this becomes a question of producing a new map by combining two or more previous maps according to logical rules (see for example (ESRI, 1991; Intergraph, 1991)). See Figure 13.1 for an example of the result of overlaying two simple maps.

When implemented on a GIS which stores real world objects in vector format as points, lines and polygons, overlay queries become geometrical computations. Overlay is a computationally intensive application as it is combinatorial in nature,

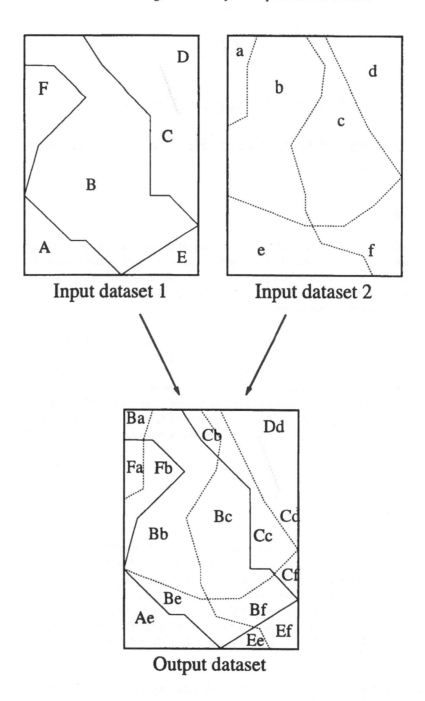

Figure 13.1: The result of overlaying two simple maps to create a third (after Burrough, 1986)

requiring, as part of the overall process, the comparison of many geometries (geoms) with many others for the purpose of detecting intersections.

13.3 Stages in Polygon Overlay

A typical polygon overlay algorithm assumes that input maps contain lists of geoms (which may or may not be fully/partially sorted) with associated edge records and that the output map requires suitably derived geom and edge records, together with associated face and area records. See Chapters 6 and 8 for details of record types.

Accordingly, the algorithms are usually divided into several stages (Wagner, 1991):

- **Transformation**: The transformation of the polygon sets to the same coordinate system.

- **Geom intersection**: Locating intersections between the geoms in the input maps and subdividing the geoms to create new geoms and edge records for the output map.

- **Polygon linking**: Using the new edges from the intersection stage, polygon boundaries are traced to generate new area records for the output map.

- **Output of new edges and new area records** (polygons).

- **Database population**: Assignment of attributes of parent polygons to child polygons by processing attdesc records.

These stages are now described in more detail.

13.3.1 Transformation

Since the transformation is not an operation that is of primary interest in terms of parallelisation, it will be assumed in this discussion that all input map datasets have already been pre-processed, if necessary, into the same coordinate system and projection.

13.3.2 Geom Intersection

This consists of:

- Location of intersections between geoms from the input maps.

- Formation of new nodes at these points of intersection.

- Formation of new geoms by subdivision of the existing geoms at the new nodes.

- Assignment of left/right face IDs to the edge records associated with the new geoms.

The geom intersection algorithm must locate all intersections between the geoms in the first and the second maps, assuming a two-map overlay problem is being considered. This can be achieved by fitting the equation $y = mx + b$ to each geom pair and simultaneously solving the resulting equations, but it is in fact better to use the form $ax + by + c = 0$ to overcome the problem of vertical geoms, since none of the terms will approach infinity. Although conceptually simple, there are many special cases that need to be considered, such as collinear geoms, or intersections which occur at the junction of a vertex of one geom and a midpoint of another. A further potential problem occurs when the gradients of the two geoms differ by a number too small for numerical representation within the precision of the machine. These factors make the production of a general and robust line geom intersection routine a considerably more complex task than it might appear at first glance (Douglas, 1974).

Testing two geoms for intersection is a compute-intensive process. Using integer arithmetic can result in greater accuracy which helps avoid problems caused by floating point imprecision; for example in a case where intersecting geoms are nearly parallel, the calculated intersection point could move slightly, thus causing it to come within the tolerance range of another node and produce a topological inconsistency where the two nodes are confused with one another. However, integer arithmetic is only really practical if the coordinates are represented as integers, as is the case for data held in NTF format. Another potential solution to this problem is to use rational arithmetic, where numbers are represented as fractions rather than decimals.

A key issue in this stage of the processing is to minimise the number of geoms in the second map that have to be compared with any geom in the first. Techniques for reducing the number of geom comparisons in geom intersection are given in section 13.5.

As each intersection is found, new entries may be placed in the node table and the two intersected geoms subdivided to form up to four new geoms, which are given new unique identifiers and whose edge records are eventually assigned new left and right face identifiers (see Figure 13.1).

13.3.3 Polygon Linking

In standard algorithms, this stage comprises the formation of new polygons by tracing their boundaries from amongst the new geoms. If node-geom relationships are present then linking is very efficient, particularly if the node-geom lists are ordered by angle in which case the next geom in tracing round the boundary can be found by taking the one that subtends the minimum angle with the current one. If such relationships do not exist, the geoms must be searched to find matching endpoints and left/right face-IDs. Some sort of hash table (Knuth, 1973) for start/end coordinates is the best method to be employed to search for the next geom in the polygon being linked.

Island boundaries can be resolved at this stage using the left-right coding of the face records and the determinant formula,

$$\text{area} = \Sigma(e.x - s.x)(s.y + e.y)$$

where s.x and s.y are the x and y coordinates at the beginning of a geom and e.x and e.y are the x and y coordinates at the end of a geom.

The determinant formula will produce a negative area if the coordinates of the geoms comprising a polygon boundary are processed in a clockwise direction. If the coordinates are processed in an anti-clockwise direction the area is positive (Nordbeck & Rystedt, 1972).

Note that two simple polygons may overlay to give complex polygons (Figure 13.2). For this project, complex polygons, that is, those polygons consisting of several non-contiguous, non-overlapping areas, are not considered valid input. Allowance must be made for this by assigning distinct IDs to the separate parts prior to resolving any islands.

In the present design, the Partial Topology Functions in TSO (see Chapter 9) are used to build chains from the new geoms and edges. These chains in turn form the basis of the output polygons. In this way, some of the algorithmic complexity inherent in the above stages is hidden within the context of the PTF library.

13.3.4 Output of New Edge and New Area Records

Output of an edge record cannot be considered complete until the new IDs of its left/right faces have been added. Output of an area record (polygon) is not possible until the new IDs of its constituent edges have all been determined and it is not complete until all contained holes have been added.

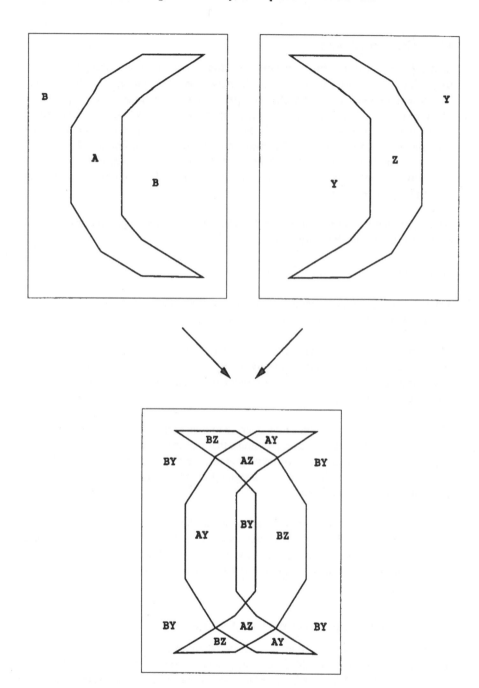

Figure 13.2: Simple polygons overlaid to produce complex polygons (after Waugh & Hopkins, 1992)

13.3.5 Database Population

Typically, each polygon in the input maps will have a number of attributes, held as a list of attribute codes linked to polygon IDs. In the NTF structure, this would correspond to ATTDESC records linked to area records (Chapter 6).

The assignment of attributes to the output polygons is essentially a database join operation. Each polygon in the output map has the two polygon IDs from its parent polygons in the input coverages. These are used as keys to query the parent polygon-attribute tables and a new attribute list is formed by concatenating the parent attribute lists. Each output polygon has an ID assigned to it in the new coverage. The parent polygon IDs can then be discarded. The new attribute list is built, based on the new IDs.

Database population is not considered to be part of the algorithm implemented here.

13.4 Non-intersecting Polygons

Certain links may not intersect with any links in the other coverage. For example, the geoms making up polygons labelled D and d in Figure 13.1 are non-intersecting.

Some polygon linking algorithms use left/right parent polygon IDs to search for the next edge belonging to a polygon being linked. The basic geom intersection algorithm will not supply both parent polygon IDs for geoms that do not intersect with any other geoms. Therefore to use such polygon linking algorithms, a modification is required to the link intersection stage.

A typical method extends an east-west or north-south line until it hits a polygon boundary whose left/right polygon ID is already known (the boundary of the 'external polygon', that is the edge of the map area, if nothing else). A geom intersection algorithm such as plane sweep (section 13.5.3) can find the parent polygon IDs of non-intersecting geoms easily.

13.5 Techniques for Efficient Geom Intersection

The 'brute force' geom intersection method entails comparing every geom in the first map with every geom in the second. For two maps with N and M geoms respectively, such a method has asymptotic time complexity $O(N \times M)$. Efficiency can be considerably improved by localisation strategies which greatly reduce the number of intersection tests (comparisons) by exploiting the simple observation that intersection between links lying in distinct areas of the coverage is impossible.

Four well known classes of localising technique are:

- Bounding boxes

- Spatial sorting

- Plane sweep

- Domain decomposition.

13.5.1 Bounding Boxes

Geoms or polygons may have their *bounding boxes* (enclosing rectangles) computed (Guevara, 1983). These might be the minimum enclosing rectangles, or minimum enclosing rectangles with sides parallel to the coordinate axes; the latter are more easily computed. The enclosing rectangle for a geom can be stored as part of the data structure and only needs re-computation when a geom is subdivided at an intersection. This method is effectively adopted in many commercial GIS but does not form part of the NTF structure *per se*.

During the geom intersection stage, a simple comparison test is applied to determine whether the bounding boxes overlap. If there is no overlap, there is no intersection. If there is an overlap, the objects (geoms, polygons) may or may not intersect, so bounding boxes must be compared at a lower level of object or a geometric intersection test performed. Clearly there is a trade-off between the level of object for which bounding boxes are stored and the reduction in intersection tests; bounding boxes can be stored for sections of links between turning points. *Turning points* are where the derivative in *x* or *y* changes sign. Comparison of bounding boxes can considerably reduce the number of line intersection tests made, but the asymptotic time complexity is only affected by a constant factor and so it remains quadratic.

A slight variation on storing the bounding box of a link is to store a straight line with the maximum deviation of the link from that line. This concept is closely related to the Douglas-Peucker line generalisation algorithm (Douglas & Peucker, 1973).

13.5.2 Spatial Sorting and Domain Decomposition

One might choose to sort links according to:

- **Full sort**:

  Geoms are sorted into ascending or descending order of lowest or highest x or y coordinate.

- **Bucket sort or domain decomposition**:

  The data space is divided into image strips (one-dimensional bucket sort) or rectangular domains (two-dimensional bucket sort). The domains may not be of equal size and may depend on object density. Each object is sent to the spatial 'bucket' that contains it; bucket sorting has linear time complexity. Geoms crossing domain boundaries are typically replicated or subdivided, but optionally the one-dimensional bucket sort may be implemented as a binary tree or perhaps a b-tree (Knuth, 1973; Sedgewick, 1988), in which case one might choose to lodge geoms at the lowest level able to contain their extent.

- **Quadtree decomposition and sort**:

  Quadtree decomposition is a two-dimensional sort technique which, unlike the sorts described above, places no preference on either the x or y coordinate, treating each equivalently. The data space is recursively divided into four to produce a tree data structure with four sub-nodes to each node (Samet, 1990). The number of levels may vary across the tree and is often made a function of object density.

  Each object may then be placed into the quadtree at the lowest level not requiring object subdivision, a method which can perhaps be considered as a quadtree sort. Alternatively, geoms may be duplicated or subdivided as necessary where they cross quadtree boundaries and then lodged at the leaf nodes, a technique which can be termed quadtree decomposition.

A geom intersection algorithm employing domain decomposition reduces the number of intersection tests performed by dividing up the study area in some fashion, for instance by using a regular grid. Intersection tests are thus only performed upon geoms that fall within the same grid cell (Franklin *et al.*, 1989). Alternatively the area may be divided into regular strips, or some more sophisticated method may be used which takes account of data distribution in some way, for instance irregular strips which vary in width according to data density or quadtree decomposition as described above. This still has quadratic asymptotic complexity but with a smaller constant factor.

The use of bounding boxes and spatial sorting is not exclusive; indeed, a spatial sort typically uses bounding box coordinates as the sort key and the resultant ordering is used to reduce the number of overlap comparisons necessary between boxes.

Importantly for this project, the localisation techniques used to improve efficiency of link intersection are naturally allied to parallelisation techniques (Hopkins & Healey, 1990; Hopkins *et al.*, 1992; Waugh & Hopkins, 1992). These are discussed in detail below.

13.5.3 Plane Sweep

A full sort lies at the heart of the plane sweep algorithm for geom intersection, which is illustrated in Figure 13.3. The geoms are pre-sorted according to, say, their minimum x coordinate. Processing proceeds in a band that moves across the data space. Since the data are sorted, it is possible to determine easily all geoms crossing the band at any time and intersection tests are only made between geoms in the band. The sorting process has O(nlog n) complexity in the number of input geoms and the reporting of all k intersections can take place in time O((n + k)log n). The plane sweep method was used by the WHIRLPOOL polygon overlay processor, a component of the ODYSSEY geographic information processing system (White, 1978; Chrisman *et al.*, 1992).

A brief summary of the plane sweep method may be useful here, even though it is covered in standard texts on computational geometry, because the topic is examined in more detail in later sections. The following description is largely based on sections 1.2.2 and 7.2.3 of Preparata & Shamos (1985).

This description assumes that the plane is being swept from left to right. Consider a vertical straight line l that partitions the plane into two half-planes, one to the left and one to the right of l. Assume that each of these half-planes contains endpoints of some of the geoms in the input datasets. Obviously the solution to the problem of finding all intersections between the geoms in the input datasets is the union of the solutions in each of the half-planes, and if the set of intersections to the left of l has already been found, then this set is not going to be affected by geoms lying to the right of l; this point may be illustrated by considering Figure 13.3, choosing any one of the lines *l1* to *l13* as the line l that divides the plane.

Moreover an intersection may only occur between two geoms whose intersections with some vertical line are adjacent; e.g. in Figure 13.3 the geoms *s4* and *s5* have adjacent intersections with the line *l8*, whereas there can be no vertical line with which the geoms *s3* and *s6* have adjacent intersections. The impossible task of generating all possible vertical cuts is avoided by realising that the vertical ordering of the geoms' intersections with a vertical line moving across the plane in the sweep direction only changes either when a geom's end-point is encountered (in which case the geom either enters or leaves the ordering, depending on which of its end-points is encountered) or when an intersection between geoms is encountered (in which case the geoms become transposed in the ordering). Thus all that is required is to jump from the left boundary of one strip to its right boundary, update the order

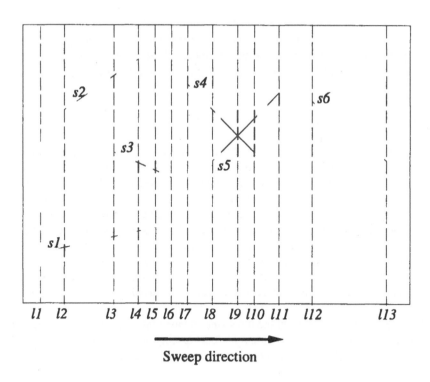

Figure 13.3: Event points in plane sweep geom intersection

of the intercepts and test for any new intersection among 'adjacent' geoms. These strip boundaries correspond to the vertical lines *l1* to *l13* in Figure 13.3.

So essentially there is a vertical line (called the *sweep-line*) that sweeps the plane from (in this case) left to right, halting at special points called *event points*. In the case of the geom intersection problem, these *event points* occur at the end-points of geoms (so each geom necessarily has two event points associated with it) and at points of intersection between geoms. The intersection of the sweep-line with the problem data contains all the information relevant to the continuation of the sweep. (The 'band' mentioned above is formed by the extent of the geoms in the sweep-line status.) Thus two basic structures are required:

The event point schedule:

This is a sequence of vertical straight lines, called abscissae, ordered from left to right, which define the halting positions of the sweep-line. The event point schedule is not necessarily entirely extracted from the input data, but may be dynamically updated during the execution of the plane sweep. Thus the endpoints of geoms form the part of the event point schedule which can be extracted from the input data, whereas the intersections between geoms form the part which is dynamically created during the plane sweep. The abscissae *l1* to *l13* in Figure 13.3 form the event point schedule for the input data geoms *s1* to *s6* .

The sweep-line status:

This is an adequate description of the intersection of the sweep-line with the geometric structure being swept. 'Adequate' means that this intersection contains the information that is relevant to the specific application. The sweep-line status is updated at each event point. In this case the sweep-line status comprises an abscissa located at an event point and the ordered list of geoms intersected by this abscissa; for instance at *l8* the geoms *s4* and *s5* are adjacent in the sweep-line status, thus it is at this point that they are tested for intersection and so the intersection at *l9* is discovered. Only geoms that are adjacent in the sweep-line status at some time can intersect, as stated above; thus comparisons between geoms are further reduced.

In the WHIRLPOOL system the links were divided into *monotonic sections* prior to sorting; monotonic sections contain no local turning points in the *x* or *y* directions, so that each can intersect a sweep-line no more than once at any stage.

13.6 Previous Work Related to Parallel Polygon Overlay

The previous sections have dealt with the major issues in standard overlay on serial processing machines and possible areas of consideration for parallel versions of overlay have been identified in the section on spatial sorting and domain decomposition. However, before proceeding to examine parallelisation strategies in more detail, it is necessary to review existing work in the field and the importance of machine architectures to an understanding of the types of implementation that are appropriate.

At the outset, care must be taken to clarify the type of target machine architecture for proposed algorithms, as this can have a significant effect on the way in which the operation can be performed. Specifically there is a need to distinguish between shared and distributed memory applications, since in the former case it is possible to design algorithms that minimise distribution of data to processors, while in the latter case the issue of data distribution is an essential component of the overall operation.

The best known approach in the GIS literature for the geom intersection stage is the uniform grid method, proposed by Franklin *et al.* (1989). As noted above, this involves a regular domain decomposition of the data for each of the map datasets to be overlaid, with individual processors assigned single or multiple grid squares on which intersection detection is to be performed. Franklin *et al.* (1989) report good speed-up using a SEQUENT shared memory multi-processor. Later development of the method for a distributed memory transputer array (Hopkins & Healey, 1990) demonstrated its feasibility for this kind of architecture, but subsequent implementation and test results (Hopkins *et al.*, 1992) indicated that, in the absence of virtual memory on the processors, determination of the appropriate amount of data to send to each processor, while avoiding memory overflow, could prove problematic.

Other approaches have favoured the use of the plane sweep algorithm, rather than a grid-based method. Early theoretical work in computational geometry (Atallah & Goodrich, 1986) proposed a strategy for parallelising based on a combination approach of building a binary tree to hold the plane sweep data, coupled with pre-processing to sort the geoms by their x coordinates. Further to this, a divide-and-conquer method of dividing the data space into equally sized sub-problems, to be solved in parallel using this method, was proposed. Detailed algorithm design or implementation results were not presented by the authors, but such an approach could be employed on either shared or distributed memory architectures. More recently, Wang (1993) has proposed a modified version of White's band algorithm for intersection detection to run in parallel. The problem is represented as a series of nested loops and parallelism is applied at successively deeper levels of nesting to yield very good speed up on a shared memory machine. However, while processing is farmed out to individual processor nodes, the data currently in the plane sweep band are held in shared memory with a locking mechanism to control access by multiple processors. Clearly, an algorithm of this kind would not, therefore, transfer to a distributed memory architecture.

13.7 Design for Parallel Polygon Overlay

In general, previous work has not always been reported in the context of either a complete implementation of all stages of the overlay algorithm (but see Franklin *et al.*, 1989 and Waugh & Hopkins, 1992) or the requirements of related operations, as described in Chapters 5 and 7. In the light of these additional considerations, the final choice of approach adopted here differs from any of those reported above, but has most in common with that of Atallah & Goodrich (1986) and subsequent related developments summarised in Akl & Lyons (1993).

The basic points guiding the development of the approach can be summarised as follows:

1. The divide-and-conquer approach of dividing the space up into strips is a first major source of parallelism.

2. Necessary sorting of geoms can be undertaken by the sort/join/GAD module (see Chapter 8).

3. Within the individual strips, a plane sweep is the most effective means of intersection detection, but is implemented sequentially. In a distributed memory machine with a large number of strips, little would be gained by trying to parallelise aspects of the sweep traversal itself, there would be added complexity for the algorithm designer and possible performance degradation rather than enhancement, as a result of the increased communication costs.

4. Further parallelism is introduced into the operation as a whole, rather than into a particular component of it, by overlapping the pipeline of tasks such that data distribution, intersection detection and new polygon linkage are happening concurrently on different processors for different sets of geoms. TSO event handling (see Chapter 9) enables this by allowing a worker to participate in stitching, interleaved with receiving data from GAD processes.

13.7.1 Partitioning the Space into Strips

The data may be divided either into strips orthogonal to the direction of the sweep or into strips parallel to the direction of sweep (Figures 13.4a and 13.4b). Geoms crossing domain boundaries may be subdivided by splitting on an existing vertex.

Strips orthogonal to sweep direction

This means that the data are divided perpendicularly to the direction of the sweep, e.g. if the sweep takes place from left to right then the strips are vertical.

Advantages:

- No predetermination of data distribution is necessary as each strip process can be fed data in turn until its limit is reached.

Disadvantages:

- The strips have a width equal to the full width or length of the area being overlaid, depending on the direction of the sweep. This means that proportionately more memory is required to hold the sweep-line status on each processor.

- If the sorted data are streamed into the worker processes then each strip process has to receive its entire data portion at once so that the next process may receive its own portion. This means that the low memory usage advantage of the plane sweep would be lost (put another way, the strips will need to be narrower than otherwise required). This problem may readily be solved by pre-sorting the data into strips so that each process may receive its own data independently.

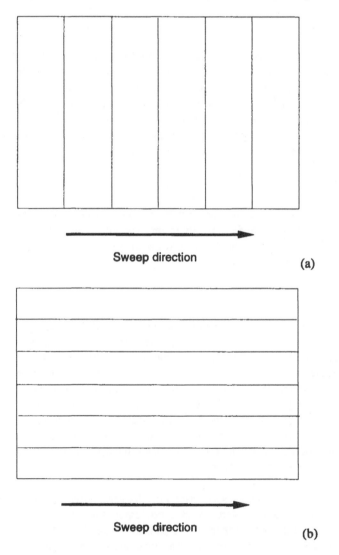

Figure 13.4: *Strips are orthogonal to sweep direction (a). Strips are parallel to sweep direction (b)*

Strips parallel to sweep direction

This means that the data are divided along the direction of the sweep, e.g. if the sweep takes place from left to right then the strips are horizontal.

Advantages:

* The strips have an extent equal to the full width (or length, depending on the direction of the sweep) of the overlay area divided by the number of strips, if a regular decomposition is used. If an irregular decomposition is used then they have variable extents depending both on the distribution of the data across the overlay area and the number of strips. Either way the strips are narrower than the full width (or length) of the map and thus the sweep-line status on each processor requires proportionately less memory.

Disadvantages:

* If the sorted data are streamed into the worker processes then a bottleneck may occur in cases where data density is very variable. Again this problem may be solved by pre-sorting the data into strips so that each process may receive its own data independently.

*

For either method, provided that the computer coverages are being processed (i.e. that a sub-area of interest has not been specified), the parent polygon IDs for non-intersecting geoms can be determined from information available to the sweep-line, since the strips extend to the edge of the dataset. In the case of strips parallel to the direction of sweep, the sweep-line starts from one end of each strip and therefore will encounter geoms which define the enclosing area at some point. The polygon IDs on either side of transitions along the sweep-line can be determined. Similarly, in the case of strips perpendicular to the direction of sweep, the sweep-line extends to the edge of the dataset and therefore must intersect a geom of the enclosing poly unless it is the universal polygon.

Since the sort and join stages, described in Chapter 8, have already converted the unsorted input data into a sorted vector topological form, the additional processing required to sort the data further into strips is not large. This allows processes to receive their strip data independently, as noted above, so the only remaining issue is that of memory usage. Here, as overlay datasets themselves may vary significantly in shape, the most appropriate combination of sweep direction and strip orientation can be chosen to optimise memory use.

13.7.2 Decomposition into Strips and Geom/Attribute Distribution

The Geom/Attribute Distribution component is described generically in Chapter 8. Its operation is summarised here, in the context of its specific application to polygon overlay. Its main role is to data-decompose the output data space over a group of Worker processes and then distribute the data to the Workers as required. The space is decomposed into strips which are rectangular and span the full width/length of the output data space. This operation has to take place for each of the datasets involved in the overlay.

The top-level control structure in the Worker process for polygon overlay is different than in the case of V2R (Chapter 11). V2R employs a more traditional task-farm approach with a loop which simply waits for a strip and then processes it. TSO, which performs vector-topological data output including re-combination of distributed vector data (Chapter 9), requires polygon overlay to use an event-driven model, because TSO has other activities to perform around Geom/Attribute distribution.

Each Worker group member receives one Geom parcel per dataset and one GeomValue parcel per dataset from each GeomAttributeServer group member for the current strip. The final sort/join of the Geomvalue and Geom records, to produce a stream of geoms with left/right attributes which are sorted by descending Ymax, is performed on the Worker. Since there is more than one input dataset, the data items are labelled with a dataset id number.

The basic task-farm model of GAD involves Workers signalling to the Source that they are ready for more work. When the Source allocates a strip to a Worker it sends a strip-descriptor to the Worker. On receipt of this strip-descriptor, the Worker can set up the necessary structures for merge-sorting the Geoms and for storing the GeomValues.

GeomAttributeServers then send data to the Worker. On arrival of these data, the Worker can process the GeomValue data and the merge-tree to produce sorted Geoms in the merge-tree output buffer. Once the merge-tree output buffer is non-empty, the application can then request geom-descriptors i.e. a GeomDesc structure comprising a pointer to a geom and its left and right attribute values. The stream of sorted GeomDescs is the output from the GAD workers. The data from the two datasets are merged into a single stream.

When the GAD structures have been emptied, typically when the strip has been processed, they can be destroyed to free up memory for the next strip. This model repeats until a 'flush' strip-descriptor is received from the Source, at which point the Worker closes down and sends a 'closing-down' message to the Source.

13.7.3 The Intersection Detection Phase

The intersection detection phase (hereafter called Intersect) of the overlay operation is based on a plane-sweep algorithm as outlined in section 13.5.3. However, whereas the basic plane-sweep simply reports intersections, Intersect, as described here, will also link to the topology stitching and output module TSO (Chapter 9) for the building of vector-topological output data (Harding *et al.*, 1994). The overall process involves:

- the creation of output geoms from the merged stream of input geoms, splitting input geoms at intersections when necessary

- the creation of output nodes from input nodes and from intersections

While Intersect proceeds with the plane sweep algorithm, it passes processed geom and node data to TSO, which undertakes:

- the building of the output polygons, with attribute values taken from the left and right attribute values of the input geoms.

Therefore Intersect only requires geoms (plus left/right attribute data to ascertain the attributes for output polygons). Assuming a plane-sweep algorithm with a north-to-south sweep of the geom data we can now expand on the earlier description of the two basic structures:

Sweep-Line Status (SLS)

Geometrically, the sweep-line is a (not necessarily straight) line which bisects the plane - more specifically it bisects the set of event-points (see below) into two subsets: those which have been processed (and lie to the north of the sweep-line) and those which are yet to be processed (and lie to the south of the sweep-line). The sweep-line therefore progresses from north to south as the algorithm progresses. The sweep-line does not have a defined position in terms of coordinates.

The *sweep-line status* (SLS) is conceptually a list of the geoms which cross the sweep-line at a given time, ordered in west→east order. Each element of the SLS is a sweep-line element. Every sweep-line element also points to a transition marking an edge (and its chains) which crosses the sweep-line.

The Event-Point Schedule (EPS)

This is a list of points at which the sweep-line status is to be modified, and is ordered in decreasing y-order. It consists essentially of a list of (a) start- and end-vertices for the geoms, and (b) intersections. It is updated dynamically as the algorithm progresses - event-points to the north of the sweep-line are deleted and new ones to the south of the sweep-line are inserted.

Event-points are said to be coincident if they lie at the same *xy*-position. The set of event-points coinciding at any *xy*-point is termed a coincident event point group (CEPG) and each CEPG forms the basis of an output node or output vertex. The transitions pointed to (via sweep-line elements) by the event points in each CEPG correspond to the edges (and chains) entering or leaving the output node or vertex.

The fundamental form of the algorithm is:

```
for(start at top of EPS;
    move North to South through EPS;
    EPS not empty)
{
    process next coincident event-point group;
}
```

Where the processing of each CEPG consists essentially of:

```
modify SLS by inserting and deleting geoms
    according to EPS;
create output node or vertex and build topology
    around it;
compare new geoms with neighbours to find
    intersections;
if(intersections exist)
{
        insert intersections into EPS;
}
```

The building of topology around an output node or vertex also modifies the SLS because it modifies the transitions therein. The results from each strip are handled by TSO, which writes out vector-topological records which are fully determined within the strip. The interaction between Intersect and TSO is examined in more detail in the next sub-section.

Thus the EPS determines the modification of the SLS and the SLS is used to find new intersections for insertion into the EPS. The number of intersection tests can be minimised by bounding-box tests and appropriate ordering of geoms around CEPGs.

The EPS and SLS are illustrated in Figure 13.5.

When Intersect has found all the intersections between input geoms and output nodes have been formed, new output polygons can be built. Each output polygon comprises the intersection of two input polygons and thus it is given two attribute values, one from each parent, in the usual manner.

The appendix to the chapter contains a more detailed worked example of this version of the plane sweep algorithm.

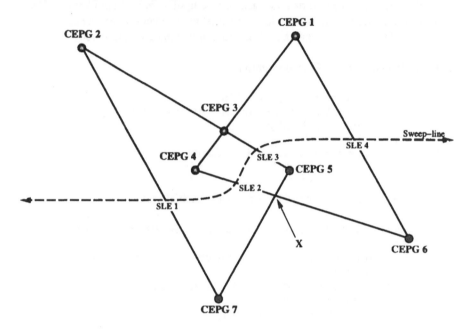

Figure 13.5: The sweep-line status and event-point schedule at a particular point in the plane-sweep algorithm. The numbering reflects the ordering in the EPS and SLS. CEPGs 1-4 have been processed and deleted from the EPS whilst 5-7 await processing. When CEPG has been processed, geom comparison will yield an intersection at X which will in turn be entered as an event point into the EPS. The SLS currently consists of four sweep-line elements.

13.7.4   Interaction between Intersect and TSO

Among the range of functions performed by TSO, the partial topology functions are of particular importance in the overlay process. A number of the functions are concerned with the building of vector-topological structures within a strip, while others perform tasks such as handling strip border information.

As the plane sweep progresses, Intersect calls these topology-building functions to allow:

- vertices to be appended to output geoms

- output nodes to be created

- chains to be created, appended to and merged

- edges to be created, merged and terminated.

In general, these functions can also be used to create and write out a vector-topological dataset for any operation which:

- can specify the meeting-points of all output geoms in a sweep-line or scan-line type ordering (these points will form output vertices and nodes) and

- has the following information at each of these meeting-points:

  - a list of the geoms which end at the point, in clockwise (or anti-clockwise) order

  - a list of the geoms which start at the point, in clockwise (or anti-clockwise) order, with intervening output attribute values.

13.7.5 Linkage and Output of Polygons that Cross Strips

While the above interaction between the Intersect and TSO modules allows polygons that are determined within a strip to be linked and output, the situation is more complex for output polygons which cross strips. These obviously require output data from Intersect operations on multiple processes, which are handling the different strips, to be coordinated before the full topological information can be assembled and the polygon written out. The functions to perform these tasks are discussed in detail in the chapter on TSO.

13.7.6 Error Conditions in Intersect

The above description of stages in the overlay operation suggests that a number of possible error conditions could arise, which require handling. These include:

- non-monotonic geoms

- geoms not correctly in descending order of maximum y coordinate

- geoms which cross a strip border and have more than one vertex lying beyond that border

- geoms of the same dataset which are collinear or which intersect elsewhere than at a node (although this cannot be an exhaustive check since some points which were vertices in the original dataset become nodes after splitting by Sort and Geom Attribute Distribution)

- geoms consisting of fewer than two vertices or containing coincident vertices (zero-length geoms)

- 'danglers', namely input geoms which have a node where no other geom from the same input dataset has a node (except when the node lies outside the strip)

- inconsistent attribute values of two input geoms bounding the same output polygon.

13.8 Conclusions

From the foregoing it is clear that the role of parallelism within the polygon overlay operation depends on the particular stage of the processing that is being considered. The modular approach, which applies to pre- and post-processing, as well as intersection detection itself, allows appropriate parallel strategies to be deployed when required. The use of parallel sort/merge to pre-process input data is very significant, in terms of overall throughput for production GIS overlay work. The strip decomposition for parallel intersection detection is conceptually simple and allows this part of the geometric processing to be based in part, at least, on existing sequential algorithms, although the distinction between output polygons that are fully determined within a strip and those that are not adds new aspects to the problem. Finally, the stitching together of strips and the re-assembling of polygons from component geoms handled by different processes has depended on the new parallel algorithm described in Chapter 9. The fundamental importance of this has been demonstrated by its deployment as one of the main processing stages of both raster-vector conversion and polygon overlay.

13.9 References

Akl, S.G. & Lyons, K.A., 1993, *Parallel Computational Geometry*, Prentice Hall, Englewood Cliffs, N.J.

Atallah, M.J. & Goodrich, M.T., 1986, Efficient plane sweeping in parallel. *Proceedings, 2nd Symposium on Computational Geometry*, York Town Heights, NY.

Burrough, P.A., 1986, *Principles of GIS for Land Resources Assessment*, Oxford University Press, Oxford.

Chrisman, N.R., Dougenik, J.A. & White, D., 1992, Lessons for the design of polygon overlay processing from the ODYSSEY WHIRLPOOL algorithm. In Bresnahan, P., Corwin, E. & Cowen, D. (Eds), *5th International Symposium on Spatial Data Handling*, Vol. 2, Humanities and Social Sciences Computing Lab, University of South Carolina, 401-410.

Douglas, D.H. & Peucker, T.K., 1973, Algorithms for the reduction of the number of points required to represent a digitized line or its caricature. *The Canadian Cartographer*, **10** (2), 119-122.

Douglas, D., 1974, It makes me so CROSS. In Peuquet, D.J. & Marble, D.F. (Eds), 1990, *Introductory Readings in Geographic Information Systems*, Taylor & Francis, London, 303-307.

ESRI Inc., 1991, *ARC/INFO Data Model, Concepts and Key Terms. V6.0 ARC/INFO Users' Guide*, ESRI, Redlands, CA.

Franklin, W.R., Narayanaswami, C., Kankanhalli, M., Sun, D. & Wu, P.Y., 1989, Uniform grids: A technique for intersection detection on serial and parallel machines. In Anderson, E. (Ed.), *AutoCarto 9: Proceedings of the 9th International Symposium on Computer-Assisted Cartography* (Baltimore, MD, April 1989), The American Society for Photogrammetry and Remote Sensing and The American Congress on Surveying and Mapping, Falls Church, VA, 100-109.

Guevara, J.A., 1983, *A Framework for the Analysis of Geographic Information Systems Procedures: the Polygon Overlay Problem, Computational Complexity and Polyline Intersection*. PhD thesis, State University of New York at Buffalo.

Harding, T.J., Hopkins, S. & Dowers, S., 1994, *Intersect Component of Polygon Overlay: Detailed Design*. Document ID: EPCC-PAP-GIS-PO-DD-INTERSECT, EPCC and The University of Edinburgh, Department of Geography, Edinburgh.

Hopkins, S. & Healey, R.G., 1990, A parallel implementation of Franklin's uniform grid technique for line intersection detection on a large transputer array. In Brassel, K. & Kishimoto, H. (Eds), *Proceedings of the 4th International Symposium on Spatial Data Handling* (Vol. 1), Zurich, Switzerland.

Hopkins, S., Healey, R.G. & Waugh, T.C., 1992, Algorithm scalability for line intersection detection in parallel polygon overlay. In Corwin, E., Bresnahan, P. & Cowen, D. (Eds), *Proceedings of the Fifth International Symposium on Spatial Data Handling*, Vol. 1, Humanities and Social Sciences Computing Lab, University of South Carolina, 210-218.

Intergraph Corporation, 1991, *MGE Analyst Reference Manual.* Intergraph Corporation, Huntsville, AL.

Knuth, D.E., 1973, *Sorting and Searching, Volume 3 of The Art of Computer Programming*, Addison-Wesley, Reading, MA.

Preparata, F.P. & Shamos, M.I., 1985, *Computational Geometry: An Introduction*, Springer-Verlag, NY.

Nordbeck, S. & Rystedt, B., 1972, *Computer Cartography*, Studentlitteratur, Lund.

Samet, H., 1990, *Applications of Spatial Data Structures: Computer Graphics, Image Processing and GIS*, Addison-Wesley, Reading, MA.

Sedgewick, R., 1988, *Algorithms*, Addison-Wesley, Reading, MA.

Wagner, D.F., 1991, *Development and Proof-of-Concept of a Comprehensive Performance Evaluation Methodology for Geographic Information Systems*. PhD thesis, Ohio State University.

Wang, F., 1993, A parallel intersection algorithm for vector polygon overlay. *IEEE Computer Graphics and Applications*, March, 74-80.

Waugh, T.C. & Hopkins, S., 1992, An algorithm for polygon overlay using cooperative parallel processing. *International Journal of GIS*, 6 (6), 457-468.

White, D., 1978, A design for polygon overlay. In Dutton, G. (Ed.), *Harvard Papers on Geographic Information Systems: Volume 6, Spatial Algorithms: Efficiency in Theory and Practice*, Laboratory for Computer Graphics and Spatial Analysis, Harvard University, Cambridge, MA.

13 Appendix

Example Plane Sweep Overlay

13 A.1 Introduction

As an illustrative example of the working of the plane-sweep algorithm, the steps involved in overlaying the two triangles ACF and BDG (see Figure 13A.1) are now shown. ACF comprises dataset 0 and BDG comprises dataset 1. The example assumes that in the input datasets the triangles were formed from one geom each, that each triangle had attribute 1 and that the universal polygon in each dataset had attribute 0.

In the diagrams used in this example, no attempt is made to differentiate between nodes and vertices, by shading for example. The reason for this is that a node in an input dataset may become a vertex in the output dataset and *vice versa*. Output nodes are however labelled N0, N1 etc. in these diagrams.

The relevant data in the input dataset is therefore:

Vertices	Dataset	Left attribute	Right attribute
ACFA	0	0	1
BDGB	1	0	1

After splitting into monotonic sections by Sort and Join the data is as follows (the precise order of geoms starting at a coincident point, such as ACF and AF, is unspecified and arbitrary):

Vertices	Dataset	Left attribute	Right attribute
AF	0	1	0
ACF	0	0	1
BG	1	1	0
BDG	1	0	1

For illustrative purposes, the example further assumes that the strip decomposition is such that the strip boundary runs South of F but North of G. The Northern strip is strip 0, the Southern strip is strip 1.

Input dataset 0   Input dataset 1

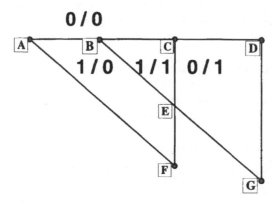

The datasets overlaid with output attribute pairs shown (the segment BC is collinear)

Figure 13A.1: Example overlay

In the transitions shown in the SLS tables below, pointers to the output chains, edges and geoms are replaced by integers for illustrative purposes. These numbers are *not* the same as the IDs which TSO grants to closed chains, edges and geoms.

The description is organised in terms of named procedures that perform different stages of the operation. These encapsulate sequences of events that link changes in the sweep line status and event point schedules.

In the SLS and EPS tables below, the symbol ^implies that the element is collinear with the previous element.

13A.2 Processing on strip 0

Strip 0 receives the following data:

Vertices	Dataset	Left attribute	Right attribute
AF	0	1	0
ACF	0	0	1
BDG	1	0	1
BG	1	1	0

13A.2.1 Initial SLS and EPS

The SLS starts off null (Figure 13A.2). The initial EPS has the raw-start nodes at A (in anticlockwise order).

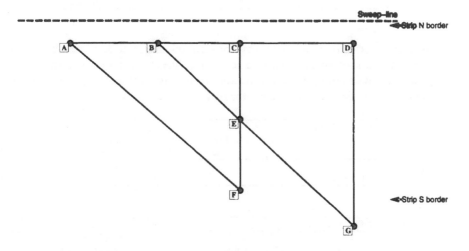

Figure 13A.2: Initial sweep-line and topology on strip 0

SLS	NULL

EPS	Position	Type	Points to
	A	RAW-START-NODE-EVENT	Geom AF
	A	RAW-START-NODE-EVENT	Geom ACF

13A.2.2 Process event-points at A

`EventPointCoincidentGroupProcessRawStartNodes` processes the first raw start-node. It inserts a new sweep-line element for the geom AF into the SLS, converts the raw start-node into a start-node and inserts an end-node event-point for the sweep-line element AF at F into the EPS.

SLS	Geom	Segment	Transition				
			Edge	Left Chain	Right Chain	Right Attribute 0	Right Attribute 1
	AF	0	---	---	---	---	---

EPS	Position	Type	Points to
	A	START-NODE-EVENT	SLS-element AF
	A	RAW-START-NODE-EVENT	Geom ACF
	F	END-NODE-EVENT	SLS-element AF

`EventPointCoincidentGroupProcessRawStartNodes` then processes the second raw start-node. It inserts a new sweep-line element for the geom ACF (in anti-clockwise order) into the SLS, converts the raw start-node into a start-node and inserts an end-vertex event-point for ACF at C into the EPS.

SLS	Geom	Segment	Transition				
			Edge	Left Chain	Right Chain	Right Attribute 0	Right Attribute 1
	AF	0	---	---	---	---	---
	ACF	0	---	---	---	---	---

EPS	Position	Type	Points to
	A	START-NODE-EVENT	SLS-element AF
	A	START-NODE-EVENT	SLS-element ACF
	C	END-VERTEX-EVENT	SLS-element ACF
	F	END-NODE-EVENT	SLS-element AF

`EventPointCoincidentGroupBuildEnds` SKIP

`EventPointCoincidentGroupProcessEnds` SKIP

`EventPointCoincidentGroupCompareCollinearStarts` finds no collinear segments starting at A.

`EventPointCoincidentGroupBuildStarts` finds two start-type event-points. The attribute pairs are obtained from the left/right attributes of the geoms AF and ACF.

SweepBuild does not create a node since there are only two transitions. It creates an edge/geom 0 with vertex A and also chains 0 and 1, and enters these into the transition data of the two start-type sweep-line elements. Chain 0 has attribute pair 0/0 and chain 1 has 1/0.

SLS	Geom	Segment	Transition				
			Edge	Left Chain	Right Chain	Right Attribute 0	Right Attribute 1
	AF	0	0	0	1	1	0
	ACF	0	0	1	0	0	0

EPS	Position	Type	Points to
	A	START-NODE-EVENT	SLS-element AF
	A	START-NODE-EVENT	SLS-element ACF
	C	END-VERTEX-EVENT	SLS-element ACF

`EventPointCoincidentGroupCompare` seeks to compares AF with its left-neighbour and AC of ACF with its right-neighbour, but no neighbours are found.

Finally the coincident event-points at A are removed from the EPS.

SLS	Geom	Segment	Transition				
			Edge	Left Chain	Right Chain	Right Attribute 0	Right Attribute 1
	AF	0	0	0	1	1	0
	ACF	0	0	1	0	0	0

EPS	Position	Type	Points to
	C	END-VERTEX-EVENT	SLS-element ACF
	F	END-NODE-EVENT	SLS-element AF

The sweep-line at this stage is illustrated in **Figure 13A.3**.

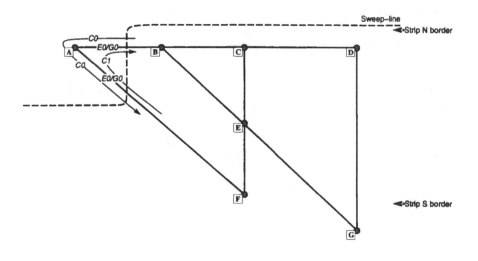

Figure 13A.3: Sweep-line and topology on strip 0 after processing of A

13A.2.3 Process event-points at B

Two new raw start nodes are introduced. The EPS tables for the processing of B are accompanied by a diagrammatic representation of the EPS elements around B to illustrate how end-type event-points may be converted to start-type and how each is used in building topology.

SLS	Geom	Segment			Transition		
			Edge	Left Chain	Right Chain	Right Attribute 0	Right Attribute 1
	AF	0	0	0	1	1	0
	ACF	0	0	1	0	0	0

`EventPointCoincidentGroupProcessRawStartNodes` processes the first raw start-node. It inserts a new sweep-line element for the geom BG into the SLS and converts the raw start-node into a start-node. It does *not* insert an end-node event-point for the sweep-line element BG at G into the EPS, because G is below the Southern strip boundary. During the insertion of the sweep-line element, it recognises that BG intersects with an existing sweep-line element AC at B and inserts an end-intersection event at B for the sweep-line element AC to mark this. A flag is set to prevent this intersection being re-discovered when the next raw start-node is processed.

SLS	Geom	Segment		Transition			
			Edge	Left Chain	Right Chain	Right Attribute 0	Right Attribute 1
	AF	0	0	0	1	1	0
	BG	0	---	---	---	---	---
	ACF	0	0	1	0	0	0

`EventPointCoincidentGroupProcessRawStartNodes` then processes the second raw start-node. It inserts a new sweep-line element for the geom BDG (in anti-clockwise order) into the SLS, converts the raw start-node into a start-node and inserts an end-vertex event-point for BDG at D into the EPS.

SLS	Geom	Segment		Transition			
			Edge	Left Chain	Right Chain	Right Attribute 0	Right Attribute 1
	AF	0	0	0	1	1	0
	BG	0	---	---	---	---	---
	BDG	0	---	---	---	---	---
	ACF	0 ^	0	1	0	0	0

`EventPointCoincidentGroupBuildEnds` finds a single end-type event-point at B.

`EventPointCoincidentGroupProcessEnds` processes the end-intersection at B by converting it to a start-intersection event and re-inserting it, and also by re-inserting the sweep-line element BCF in the anticlockwise order of its lower part (beyond the intersection B). The geom ACF is updated to become BCF.

SLS	Geom	Segment		Transition			
			Edge	Left Chain	Right Chain	Right Attribute 0	Right Attribute 1
	AF	0	0	0	1	1	0
	BG	0	---	---	---	---	---
	BCF	0	0	1	0	0	0
	BDG	0 ^	---	---	---	---	---

`EventPointCoincidentGroupCompareCollinearStarts` finds the collinear output segment starting at B and ending at C. An end-intersection event is inserted at C for the sweep-line element BDG to mark this.

SLS	Geom	Segment	Transition				
			Edge	Left Chain	Right Chain	Right Attribute 0	Right Attribute 1
	AF	0	0	0	1	1	0
	BG	0	---	---	---	---	---
	BCF	0	0	1	0	0	0
	BDG	0 ^	---	---	---	---	---

`EventPointCoincidentGroupBuildStarts` finds three start-type events at B but two of these are collinear, so there are only two transitions.

SweepBuild was told of three event-points at B so it creates a node 0. It terminates the corresponding edge 0 with the node 0. It creates edge/geom 1 and edge/geom 2 from node 0 and starts a chain 2 between them with attribute pair 1/1. This is entered into the transition data of the new sweep-line elements. It also extends chain 1 by adding edge 1 to it and extends chain 0 by adding edge 2.

SLS	Geom	Segment	Transition				
			Edge	Left Chain	Right Chain	Right Attribute 0	Right Attribute 1
	AF	0	0	0	1	1	0
	BG	0	1	1	2	1	1
	BCF	0	2	2	0	0	0
	CDG	0 ^	2	2	0	0	0

`EventPointCoincidentGroupCompare` compares BG with its left-neighbour AF but finds no intersect. It does not need to compare BC because it is collinear with AC, which is not a new segment.

Finally the coincident event-points at B are removed from the EPS.

SLS	Geom	Segment	Transition				
			Edge	Left Chain	Right Chain	Right Attribute 0	Right Attribute 1
	AF	0	0	0	1	1	0
	BG	0	1	1	2	1	1
	BCF	0	2	2	0	0	0
	CDG	0 ^	2	2	0	0	0

EPS	Position	Type	Points to
	C	END-INTERSECTION-EVENT	SLS-element CDG
	C	END-VERTEX-EVENT ^	SLS-element BCF
	D	END-VERTEX-EVENT	SLS-element CDG
	F	END-NODE-EVENT	SLS-element AF

The sweep-line is illustrated in Figure 13A.4.

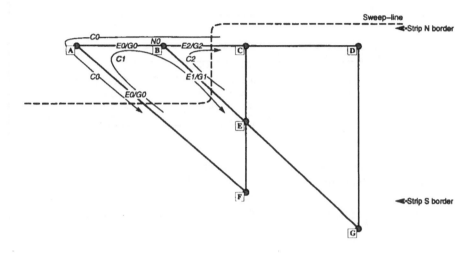

Figure 13A.4: Sweep-line and topology on strip 0 after processing of B

13A.2.4 Process event-points at C

EventPointCoincidentGroupProcessRawStartNodes: SKIP.

EventPointCoincidentGroupBuildEnds finds two collinear end-type
event-points implying a single end-type transition.

EventPointCoincidentGroupProcessEnds processes the end-
intersection at C in a similar way to the one at B, by converting it to a start-
intersection event and re-inserting it, and also by re-inserting the sweep-line
element CDG in the anti-clockwise order of its lower part (beyond the intersection
C). The geom BDG is updated to become CDG.

It processes the end-vertex event similarly, by converting it to a start-vertex event
and re-inserting it, and also by re-inserting the sweep-line element BCF in the anti-
clockwise order of the next segment and incrementing the segment number and
finally by inserting an end-node event-point at F for BCF.

SLS	Geom	Segment	Transition				
			Edge	Left Chain	Right Chain	Right Attribute 0	Right Attribute 1
	AF	0	0	0	1	1	0
	BG	0	1	1	2	1	1
	BCF	1	2	2	0	0	0
	CDG	0	2	2	0	0	0

EPS	Position	Type	Points to
	C	START-VERTEX-EVENT	SLS-element BCF
	C	START-INTERSECTION-EVENT	SLS-element CDG
	D	END-VERTEX-EVENT	SLS-element CDG
	F	END-NODE-EVENT	SLS-element BCF
	F	END-NODE-EVENT	SLS-element AF

`EventPointCoincidentGroupCompareCollinearStarts` finds no collinear segments starting at C.

`EventPointCoincidentGroupBuildStarts` finds two start-type events at C.

SweepBuild creates a node 1 and terminates edge 2 on node 1. It creates edge 3 and edge 4 from node 1 and starts a chain 3 between them with attribute pair 0/1. This is entered into the transition data of the new sweep-line elements. It also extends chain 2 by adding edge 3 to it and extends chain 0 by adding edge 4.

`EventPointCoincidentGroupCompare` compares CF of BCF with its left-neighbour BG and finds an intersect at E. It inserts a cross-intersection event for BCF and BG at E.

Finally the coincident event-points at C are removed from the EPS.

SLS	Geom	Segment	Transition				
			Edge	Left Chain	Right Chain	Right Attribute 0	Right Attribute 1
	AF	0	0	0	1	1	0
	BG	0	1	1	2	1	1
	BCF	1	3	2	3	0	1
	CDG	0	4	3	0	0	0

EPS	Position	Type	Points to
	D	END-VERTEX-EVENT	SLS-element CDG
	E	CROSS-INTERSECTION-EVENT	SLS-element BCF, BG
	F	END-NODE-EVENT	SLS-element BCF
	F	END-NODE-EVENT	SLS-element AF

The sweep-line is illustrated in Figure 13A.5.

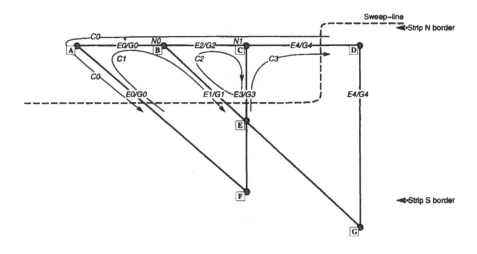

Figure 13A.5: Sweep-line and topology on strip 0 after processing of C

13A.2.5 Process event-points at D

`EventPointCoincidentGroupProcessRawStartNodes`: SKIP.

`EventPointCoincidentGroupBuildEnds`: finds one end-type event-point at D.

`EventPointCoincidentGroupProcessEnds` processes the end-vertex in a similar way to the one at C, by converting it to a start-vertex event and re-inserting it (actually a nullop in this case), and also by re-inserting the sweep-line element CDG in the anticlockwise order of the next segment (also a nullop) and incrementing the segment number. It does *not* insert an end-node event-point for the sweep-line element CDG at G into the EPS, because G is below the Southern strip boundary.

SLS	Geom	Segment	Transition				
			Edge	Left Chain	Right Chain	Right Attribute 0	Right Attribute 1
	AF	0	0	0	1	1	0
	BG	0	1	1	2	1	1
	BCF	1	3	2	3	0	1
	CDG	1	4	3	0	0	0

EPS	Position	Type	Points to
	D	START-VERTEX-EVENT	SLS-element CDG
	E	CROSS-INTERSECTION-EVENT	SLS-element BCF, BG
	F	END-NODE-EVENT	SLS-element BCF
	F	END-NODE-EVENT	SLS-element AF

`EventPointCoincidentGroupCompareCollinearStarts`: SKIP.

`EventPointCoincidentGroupBuildStarts`: finds one start-type event-point at D.

SweepBuild does not create a node but just adds a vertex at D to geom 4.

`EventPointCoincidentGroupCompare` compares DG of CDG with its left-neighbour CF of BCF and finds no intersect. It seeks to compare DG of CDG with a right-neighbour but none is found.

Finally the event-point at D is removed from the EPS.

SLS	Geom	Segment	Edge	Left Chain	Right Chain	Right Attribute 0	Right Attribute 1
					Transition		
	AF	0	0	0	1	1	0
	BG	0	1	1	2	1	1
	BCF	1	3	2	3	0	1
	CDG	1	4	3	0	0	0

EPS	Position	Type	Points to
	E	CROSS-INTERSECTION-EVENT	SLS-element BCF, BG
	F	END-NODE-EVENT	SLS-element BCF
	F	END-NODE-EVENT	SLS-element AF

The sweep-line is illustrated in Figure 13A.6.

13A.2.6 Process event-points at E

`EventPointCoincidentGroupProcessRawStartNodes`: SKIP.

`EventPointCoincidentGroupBuildEnds` finds a cross-intersection event i.e. two end-type transitions.

`EventPointCoincidentGroupProcessEnds` processes the cross-intersection event by updating the geoms BG and BCF to EG and BEF. The sweep-line elements are swapped in the SLS.

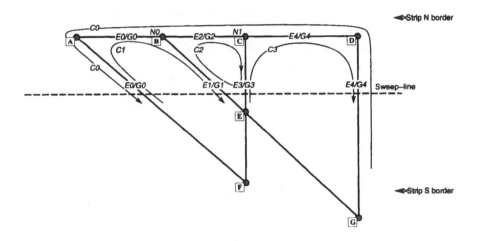

Figure 13A.6: Sweep-line and topology on strip 0 after processing of D

SLS	Geom	Segment	Transition				
			Edge	Left Chain	Right Chain	Right Attribute 0	Right Attribute 1
	AF	0	0	0	1	1	0
	BEF	1	3	2	3	0	1
	EG	0	1	1	2	1	1
	CDG	1	4	3	0	0	0

EPS	Position	Type	Points to
	E	CROSS-INTERSECTION-EVENT	SLS-element BEF, EG
	F	END-NODE-EVENT	SLS-element BEF
	F	END-NODE-EVENT	SLS-element AF

`EventPointCoincidentGroupCompareCollinearStarts` SKIP.

`EventPointCoincidentGroupBuildStarts` finds a cross-intersection event i.e. two start-type transitions.

SweepBuild creates a node 2 at E. It terminates edge 1 and edge 3 at node 2 and joins the two branches of chain 2. It also creates edge 5 and edge 6 from node 2 with a new chain 4 between them with attribute pair 0/0. Finally it extends chain 1 to the left of edge 5 and chain 3 to the right of edge 6.

`EventPointCoincidentGroupCompare` compares EF of BEF with its left-neighbour AF and finds the intersect at F. It does not insert it into the EPS however, because the intersection occurs at a vertex for both segments. It also compares EG with its right-neighbour DG of CDG and treats it in the same way.

Finally the event-point at E is removed from the EPS.

SLS	Geom	Segment		Transition			
			Edge	Left Chain	Right Chain	Right Attribute 0	Right Attribute 1
	AF	0	0	0	1	1	0
	BEF	1	5	1	4	0	0
	EG	0	6	4	3	0	1
	CDG	1	4	3	0	0	0

EPS	Position	Type	Points to
	F	END-NODE-EVENT	SLS-element BEF
	F	END-NODE-EVENT	SLS-element AF

The sweep-line is illustrated in Figure 13A.7.

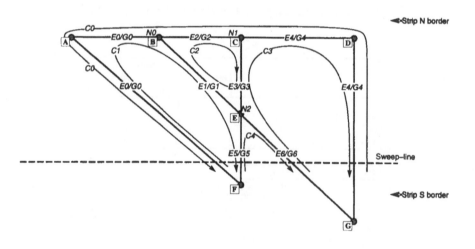

Figure 13A.7: Sweep-line and topology on strip 0 after processing of E

13A.2.7 Process event-points at F

EventPointCoincidentGroupProcessRawStartNodes: SKIP.

`EventPointCoincidentGroupBuildEnds` finds two end-type Event-Point Schedules.

`EventPointCoincidentGroupProcessEnds` processes the end-nodes by deleting the sweep-line elements they reference.

SLS	Geom	Segment			Transition		
			Edge	Left Chain	Right Chain	Right Attribute 0	Right Attribute 1
	EG	0	6	4	3	0	1
	CDG	1	4	3	0	0	0

EPS	Position	Type	Points to
	F	END-NODE-EVENT	SLS-element null
	F	END-NODE-EVENT	SLS-element null

`EventPointCoincidentGroupCompareCollinearStarts`: SKIP.

`EventPointCoincidentGroupBuildStarts` SKIP.

SweepBuild does not create a node, but joins edge/geom 0 with edge/geom 5 at F, and joins the branches of chain 1 and joins chain 0 and chain 4.

`EventPointCoincidentGroupCompare` seeks to compare EG with a left-neighbour but none is found.

Finally the coincident event-points at F are removed from the EPS.

SLS	Geom	Segment			Transition		
			Edge	Left Chain	Right Chain	Right Attribute 0	Right Attribute 1
	EG	0	6	4	3	0	1
	CDG	1	4	3	0	0	0

EPS	NULL

The sweep-line is illustrated in Figure 13A.8.

The EPS is now empty so the plane-sweep algorithm stops here after converting the current SLS into the Southern strip border data for TSO.

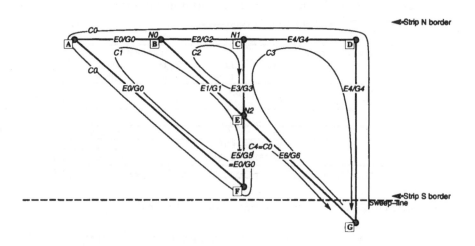

Figure 13A.8: Final sweep-line and topology on strip 0 after processing of F

13A.3Processing on strip 1

Strip 1 receives the following data:

Vertices	Dataset	Left attribute	Right attribute
DG	1	0	1
BG	1	1	0

13A.3.1 Initial SLS and EPS

The SLS starts off null (Figure 13A.9). The initial EPS has a raw-start node at B.

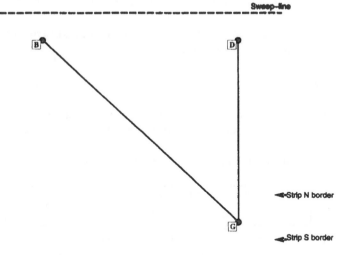

Figure 13A.9: Initial sweep-line and topology on strip 1

SLS	NULL

EPS	Position	Type	Points to
	B	RAW-START-NODE-EVENT	Geom BG

13A.3.2 Process event-point at B

`EventPointCoincidentGroupProcessRawStartNodes` processes the raw start-node in the usual way.

SLS	Geom	Segment			Transition		
			Edge	Left Chain	Right Chain	Right Attribute 0	Right Attribute 1
	BG	0	---	---	---	---	---

EPS	Position	Type	Points to
	B	START-NODE-EVENT	SLS-element BG
	G	END-NODE-EVENT	SLS-element BG

`EventPointCoincidentGroupBuildEnds`: SKIP.

`EventPointCoincidentGroupProcessEnds`: SKIP

`EventPointCoincidentGroupCompareCollinearStarts` finds no collinear segments starting at B.

`EventPointCoincidentGroupBuildStarts`: SKIP.

`EventPointCoincidentGroupCompare` seeks to compare BG with its left- and right-neighbours, but it finds no neighbours.

Finally the event-point at B is removed from the EPS.

SLS	Geom	Segment			Transition		
			Edge	Left Chain	Right Chain	Right Attribute 0	Right Attribute 1
	BG	0	---	---	---	---	---

EPS	Position	Type	Points to
	G	END-NODE-EVENT	SLS-element BG

The sweep-line is illustrated in Figure 13A.10.

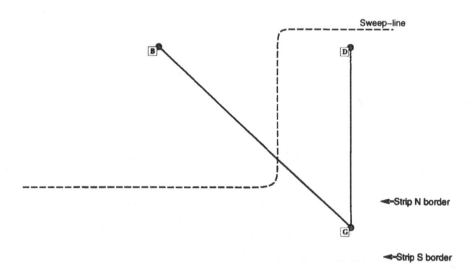

Figure 13A.10: Sweep-line and topology on strip 1 after processing of B

13A.3.3 Process event-point at D

A new raw start-node is introduced.

SLS	Geom	Segment			Transition		
			Edge	Left Chain	Right Chain	Right Attribute 0	Right Attribute 1
	BG	0	0	0	1	---	---

EPS	Position	Type	Points to
	D	RAW-START-NODE-EVENT	Geom DG
	G	END-NODE-EVENT	SLS-element BG

`EventPointCoincidentGroupProcessRawStartNodes` processes the raw start-node in the usual way.

SLS	Geom	Segment			Transition		
			Edge	Left Chain	Right Chain	Right Attribute 0	Right Attribute 1
	BG	0	---	---	---	---	---
	DG	0	---	---	---	---	---

EPS	Position	Type	Points to
	D	START-NODE-EVENT	SLS-element DG
	G	END-NODE-EVENT	SLS-element DG
	G	END-NODE-EVENT	SLS-element BG

`EventPointCoincidentGroupBuildEnds`: SKIP.

`EventPointCoincidentGroupProcessEnds`: SKIP.

`EventPointCoincidentGroupCompareCollinearStarts` finds no collinear segments starting at D.

`EventPointCoincidentGroupBuildStarts`: SKIP.

`EventPointCoincidentGroupCompare` compares DG with its left-neighbour BG and finds an intersect at G, which is not inserted as it is at an end-vertex for both segments.

Finally the event-point at D is removed from the EPS.

SLS	Geom	Segment			Transition		
			Edge	Left Chain	Right Chain	Right Attribute 0	Right Attribute 1
	BG	0	---	---	---	---	---
	DG	0	---	---	---	---	---

EPS	Position	Type	Points to
	G	END-NODE-EVENT	SLS-element DG
	G	END-NODE-EVENT	SLS-element BG

The sweep-line is illustrated in Figure 13A.11.

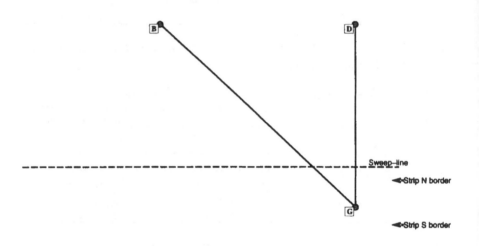

Figure 13A.11: Sweep-line and topology on strip 1 after processing of D

13A.3.4 Process event-points at G

The event-point G is the first event-point below the Northern boundary of strip 1 so transitions are created with new edges and chains and the current SLS is converted into the Northern strip border data for TSO.

SLS	Geom	Segment	Transition				
			Edge	Left Chain	Right Chain	Right Attribute 0	Right Attribute 1
	BG	0	0	0	1	---	---
	DG	0	1	2	3	---	---

EPS	Position	Type	Points to
	G	END-NODE-EVENT	SLS-element DG
	G	END-NODE-EVENT	SLS-element BG

The sweep-line is illustrated in Figure 13A.12.

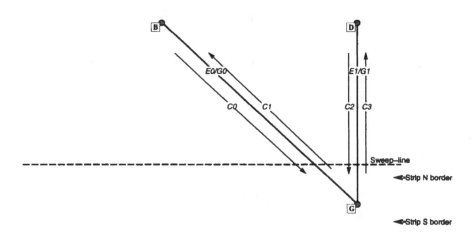

Figure 13A.12: Sweep-line and topology on strip 1 before processing of G

`EventPointCoincidentGroupProcessRawStartNodes`: SKIP.

`EventPointCoincidentGroupBuildEnds` finds two end-type Event-Point Schedules.

`EventPointCoincidentGroupProcessEnds` processes the end-nodes by deleting the sweep-line elements they reference.

SLS	NULL

EPS	Position	Type	Points to
	G	END-NODE-EVENT	SLS-element DG
	G	END-NODE-EVENT	SLS-element BG

`EventPointCoincidentGroupCompareCollinearStarts`: SKIP.

`EventPointCoincidentGroupBuildStarts`: SKIP.

SweepBuild joins edge 0 with edge 1, joins chain 0 and chain 3 and also joins chain 1 and chain 2.

Finally the coincident event-points at G are removed from the EPS.

SLS	NULL

EPS	NULL

The final SLS in this case is null (Figure 13A.13) and so no Southern-border information is passed to TSO.

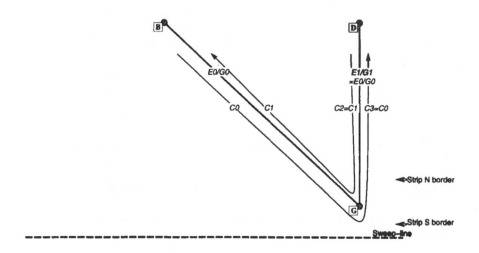

Figure 13A.13: Final sweep-line and topology on strip 1

Implementation Case Study: Generalisation of Raster Data

M.J. Mineter, G.G. Wilkinson, S. Dowers and R.G. Healey

14.1 Introduction

This chapter describes the design and implementation of parallel software to perform moving-window statistical filtering of remotely-sensed images. Its purpose is to develop a number of themes mentioned earlier in the book. Prominent amongst these are the reasons why N-fold speed-up is not usually achieved with N processors, techniques for reducing that gap, and the benefits of using libraries of parallel software.

The filtering algorithms were developed by one of the authors (Wilkinson, 1992) at the Space Applications Institute (SAI), within the Joint Research Centre of the European Union. The sequential implementation was sufficiently CPU-intensive that its exploitation was limited. The parallel software was developed during a collaboration between the SAI and the University of Edinburgh (Mineter, 1996). Regular decomposition was used (Chapter 4) so that each processor filtered one block of the image.

Section 14.2 describes the filtering algorithms; section 14.3 explores the issues affecting the parallelisation. Facets of the implementation are described in section 14.4, and the performance of the software is discussed in section 14.5.

14.2 Image Generalisation

Image generalisation reduces the number of regions in a remotely-sensed image with the purpose of facilitating further analysis, such as the conversion of the image to a vector-topological dataset of manageable size. In this study two algorithms were implemented with the goal of retaining relative class sizes at the cost of losing some locational accuracy (Wilkinson, 1992). The first algorithm, 'majority filtering', seeks to remove small regions to reduce the 'noise' in the image. The second algorithm then re-grows the under-populated classes by extending the surviving regions of those classes at the cost of over-populated classes. The resultant effect is that areas in which the different classes dominate will be distinguished more clearly, so, for example, pixels representing isolated buildings in a forest would be removed and the size of more prominent built-up areas exaggerated.

Both algorithms are iterative, and use a 'moving-window'. In each iteration, the window (a square, typically of sides of between 3 and 15 pixels) is passed over the image, and in each position the central pixel in the window is modified if certain criteria are met by the contents of the moving-window. These criteria concern the populations of the classes within the moving-window.

At each position of the moving-window the following must occur:

1.   Build statistics relevant to the filtering algorithm from the contents of the moving-window.

2.   Sort the statistics according to the criteria of the filtering algorithm.

3.   If necessary, overwrite the central pixel.

After each iteration, the number of changed pixels and other statistics specific to each algorithm are examined. The filtering is stopped when the image has stabilised or alternatively when a user-specified maximum number of iterations has been performed.

During majority filtering, at each pixel the moving-window is analysed to seek a class which exists in the majority of pixels within the window. If such a class exists then it is used to overwrite the central pixel. ('Majority' defaults to mean more than half the number of pixels in the window. This can be reset by a user-specified parameter if required.) Two additional options can be selected at run-time:

1.   Central pixels in the second most frequent class are preserved.

2.   If no majority exists, then the most frequent class is used to overwrite the central pixel unless that pixel has the first or second most frequent class.

The re-growing algorithm expands regions of connected pixels belonging to classes which have been depopulated by the majority filtering. Before each iteration, statistics of class populations across the complete image are derived and compared with those of the initial image before majority filtering. For each position of the moving-window, if the central pixel is over-populated in the complete image and the window includes classes which, across the complete image, are currently under-populated, then the central pixel is overwritten. The new value is that of the most under-populated class which is present in the window. Two thresholds are used. Both are percentages defining limits based upon variations between current and initial image statistics. The first determines when a class is judged to be over- or under-populated. When the populations of all classes fall within the second limit, convergence is judged to have occurred. These are necessary to prevent oscillations between two states of the image.

The re-growing algorithm is a variation on Wilkinson (1992), in which any over-populated pixel within the filtering window was overwritten by the value in the central pixel if that value was in an under-populated class. Once the issues of parallelisation have been described, the need for modifying that approach will be seen.

14.3 Parallelisation Issues

14.3.1 Languages, Libraries and Layers

The two basic approaches to MIMD processing were discussed in Chapter 2. The first is to use a parallel language such as High Performance FORTRAN or C which incorporates special parallel constructs. Communication between the processors is a consequence of the compilation of these constructs. The second is to use a standard language with calls to functions in a message-passing library. It was adopted here, using ANSI-C and MPI, because the resultant software is portable and capable of optimisation to a greater extent than a parallel language would allow at present. There is also greater control over the frequency and circumstances of process synchronisation and messaging: with a parallel language these are determined by the compiler. The effort inherent in the chosen approach was reduced by use of a library comprising a parallel utility supporting suitable data decomposition and messaging.

Figure 14.1 compares the resulting layers of software with those of a typical parallel and typical sequential application. In a sequential environment, software can often be viewed as three layers: the operating system, some generic functions (such as database reading/writing), and then the application-specific code. As the number of related applications grows, so another layer can develop below the application code, comprising functions related to a group of applications. In geographical applications this corresponds to data conversions, overlays or other standard operations found in a commercial GIS package. The resulting schematic is shown in Figure 14.1a. Parallel processing requires additional layers for message-passing and for data decomposition and creation (Figure 14.1b). The I/O functionality either is unchanged, using one processor only, or else needs redevelopment to exploit the multiple processors. The 'application-related utilities' typically need redevelopment to take account of the distributed data. In many cases this can hide the consequences of the parallelism from the application code. This pattern is present in each of the layers: each encapsulates a number of functions of use to the higher layer, reducing the complexity of that higher layer. Bringing those functions together into a software library facilitates reuse of the functions.

Chapter 4 described how parallel libraries have been developed to support commonly used approaches to parallel software design. In the example of image generalisation each process filters a subset of the image. The PUL-RD library (Clarke *et al.*, 1995), also discussed in Chapter 4 and Chapter 10, supports data I/O and the messaging which is a consequence of the iterative processing of distributed data. In this case the 'application-related utilities' comprise software for controlling the scan of the moving-window in each process, for ensuring that the multiple processors iterate in step, and for controlling the invocation and termination of a filtering algorithm. The code specific to the two filtering algorithms comprised a small number of functions, represented in the schematic (Figure 14.1c) as the top layer.

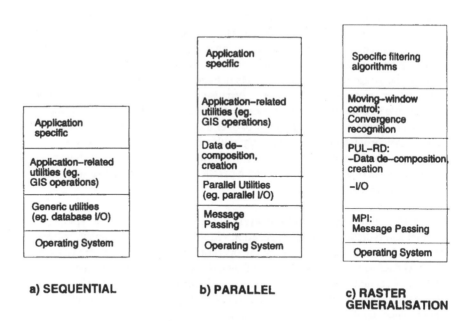

Figure 14.1: Comparison of software layers in sequential and parallel applications

14.3.2 Data Decomposition

Options for decomposing a raster in an attempt to load balance across a number of processors were discussed in Chapter 10. In regular decomposition, each processor receives one block of the raster image, the blocks are of similar size, and the assumption is made that equal sized blocks take equal amounts of processing time. Regular decomposition was considered sufficient for this application because the amount of processing entailed in the filtering is only weakly dependent on the contents of the image. The overheads in both run-time and development time of a more complex approach were considered to outweigh any possible benefits.

The moving-window filtering changes the class of the central pixel in the window according to the classes of neighbouring pixels. In order that pixels at the edges of the blocks can be filtered, some data are held on two processors. The data held within each processor thus comprises a block surrounded by a 'halo' of pixels (Figure 10.1). The halo width is the half-width of the moving-window so that data are available for filtering the edges of the block. After each iteration each halo is refreshed by messages from those processes which hold the adjoining blocks.

In section 14.2 it was mentioned that the initial sequential code for the re-growing algorithm, as described in Wilkinson (1992), used tests of the central pixel to control overwriting of the other pixels in the moving-window. This is much more difficult to implement in parallel as there would be competition between two processes to overwrite pixels close to block borders: the pixels bordering the edge of one block would also be overwritten in the halo of the block held by another process.

Both the initial and the resultant data are images held on disk. This project was constrained to using the standard UNIX disk I/O functions supported by PUL-RD, rather than exploiting parallel streams (Chapter 3). The constraints were in time available for development and in the need for existing software packages to be able to access the images. Consequently, at initialisation one processor reads the data and distributes them, and at completion of filtering, the final dataset is collated on one processor and written to disk. It can be seen that as the number of processes is increased so the effort entailed in distributing and collating the images will also rise. The I/O itself remains a sequential operation, so that the implications for scalability are clear.

14.3.3 Recognition of Convergence

Statistics for the complete image must be generated in order that convergence of an algorithm can be recognised. To do this, one processor is given the role of Coordinator. Each process sends to the Coordinator statistics about its block. In majority filtering the statistics comprise the number of pixels to have changed value in the last iteration. Convergence occurs when no pixels change in an iteration. In re-growing, the extent to which each class population deviates from the original image is required. The Coordinator collates the statistics, and determines whether convergence has occurred, and the operation should stop, or, in re-growing, which classes in the complete image are still under-populated. Appropriate information must then be communicated to all processes, using techniques described further in section 14.4.2.

The Coordinator is selected by using the identifier allocated by MPI (MPIF, 1994) to each process. These are used to specify the destination of messages. The process with identifier 0 is typically chosen to be the Coordinator, otherwise in tests with only one process these functions are not performed.

14.4 Implementation

14.4.1 Overview

Figure 14.2 schematically shows the processing required for a sequential implementation of the filtering algorithms. The image from one iteration is used to

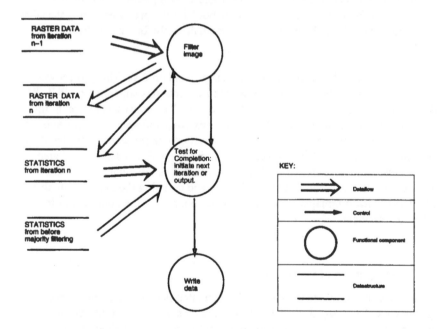

Figure 14.2:Sequential raster generalisation: processing in the nth iteration

derive the next image - hence two arrays of data are required. After each iteration, image statistics are derived, and used to determine whether filtering should cease and, in the case of re-growing, which classes are outside the desired limits.

Figure 14.3 illustrates the parallel implementation. The code that manipulates the moving-window and alters the central pixel can be identical in parallel and sequential implementations. Instead of processing the complete image, as in the sequential case, each Worker process will filter one block of that image and

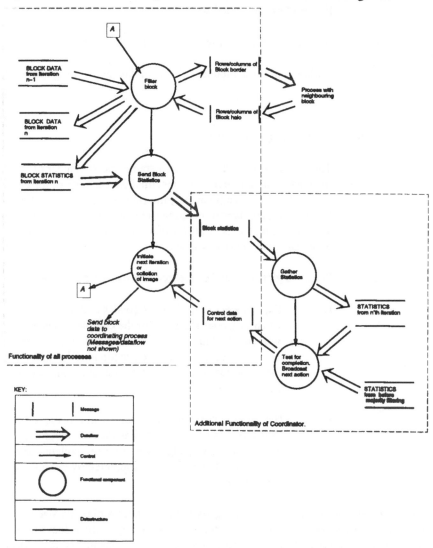

Figure 14.3: Parallel raster generalisation: processing in the nth iteration

generate statistics for its own block. Additionally, one process is responsible for coordinating the filtering, as described in section 14.4.3.

This filtering code is invoked by a framework supporting the coordination and communication required for parallelism, and comprising:

1. The PUL-RD library (Clarke *et al.*, 1995) to provide:

    a) File reading and data distribution

    b) Boundary (i.e. halo) swapping between processes.

    c) Collation of the component blocks of data and writing to file.

2. Additional messaging using MPI:

    a) The broadcast of initial run-time parameters. A number of parameters are required to be specified for each run of the code. These specify image filenames, filtering window sizes, options for filtering, numbers of iterations, etc. and are held in a file. This is read by the Coordinator into a structure which is then broadcast to all other processes.

    b) Statistics gathering and convergence tests. After each iteration:

        i) Each process sends block statistics to the Coordinator, which generates statistics for the complete image by summing the block statistics. (The gathering of the statistics is supported by the function *MPI_Reduce*.) The Coordinator can then test for convergence of the filtering algorithm.

        ii) The Coordinator then broadcasts to all processes the necessary control data to initiate the next iteration, change the filtering algorithm, or cause collation and writing out of data. (This is supported by the function *MPI_Bcast*.)

In the typical case when the two filtering algorithms require differently sized moving-windows, the larger is used for determining the size of the haloes around the data blocks created during the regular decomposition. This incurs an overhead of exchanging superfluous data at each iteration of one of the algorithms. The alternative is to reorganise the data distribution between running the two algorithms. In the present case the overhead would be even greater, as PUL-RD would require that the data be collated and written to disk, and then read and redistributed. In most runs the majority filtering uses the larger window, and the Re-growing requires fewer iterations.

After the filtering is completed a margin will usually be created by overwriting the border pixels of the image with a distinctive class. This is very trivial in sequential processing; in parallel processing it is necessary to distinguish between block borders which are internal to the image (where the margin is not required), and those which border the image. Although still simple to accomplish, it is indicative of the nature of parallel processing: care must always be taken to ensure that the appropriate information is available to each process. In this case, PUL-RD provides a function which allows the location within the image of a pixel within a block to be determined.

The code was written in C, and linked with the PUL-RD and MPI libraries. Commands provided as a part of an MPI implementation were used to execute the image on the multiple processors.

14.4.2 Non-blocking Communication

When a message is sent to another process one of two main modes can be used. For current purposes, *blocking* is defined to entail the sender process waiting until the data have been read; *non-blocking* entails the sender initiating the sending of the message, and then continuing with other application processing. At a later stage, and in particular before the memory holding the message is reused, the sending process must check that the message has been sent successfully. Similarly, non-blocking reads entail specifying a region of memory to receive a message, and at a later stage checking that the read has completed, so that the data can be used. Non-blocking is generally preferred due to the overlap of communication and computation. Non-blocking communication is used by PUL-RD. (The MPI specification (MPIF, 1994) gives a more detailed classification of communication modes.)

In this application there are two benefits from using non-blocking messaging. The first is in avoiding deadlock during the halo exchange, in which each process sends messages containing data bordering its block, and receives messages containing the halo of its block. Were this to be coded using blocking messaging, then the danger is that each process would attempt to send a message, and all processes would then be in the same blocked state. The use of non-blocking allows each process to prepare and send its second and subsequent messages without waiting for each to be read. The second benefit arises from the fact that each data block is much larger than the size of the moving-window. The consequence is that most processing is associated with filtering the central region of each block which is uninfluenced by the contents of the halo around that block. Hence, at the start of each iteration, the haloes are sent in non-blocking messages, and then the central region of each block can be filtered whilst the communication is in progress, prior to receiving the messages which enable filtering of the borders of the block. It is worth noting that a processor which is reading a number of messages in non-blocked mode has no control over the order in which the messages are received. This is determined by the communications network and by the timing of events on other processors.

14.4.3 Collective Communications

As described in section 14.3.3, checks for convergence of an algorithm need to be made. These entail gathering statistics, and broadcasting control data. The messaging is implemented by calls to the *MPI_Reduce* and *MPI_Bcast* functions which support many-to-one and one-to-many messaging respectively. *MPI_Reduce* allows each process to send an array to the Coordinator which sums the corresponding array elements, reducing the N arrays to one array. This is used to collate class populations and a count of the number of pixels to have changed in an iteration.

The *MPI_Bcast* function allows the Coordinator to send a message to all processes. The broadcast data comprises a number of items. Rather than issue a number of broadcasts, these items are held as one C structure, and so can be broadcast more efficiently in the same message. The broadcast data specifies whether another iteration is required and, in re-growing, which classes are under- or over-populated.

MPI collective functions block until completion: it is not possible to overlap useful application work with these. Furthermore, during the sequence of data reduction and broadcasting, there is a phase in which only the Coordinator is active: the other processes are awaiting receipt of the broadcast. The consequence of the enforced waiting time and synchronisation across all processes is that the scalability of the application will deteriorate, an effect which will be more evident as the number of processors is increased.

At the time of writing, non-blocking versions are being considered for inclusion in the MPI-2 standard (L.J. Clarke, private communication); as with simple point-to-point messaging, this would allow a call to be made to these collective functions, and processing to continue while the communications took place. In the case of majority filtering (where convergence requires only that no pixels be updated) a process would continue with little if any delay due to these functions until its own block had stabilised. In the case of re-growing, the new global statistics are required before a new iteration can proceed so that the gain would be slight at best.

One optimisation in this area was made in the present code. In the case of majority filtering, experience has shown that if in one iteration a 'large' number of pixels are amended, then convergence is unlikely to occur in the next 'few' iterations. Consequently, use of parameters allows a number of iterations to be processed with no collective functions being invoked.

14.4.4 Pseudo-code of the Parallel Framework

The following pseudo-code indicates the use of MPI functions (MPI_...) and PUL-RD functions (rd_...):

```
int doLocPos[2], doGloPos[2], doSize[2];    /* arrays used by PUL-RD to
                                               specify regions of the
                                               locally held block which can be
                                               filtered. */

byte    *array1, *array2;                   /* pointers to arrays which hold
                                               local block and surrounding haloes
                                               of data from neighbours */

byte    *fromBlk, *tmpBlk, *toBlk;          /* allow switching of roles for
                                               the two arrays ie. filter from
                                               array1 to array2, then from array2
                                               to array1. */

main(int argc, char *argv[])
{
    MPI_Init(argc, argv);                   /* initialise MPI */
    rd_initMPI(....);                       /* initialise the version of PUL-
                                               RD built upon MPI */

    procId = rd_groupNum(...);              /* Get the process id from MPI via
                                               PUL-RD */

    coordinator = (procId == 0);            /* Flag if this process is the
                                               coordinator */

    if (coordinator)
    {
        readParams();                       /* read in run-time parameters
                                               from a file*/
        readImageHeader();                  /* obtains image size from the
                                               header of the image file */
    }

    MPI_Bcast(...);                         /* From process 0 broadcast run-
                                               time parameters and image size to
                                               all processes */

    rd_open(..);                            /* specify image and halo size; RD
                                               determines data decomposition. */
    array1 = rd_makeBlock(..);              /* allocate first array for the
                                               block */
    array2 = rd_makeBlock(..);              /* allocate second array for the
                                               block */
```

322   *M.J. Mineter, G.G. Wilkinson, S. Dowers and R.G. Healey*

```
rd_blockSize(...)                       /* find the size of the block
                                        allocated by PUL-RD to this
                                        process */
rd_blockOff(...)                        /* find the location of this block
                                        in the image */

fromBlk = array2, toBlk = array1;       /* initialise pointers to data
                                        blocks */

rd_read(... *toBlk);                    /* read and distribute the image
                                        */

generate initial class statistics, for use in controlling re-growing.

while(moreFiltering)                    /* do next iteration */
{
        tmpBlk =toBlk, toBlk= fromBlk, fromBlk=tmpBlk;
                                        /* exchange source/destination
                                        arrays */

        rd_startSwap(....fromBlk);      /* initiate halo exchange using
                                        non-blocking messages */

           /*----------------------------------------------*/
           /* filter the central part of the block    */
           /*----------------------------------------------*/

        if (rd_nextUpdate(... , doSize, doLocPos, doGloPos) == RD_OK)
                                        /* then data of the size returned
                                        in doSize is available for
                                        processing from coordinates
                                        doLocPos in the local block,
                                        corresponding to coordinates
                                        doGloPos in the total image. */
        {
            FilterPixels( fromBlk,toBlk,doSize, doLocPos, doGloPos);
        }
           /*----------------------------------------------*/
           /*  Process remaining parts of the block       */
           /*----------------------------------------------*/

        rd_endSwap(........fromBlk);    /* test for receipt of halo;
                                        Wait if some messages have yet to
                                        be received: */

        while (rd_nextUpdate(... doSize, doLocPos, doGloPos) = = RD_OK)
                                        /* there is another region to
                                        filter */
        {
            FilterPixels(..., fromBlk,toBlk,doSize, doLocPos, doGloPos);
        }
           /*----------------------------------------------*/
           /* Generate and collate statistics; check for convergence */
           /*----------------------------------------------*/
```

```
        makeBlockStatistics(toBlk);        /* build statistics for the newly
                                              filtered block */


/* gather statistics from each block, summing them in the Coordinator...*/

        MPI_Reduce(blockStats,             /* array to send            */
                   imageStats,             /* destination for result   */
                   nclasses,               /* no. of items sent        */
                   MPI_INT,                /* type of item             */
                   MPI_SUM,                /* operation to perform     */
                   0,                      /* process to receive data  */
                   MPI_COMM_WORLD);        /* All processes send data  */
        if (coordinator)
        {
                check if filtering has completed by testing statistics in
                                    imageStats
                reset control data structure for broadcast

        }
        MPI_Bcast(controlData,             /* This structure is being
                                              broadcast from process 0;
                                              It receives data in other
                                              processes */

                ...
                0,                         /* Identifies process sending the
                                              data */
                MPI_COMM_WORLD);           /* All processes receive data*/
        if ((status held in controlData = = algorithm has converged) ||
                                 (iteration limit is reached))
                        moreFiltering = setupNewAlgorithm();    /*
                                              prepares the next filtering
                                              algorithm */
}

create a margin around the complete image. For each border of the block,
reset pixel classes if that border is along the edge of the image.

rd_write(..... toBlk);                     /* write header and new image to
                                              file */
rd_close(gridId);                          /* Terminate PUL-RD, free memory
                                              used within RD */
MPI_Finalize();                            /* Terminate MPI */
}
```

14.4.5 Single Processor Optimisation of Moving-Window Filtering

The performance of parallel software is a consequence of many factors. In addition
to load-balancing, communications and synchronisation, discussed in the previous
sections, the efficiency of the application algorithms and of the more intensively
used code is also key to minimising run-times. 'Single processor' optimisation, i.e.
the optimisation of software which is executed independently on each processor, in
between communication or synchronisation, is achieved for parallel software in the

same way as is the case for sequential processing. As Booth & MacDonald (1994) describe, well-known techniques include ensuring effective use of:

- the memory cache by judicious arrangement of data so that the rate of cache refreshes is low

- code structuring, to minimise processing within loops, and to maximise pipelining of instructions

- optimised library functions (e.g. for mathematics), not relevant in this case

- the compiler optimisation switches

- tabulated data to avoid multiple calculations of CPU-intensive expressions (also not relevant here).

Optimisation also requires that the filtering itself should be implemented using efficient algorithms. Such optimisation becomes relevant once a prototype has demonstrated the correctness of the filtering. In this case a sequential prototype was developed using an array in which each element corresponds to a different class. The array contents are recreated from scratch for each position of the window in turn so that class i has its population within the moving-window held in element i. In the case of a 15 by 15 pixel window 225 pixels must be read and the corresponding array elements incremented. Recognition of the most populated class within the window entails inspecting the complete array. The data structures are simple; the code is short, easily tested and simple to modify.

In developing the parallel version, the following facts were used:

1. The windows around neighbouring pixels overlap substantially. This has two consequences:

    a) The statistics for the window surrounding one pixel can be derived by modifying the statistics pertaining to its neighbour.

    b) The statistics of their moving-window populations will often be similar.

2. The number of classes in the window will often be much smaller than the number of classes in the complete image.

The scanning of the image block is shown in Figure 14.4 and is similar to that described by Mather (1987). Moving down columns of the image by dropping one row from the top of the current window, and adding a new row to complete the bottom of the next window, reduces the number of pixels which must be accessed.

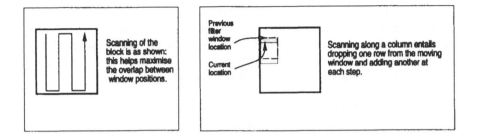

Figure 14.4: Raster scanning: arranged for efficiency

In contrast to the 225 of the prototype, only 30 pixels need to be read. The fact that adjacent pixels in a row will be held adjacent in memory will, on some processors, result in better use of cache than if the window were scanned along rows, dropping and adding columns.

In place of the simple array of the prototype, the generation of the statistics within the moving-window uses an array of structures, one for each possible class within the image. These class-statistics structures include a count of the number of occurrences within the window of that class. They also include a pointer, by means of which a linked list of structures can be built. Only those classes within the window are held on the linked list; the selection of the new class for the central pixel can then use only the linked structures. In majority filtering, by resetting pointers to move the class with highest population to the head of the linked list, the majority class for the next window position is then often recognised simply by checking that the count at the head of the list is greater than the majority threshold. In the re-growing algorithm only those classes which are under-populated in the complete image can be used to overwrite the central pixel. Only the structures of these classes need be held in the linked list, which is ordered by the extent of their under-population. Hence whenever a central pixel belongs to a globally over-populated class, the class at the top of the linked list is used to overwrite that central pixel. To allow recognition of under-populated classes, the structures thus include the amount (as a percentage) by which their classes are over- or under-populated. This is set up prior to each iteration on receipt of data broadcast by the Coordinator.

As was shown by the pseudo-code in the previous section, the filtering is done from array 1 to array 2, then from array 2 to array 1. Using two arrays, rather than one avoids results being dependent on the direction of scan. Alternating the role of the arrays avoids the need for copying data between iterations.

The relationship between scalability and single-processor optimisation is worth noting. Improving the latter has the effect of reducing both run-times and scalability, because the filtering time (in this case), but not the parallel overhead, is reduced. The goal of high performance computing is to minimise run-times, within constraints of time and budget available for development. Monitoring scalability allows inferences to be drawn as to whether using additional processors requires different approaches, but scalability is not the primary goal.

The cumulative effect of these optimisations was tested by comparing performance of the prototype with that of an implementation on 4 processors, for an image of 512 rows of 480 pixels, comprising 9 classes. Majority Filtering was performed with a moving-window 9 pixels wide. The sequential prototype ran 44 times more slowly than the parallel version, showing that a factor of approximately 11 was gained by the optimisations. This figure is given here to demonstrate the general point that both single-processor optimisation and efficient parallelism are needed for high performance of the parallel algorithms.

14.4.6 Stages in Implementation

This section summarises the approach taken in implementing the software. This is a typical approach for parallel programs, using techniques common in sequential software implementation, together with additional steps intended to help avoid or recognise problems with the use of multiple processors. The testing stages were as follows:

1.   Stand-alone tests of individual functions.

2.   Single processor tests of the integrated filtering functions using small datasets.

3.   Multiple-processor tests of MPI calls, PUL-RD and the high-level functions using both trace and the usual debuggers. (No filtering was included.)

4.   Tests of data reading, distribution, collation and writing, to ensure that the data could be correctly distributed and gathered together.

5.   Multiple-processor tests of the complete system using small datasets.

6.   Using large datasets, results with different numbers of processors were compared to ensure that there was no dependency on decomposition. A utility to compare images was written to allow comparisons to be made between the results of a sequential prototype (which implemented the filtering in a more simple and less optimised manner) and those of the parallel software. The filtering algorithms were tested both for one and for many iterations.

7.   Optimised compilation was used once all known errors were resolved. Results were checked.

8.   Performance profiling was enabled by including calls to timing functions, bracketing the significant PUL-RD, MPI and filtering functions. Some tuning of the software was done as a result, in particular eliminating some calls to collective MPI functions as described in section 14.4.3.

14.5 Performance

This section summarises the performance of the software on a Meiko Computing Surface using up to 6 Sparc2 processors. The performance characteristics are sensitive to filtering parameters and image size. The figures reported here were obtained by processing a classified land-cover image comprising 2401 rows of 2401 pixels, and 17 classes. The parameters used for the filtering are shown in Table 14.1. (They were explained in section 14.2.) Timing data were obtained

Table 14.1 Parameters used in tests of performance

Majority Filtering	
Window size	9
Number of iterations to reach convergence	20
Filter option 1	Do not overwrite 2nd most frequent class in the window
Filter option 2	Use the most popular class in the window even if it is not a majority class.
Re-growing	
Window size	3
Number of iterations to reach convergence	4
Convergence limit - stop when all classes are within this limit	10%
Class limit: outside this limit a class is under-/over-populated	2%

from calls to an MPI function which returns 'wall-clock' times. (CPU times do not include all inactive periods on a processor, for example, when a process is waiting for a message.) The tests were carried out on a dedicated system and the timings were found to be reproducible to within approximately 5 percent.

The purpose was to determine the overall scalability of the performance. Table 14.2 shows the total elapsed times, the times taken in each of the two filtering algorithms, in reading and distributing the data into the blocks, and in gathering and writing out the resultant dataset. The overall speed-up and the speed-up achieved within the filtering algorithms is also shown. It is clear that total speed-up is compromised by the time taken for reading and distributing, and then collating and writing the images. The data distribution and collation times were obtained using an early MPI implementation, and were surprisingly large. Figure 10.3 shows that, for 5 processors, each scan-line is split into 5 messages by PUL-RD; this might overstretch that MPI implementation, causing the particularly excessive time in Table 14.2. The I/O figures are erratic beyond expectation, including that for 1 processor where no messaging should occur. Reasons for this were not investigated. The software was also installed at the SAI on a 10MByte/sec network using 3 Sparc5 and one Sparc10 workstations. The I/O figures were again relatively poor, but after linking with a different MPI library, reading/distribution and collation/writing were performed in around 6 seconds each, demonstrating that the merits of using a standard message-passing library are not only that the application software is portable, but also that different MPI implementations can be exploited as available.

The scalability of the filtering algorithms imply that only small overheads are

Table 14.2. Elapsed times and speed-up ratios using 1-6 Sparc2 processors

Number of processors	1	2	3	4	5	6
Majority Filtering (seconds)	1965	1000	665	502	401	334
Majority Filtering: speed-up		1.96	2.95	3.91	4.90	5.88
Re-growing (seconds)	147	75	51	38	31	26
Re-growing: speed-up		1.96	2.88	3.87	4.74	5.65
Read/distribute (seconds)	4	19	19	31	54	24
Write results (seconds)	45	63	66	68	66	68
Total speed-up		1.87	2.7	3.38	3.91	4.78

associated with the non-blocking halo exchange. The difference in speed-up between the two filtering algorithms is primarily due to re-growing being less CPU-intensive. To some extent it may also be due to the different numbers of calls to collective functions. These were called on each of the four iterations during re-growing, whereas an optimisation (section 14.4.3) reduced this requirement for majority filtering to 3 times in 20 iterations. Scalability suffers slightly as additional processors are employed because the parallel overhead increases, but the run-time decreases: any slight imbalance in loading and any sequential component in the processing (such as in the testing of completion of filtering by the Coordinator) result in wastage of CPU-time on more processors.

14.6 Summary

The chapter described an implementation of parallel software for raster generalisation, noting that in this case algorithm optimisation as well as parallelisation is necessary to achieve high performance. Tests demonstrated that the overhead of non-blocking messaging to achieve halo exchange was slight, and that overall speedup was constrained by I/O. The use of MPI and of PUL-RD greatly reduced the development effort. This indicates the benefits which would result from having libraries available to manipulate other formats of GIS data, as investigated elsewhere in this book.

14.7 Acknowledgements

It is a pleasure to acknowledge assistance from Dr. Mark Parsons, Dr. Lyndon Clarke and Mr. Scott Telford of the Edinburgh Parallel Computing Centre and from Mr. Dieter Flunkert of the Joint Research Centre.

14.8 References

Booth, S. & MacDonald, N., 1994, *Performance Optimisation on the Cray T3D*, Course Notes, Edinburgh Parallel Computing Centre, University of Edinburgh.

Clarke, L.J., Chapple, S.R. & Trewin, S.M., 1995, *PUL-RD Prototype User Guide*, Edinburgh Parallel Computing Centre, Document EPCC-KTP-PUL-RD-PROT-UG 1.7.

Mather, P., 1987, *Computer Processing of Remotely Sensed Images*, Wiley, 247-249

Message Passing Interface Forum, 1994, *MPI: A Message-Passing Interface Standard*, University of Tennessee, Knoxville, TN.

Mineter, M.J., 1996, *Pilot Study for a Parallel Processing Approach to Raster Image Generalisation with Potential Application to Forest Mapping*, Project report, Contract No. 10467-94-10 F1ED ISP GB, for the Joint Research Centre, Ispra, Italy. Department of Geography, University of Edinburgh.

Wilkinson, G.G., 1992, The generalisation of satellite-derived raster thematic maps for GIS input. *Geo-Informations-Systeme*, **6** (5), 24-29.

Part Four

Application of Parallel Processing

15

Parallel Database Management Systems for GIS

R.G. Healey, S. Dowers, B.M. Gittings and M.J. Tranter

15.1 Introduction

The quest for novel hardware solutions to the computational and I/O requirements of database management applications has a long history. From the early days of the database computer (Banerjee & Hsiao, 1978), the more esoteric and special-purpose types of hardware, as in other areas of computing technology, have progressively been replaced by commodity processors and disks. At the same time, with the rapid adoption of relational database technology during the 1980s (Date, 1986), there was increasing recognition both that database operations in general could be accelerated by means of parallel processing techniques and that relational database operations, in particular, were very well suited to parallelisation (DeWitt & Hawthorn, 1981; Bitton *et al.*, 1983).

However, through much of this period the applications of parallel computing to database management comprised a rather distinct thread of development from the mainstream uses of parallel processing in the scientific computing field. With the advent of more general purpose parallel platforms (Hack, 1989) and the abandonment of special purpose operating systems in favour of versions of UNIX, the 1990s have brought a convergence of hardware requirements for the two types of application. As a result, both high performance SMP machines and MIMD-DM machines from vendors such as Meiko and nCube are now used extensively for scientific computation, but they also support parallel versions of commercial database software systems (Statler, 1993).

The 1980s also saw the very rapid development of GIS from modest beginnings into a major sector of the IT industry (Maguire *et al.*, 1991), drawing together concepts and technologies from the fields of digital cartography and database management. Given the timing, it is not surprising that relational methods were adopted at an early stage in this process. They continue to be the dominant force for database management in GIS (Healey, 1991), despite the recent growth of interest in the use of object-oriented data structures (Worboys, 1992) and the availability of a new generation of systems utilising object technology, such as Smallworld and Laserscan Gothic (Chance *et al.*, 1990).

Initially, most GIS implementations were effectively pilot projects with limited size datasets, and were often developed without detailed consideration of their potential linkages to wider management information systems requirements within an organisation. Developments of this type were not of a scale that invited consideration of the specialised technology of parallel processing. Subsequently, it became apparent that for many large organisations, such as utility companies and regional government or planning authorities, spatially referenced data and, by implication, GIS technology, underpinned the vast majority of their information requirements for both strategic decision-making and operational planning. Unfortunately, while these organisations had accumulated large mainframe databases to support their accounting-related requirements, the equivalent map databases showing the disposition of their assets (e.g. pipes or cable networks) did not exist in digital form and could not therefore be linked into an overall information system. Also, the traditional mainframe was ill-suited for serving the graphics intensive requirements of GIS-based applications. As a result, most GIS developments took advantage of the rapid improvements in workstation technology as a preferred route to implementation, although this often left the question of efficient interfacing to existing mainframe databases unanswered.

In addition to the 'consumers' of digital cartographic data, digital map producers, many of them national mapping agencies, now also have very large data management and processing requirements. Prominent among these organisations are the UK Ordnance Survey and the US Geological Survey. The Ordnance Survey is approaching completion of the digitisation of mapping scales from 1:1,250 to 1:625,000 for the entire country, while a slightly longer timescale is necessitated by the magnitude of the task facing the US Geological Survey. However, in collaboration with the Bureau of the Census, the latter organisation completed the digitising of all 1:100,000 scale maps, together with over 350,000 census enumeration maps, in the course of the TIGER project for automation of the 1990 census (Marx, 1986). GIS databases of this kind, with national coverage, already extend into the 100 GB range, and present significant problems for efficient storage, management and update. While most customers would only require a small subset of these data for a specific area, it should also be borne in mind that larger utilities, such as gas and telecommunications, may have an overall requirement for tens of thousands of 'sheets' of large scale data, as background referencing for the plotting of their pipe and cable networks.

Prior to the widespread adoption of relational databases and entity-relationship data modelling approaches (Chen, 1976), a variety of special purpose file formats were used to store digitised cartographic datasets, although frequently the topological relationships between cartographic elements were not stored in conjunction with the coordinate data. However, in 1984 it was demonstrated that a complete description of digital cartographic data could be provided within a relational database framework (Van Roessel & Fosnight, 1984), such that both coordinate and attribute data could be handled within the same environment. This *integrated model* approach, as it came to be known (McLaren & Healey, 1992), is in contrast to the *hybrid* approach (Morehouse, 1985), where digital cartographic data are handled using the standard file system for speed, while attribute data for map features are managed within a relational framework, for the convenience of the user. In the early and mid 1980s there was strong pragmatic justification for using a hybrid approach, despite the additional software layers required to manage the linkages between the digital cartographic and associated attribute data, because of the relatively poor performance of relational systems at that time. This was exacerbated by the fact that even relatively modest sized areas of digital map coverage could involve millions of *xy* coordinate pairs. This would result in extremely large database tables, whose rows could not be retrieved to the graphics screen for map display in any reasonable time (Bundock, 1987).

The theoretical attractions of the integrated model approach for data storage and retrieval in a standardised manner led the Ordnance Survey to address the problems of relational database performance by means of Britton-Lee database machine technology (Smith, 1987). Similar enthusiasm for the integrated relational approach was demonstrated by the US Geological Survey (Guptill, 1986, 1987), although a database machine solution was not used for production purposes. However, it is interesting to note that the approach was not adopted for the TIGER project, which employed specially designed file structures and a data processing strategy that utilised the combined resources of several large mainframes (R.W. Marx, pers. comm.)

From this overview of database developments in GIS, it is apparent that both the integrated and hybrid model approaches to data management offer important opportunities for the deployment of parallel database technology. For systems based on the integrated model, they offer a more flexible version of the hardware database engine solution. For hybrid systems, they offer the potential to meet both the transaction processing and ad hoc query requirements of centralised MIS systems. At the same time, they can also provide high performance linkages between the DBMS and the GIS, and act as digital cartographic data servers in a networked client/server environment.

Developments in parallel database technology will now be reviewed in the context of these different requirements for GIS processing, highlighting the areas of specific benefit, as well as issues and problems that currently remain.

Since a number of reviews of parallel database technology have recently been assembled into an edited collection of papers (Lu *et al.*, 1994), it is not necessary to rehearse in detail all the major types of development here. Instead, some of the key trends in hardware and software will be summarised, to provide the necessary context for examination of their implications for GIS related work.

15.2 Hardware

Early enthusiasm for special purpose hardware solutions to database problems derived from the limitations of general purpose hardware and software during the 1970s. Indeed, two of the particular advantages claimed for such solutions were that the hardware database computer was not a general purpose machine and that it was 'almost entirely devoid of software' (Banerjee & Hsiao, 1978). While excessive concern with disk head technology and novel hardware such as bubble memory may now seem misplaced, the early database computers, by introducing the notion of back-end database processing, opened the way for wider acceptance of the concept of database servers attached to client-server networks. This market niche is now of growing importance in relation to the adoption of parallel database servers.

From the software perspective, even before the innovation wave of relational database technology had diffused through the data processing departments of major organisations, the potential role of parallel processing for implementing relational database operations had been identified and explored in a preliminary way (Bitton *et al.*, 1983). However, serial processing was the only available technology for general purpose computing, throughout the mini-computer to mainframe range, at that time. Therefore, the special purpose database machine was the only type of platform where the potential of parallel processing for database operations could begin to be exploited. Unfortunately, this special purpose hardware could not be used for other processing tasks when not needed for database work, so its cost could only be justified in restricted market sectors, where the database load was a large proportion of the overall processing requirement.

SIMD machines have not proved appropriate for the multi-user, I/O intensive requirements of database processing, so these architectures have not gained acceptance as database computers. However, MIMD machines are much more suitable. This is particularly the case when a 'shared nothing' architecture is employed. This means that each processor, in addition to having local memory, also has a local disk or disks, for which it acts as a server. The commercially available Tandem and Teradata database machines, for example, employ this approach (DeWitt & Gray, 1992). The main advantage offered by the shared nothing architecture is its potential scalability up to large numbers (hundreds or even thousands) of processors. In contrast, shared memory multi-processing systems, with shared disks, have been perceived to suffer from significant limitations of bus contention. This arises from the overlapping traffic, between processors and disks, that is generated in the course of database operations. As noted in Chapter 2, this bus contention has the potential to limit the scalability of large systems, although former limitations are now being partly overcome with the advent of multi-bus and cross-bar switching configurations.

Hence, just as single chip designers have demonstrated a greater capacity to increase processor power than many might originally have predicted, designers of bus and disk sub-systems are also demonstrating an ability to circumvent potential bottlenecks with innovative technological solutions. This has led supporters of the shared memory multi-processing approach to stress that it offers a platform for all but the most demanding of database applications (Statler, 1993). The progressive announcement of higher and higher performance multi-processing servers by major vendors such as Digital and Sun lends support to this view at the time of writing.

While both massively parallel and multi-processing approaches are being employed, either singly or in combination, in special purpose database computers (Statler, 1993), hardware is not the only issue. A second key concern for the future of parallel database management, is the nature of the software employed and whether or not it is dedicated to the hardware on which it runs. This is the case for the 'traditional' database computer, but there is now a well-developed trend, as noted above, for suppliers of general purpose DBMS software to make versions of their systems available for both multi-processing and MIMD machines. The implications of this will be examined in the next section.

15.3 Software

Two stages can be identified in the utilisation of parallel processing technology by general purpose DBMS software. The first stage is characterised by extensions of existing facilities and interfaces to allow basic utilisation of many processors and parallel disk sub-systems. The second phase involves more extensive modifications to database kernel functions, to exploit parallelism more effectively. These modifications are particularly focused on the optimisation of database queries in parallel.

15.3.1 Basic Exploitation of Parallel Architectures

For some years now, major DBMS vendors have made versions of their software available which utilise the capabilities of symmetric multi-processing (SMP) machines, such as those available from Digital or Sequent. Since many of these packages use background processes for database writes or asynchronous read-ahead from disk, it has proved relatively easy to employ the process management tools provided by SMP operating systems, to distribute these processes across multiple processors. Until the machine is fully loaded, this approach markedly improves response times for a user process which is confined to a single processor. In addition, SMP machines have had considerable commercial appeal in the DBMS marketplace because of their ability to run standard applications on standard operating systems without the need for recoded parallel versions of the software. Using standard software is often of particular importance in the somewhat conservative DBMS marketplace.

More recently, vendors have extended the capabilities of their software to encompass massively parallel processors from suppliers such as Parsys (Parsys, 1991), nCube and Meiko. Vendors with MPP versions of their software now include Oracle, CA Ingres, Informix and Sybase (Beckett, 1995). Oracle on the Meiko very clearly exemplifies the evolutionary approach to the embracing of parallel technology. It can be examined under three headings (Holman & Barton, 1991), as follows:

- multi-instance support

- parallel lock management

- parallel file serving.

To understand these different components and their interactions, it is necessary first to outline the structure of the Meiko CS-1 platform, to which the Oracle software was originally ported, although it should be noted that more recent hardware configurations are now available.

Figure 15.1 shows that the CS1 platform is composed of a number of SPARC processor nodes, each running the Meiko version of the UNIX operating system, a set of transputers, a minimum of two assigned to each SPARC node, and a transputer-hosted parallel disk farm. The different processors and disk controllers are linked to each other by a proprietary network interconnect running at 20 Megabits/sec (Meiko, 1989).

Multi-instance support means that each SPARC processor runs its own instance of the Oracle database management software. Each instance supports its own set of users and has its own background processes. However, all instances link through the internal network to a single Oracle database. The individual instances are able to communicate with one another by means of Oracle's distributed database capability. By way of comparison, an SMP machine would normally run only a single Oracle instance.

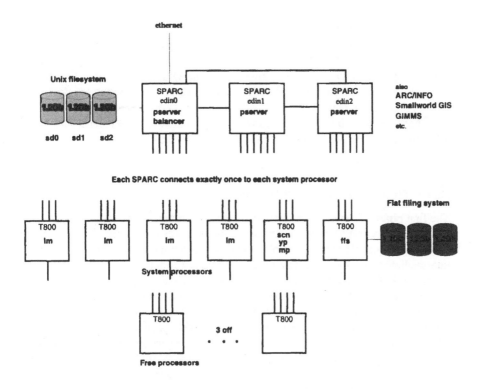

Figure 15.1: Oracle parallel implementation on Meiko CS-1

The parallel lock manager is designed to handle the locks that are acquired and released in the course of updating the same database from multiple Oracle instances. It is implemented as a number of cooperating sequential processes, running on the transputers within the machine. The processing resource required for the parallel lock manager is clearly a function of the database workload, so it is important that the number of instances can be scaled by adding new SPARC nodes. The lock manager should also be scalable by adding new transputers (Jakobek, 1990). The CS1 provides these capabilities.

Finally, the processing sub-system is complemented by a parallel file server, based on multiple disk controllers, each with a large data cache. In typical configurations this cache might be minimally 16 MB in size. The Oracle database administrator can stripe both data files and log files across multiple disks to improve performance. Log files track transactions as they are executed and allow database recovery in the event of system failure (Holman & Barton, 1991).

Vendors have claimed a number of advantages for this kind of approach, which employs standard commercial software, commodity disks and processors, and restricts the use of special purpose components to disk controllers and interconnects. These advantages include flexibility, scalability and the provision of an upward migration path from standard architectures for resource hungry database applications. Also, systems of this kind can be configured to suit either fine-grain transaction processing applications or the more complex, but less time critical, queries that tend to predominate in decision-support systems (Parsys, 1991). This latter point needs to be considered further in the context of more recent developments, because of its potential importance in the GIS context.

15.3.2 Optimisation of Parallel Queries

The first phase of database parallelisation using commercial software brought considerable benefits where the primary bottleneck was multi-user performance under heavy load. However, complex queries with numerous relational joins, or intermediate steps involving computation, could not benefit to the same extent, because all of the processing associated with a particular query was confined to a single processor. While it was possible to utilise the distributed database capability to harness the power of multiple database instances on different nodes in the processor network, this required significant additional programming by the user. Also, the extra traffic on the internal network could easily outweigh the potential performance gain. It can therefore be argued that the flexibility provided by these initial parallel configurations was biased more towards transaction processing than decision support applications.

To facilitate the processing of decision support queries, to a greater degree than that provided by data distribution across multiple disks alone, requires optimisation of the query decomposition plans adopted by the database kernel, to take account of the parallel nature of the underlying processing hardware.

This parallel query optimisation has a number of aspects. At the highest level there is the issue of whether optimisation is static or dynamic. (Pirahesh *et al.*, 1990). Static optimisation implies that the breakdown of tasks between processors is decided by the query executor at the time the query is compiled. The decision may be based on the likely 'cost' of different operations contained within the query, as with traditional query optimisation, but with the added qualifier of the relative ease with which certain operations may be parallelised. In 'shared nothing' systems, where specific disks are attached to specific processors, statistical information on the relative accessibility of data from different database tables to individual processors might also be employed. Dynamic optimisation would adapt the above plan, based on run-time information about loading on processors and/or other system parameters, such as memory availability. It is apparent that even the static optimisation approach results in a potentially very large search space to find the optimum, and the problem is even more severe in the dynamic case. This could result in significant resource use, simply to determine how to process the query.

This may be unacceptable under conditions of heavy loading. One solution is to adopt a two-stage optimisation strategy. In the first stage a query plan is determined without reference to parallelisation issues. The resulting plan is then optimised for parallel execution. While this simplifies the problem, it clearly does not guarantee an approach that is optimal overall. It is possible that operations which are expensive in a sequential environment could be decomposed to run much more cheaply in parallel, but they may not have been chosen to drive the query, having already been scored as too costly in the first stage of the optimisation.

At a more detailed level, the extent to which individual queries can be parallelised is dependent on the specific content of those queries. The SQL language is composed of a variety of clauses and conditions, some of which lend themselves to parallel execution while others do not. Relational join conditions, which specify how values in rows in one table are to be matched to those in a second table, are obvious candidates for parallel execution. This inherent parallelism of simultaneous multi-table scanning has been exploited since the early days of database machines. The situation is similar for sub-queries, where the results of one query are used to drive the search conditions of another query. However, the position is less straightforward for clauses which contain grouping operators. In general, these operators have to receive and process all of their inputs before the results stream is output. They cannot therefore proceed in parallel with other operations, if these require the output as input to their own processing.

Major commercial systems such as Oracle and Informix now support parallel data query approaches of this kind (Statler, 1993). While the specialist database machine architectures have favoured the 'shared nothing' approach underlying their parallel query optimisation, the commercial DBMS software vendors have favoured the more flexible route, whereby individual database instances on individual processors can access all the disks on the system more directly. Nevertheless, it is apparent that general purpose DBMS software vendors, by porting their packages to, and optimising them for, general purpose parallel architectures, are now able to match the performance and scalability that could only be provided previously by special-purpose database machines. This increased hardware and software flexibility, coupled with high performance, is of particular significance in the GIS context.

15.3.3 Remaining Technical Problems

Several problem areas that remain for parallel database management should also be mentioned at this stage, concentrating on those that have implications for linkage between databases and GIS. The first of these is that, depending on the locking mechanism used, transaction oriented and query oriented processing tasks may interfere with one another to degrade performance. Secondly, more generally, system related issues in parallel database optimisation have not been extensively studied. Thirdly, there are a number of possible indexing strategies and ways of partitioning databases across multiple disks in the parallel environment. However, at present, there is limited understanding of the implications of multi-dimensional indexing or distribution techniques for parallel databases.

While a detailed examination of database locking techniques is outside the scope of the present discussion, it is apparent that data records must be locked during transaction processing to prevent two users attempting to update the same record simultaneously. To maximise concurrency for a given transaction workload, these locks should be as fine as possible, i.e. at the individual record level. However, the query-oriented processing required by decision support systems is likely to access large numbers of records, while needing to retain a consistent view of the database. This may result in significant processing overhead to keep track of the transaction locks that overlap with the search sets of queries. Several approaches to this problem for sequential processors have been summarised in the literature (Pirahesh *et al.*,1990). The effectiveness of parallel systems such as the Meiko-Oracle configuration, which have a combined hardware/software solution to lock management, for a mixed transaction/query workload, will be examined in more detail below.

Following on from the specific question of lock management in relation to mixed workloads, Rys & Weikum (1994) have drawn attention to the wider issue of resource allocation in parallel systems, in relation to benefit/cost considerations, rather than speed-up or throughput alone. For example, it is possible to adopt two different approaches to handling the workload. The first is based on giving a small number of concurrent jobs large numbers of processors and other system resources. The second is the converse, namely running many jobs, each of which is allocated limited resources. Taking the example of parallel database scans and data pre-fetching techniques, the authors of the above study were able to demonstrate that a cost/benefit criterion, based on the relative costs of processors and memory, could be used to allocate the workload much more cost-effectively than an optimal speed-up criterion, while resulting in only a 30% increase in response time. Although this is only a single result for a specific type of operation, it underlines the complexity of the issues involved in effective resource management on parallel systems.

The final question to consider is that of the potential role of multi-dimensional indexing and data partitioning strategies. Until recently, partitioning has been based on the value of a single attribute in a table (DeWitt & Gray, 1992). However, if spatial or spatio-temporal data are to be stored in the database, more efficient means of partitioning to facilitate retrieval of data for contiguous blocks of space or space-time volumes need to be found. Several studies have examined the use of quadtrees or octtrees for multi-dimensional indexing of these domains on parallel machines (Bhaskar & Rosenfeld, 1988; Watson & Halsall, 1990; Mason *et al.*, 1988), but not within a relational database context. Conversely, early work by Abel & Smith (1986) and Waugh & Healey (1986) on incorporation of spatial indexes into relational databases for digital cartographic data has not been pursued in the parallel domain. It is therefore of significance that a commercial solution to this problem has now emerged. The Oracle Corporation has recently released a new module for its DBMS which utilises an indexing strategy based on a linear key derived from a multi-dimensional tree. This key is known as a helical hyperspatial code (HHCODE). The HHCODE also forms the basis of an automatic data partitioning approach that the module provides. When data are loaded, they are sub-divided automatically and dynamically into multiple tables based on the HHCODE, such that each table represents a specific region of multi-dimensional space, where the space may have up to 32 dimensions (Oracle, 1995). The module and its functions are integrated with the parallel capabilities of the Oracle system, thereby allowing multi-dimensional data management on a parallel machine, using generic commercially available software tools. As the module is relatively new at the time of writing, the authors are not aware of any published studies that evaluate the technology from the performance perspective, but it can be expected that these will be forthcoming in due course.

15.4 Implications for GIS

Following this examination of recent developments in parallel database management, it is now possible to indicate a number of likely implications for both integrated and hybrid model GIS approaches.

15.4.1 Integrated Model GIS in a Parallel Database Environment

With current levels of technology, it is now possible to store the contents of very large digital cartographic databases in fully normalised form within parallel relational databases. This can be achieved using general purpose commercial software on general purpose parallel machines. The necessity for special purpose hardware for a database computer has therefore effectively disappeared. Standard optimisations such as storage of map coordinate strings in bulk datatypes will produce much greater performance gains in a parallel rather than a sequential environment. This is particularly true if parallel processing is combined with the various array fetching mechanisms for retrieving groups of rows that are supported by leading vendors. While the hardware/software combination on the parallel machine can deliver the requisite level of I/O performance, the new helical code indexing methods can be expected to provide matching gains in search performance.

Since these technological developments have been made on the basis that users will wish to handle databases many hundreds of gigabytes in size, integrated model parallel databases are now well placed to meet the future requirements of even large national mapping agencies such as the Ordnance Survey. The more limited requirements of individual utility companies, for example, would also be readily accommodated.

A further interesting point is that integrated model GIS databases can be regarded as a step on the way towards more object-oriented DBMS (Hughes, 1991) where not only attribute and coordinate data would be stored, but also the geo-processing methods needed to manipulate the cartographic objects. Planned developments of the SQL standard, not to mention commercial pressure on relational DBMS vendors from their fully object-oriented DBMS counterparts, indicate that a trend to provide enhanced support for method handling will continue in future.

The advantage of a general purpose parallel platform hosting such a parallel GIS database is that a number of processors could be set aside for 'method processing', leaving the remainder free to concentrate on the database query and retrieval functions. This allocation of processing power could be dynamic, depending on the changing nature of the workload.

15.4.2 Hybrid Model GIS

Despite the many and growing advantages of the integrated model, hybrid approaches (for example, ESRI Arc/Info) still dominate the commercial market place at present, while major object-oriented systems such as Smallworld and Laserscan Gothic also provide connections to relational DBMS. As noted in the introduction, a parallel approach to implementation of the hybrid model can be examined under several headings. These include the overall suitability of the hardware platform, the effectiveness of the GIS/database interface and links between GIS and Management Information Systems (MIS) functions.

Of the two main types of parallel platform, the shared memory multi-processor presents no problems for installation of GIS and database software, because of the standard nature of the operating systems and the availability of versions of the widely used commercial packages. On distributed memory MIMD machines there are no commercial GIS packages that can utilise the parallel nature of the system, but a choice of relational DBMS is now available. Nevertheless, for machines in the same broad category as the Meiko CS family (as shown in Figure 15.1), since the SPARC nodes run a version of UNIX and they can individually have disks attached, there is no difficulty in running copies of GIS software on these nodes. In addition, since database instances can also run on the SPARC nodes, while harnessing the additional power of the rest of the machine, standard interfaces from the GIS software to these databases can be utilised to run the hybrid GIS, with attribute data stored in the DBMS. There are thus no technical impediments to this kind of implementation of the hybrid model which utilises a parallel database.

Some GIS, such as Arc/Info and Smallworld, have internal data management facilities, while supporting interfaces to external databases. Other packages rely entirely on external interfaces for attribute management. Owing to the nature of these interfaces, they tend to be relatively resource intensive. When spanning between systems it is difficult to provide the same level of functionality and performance as that of the relational join within a single database environment. Separation of part of the database work on to other parts of the machine, to avoid overloading the nodes running the GIS software, is therefore of particular advantage for the end user. The latter will log on to the GIS node and the parallel database facilities will provide him or her with powerful back-end attribute server capability, which is unaffected by other traffic across the external network.

The issue of the link between GIS and MIS functions relates directly to the problem of query versus transaction loads discussed in an earlier section. In the case of a utility company, for example, large parts of the MIS database would be concerned with customer and billing information, while other parts would be concerned with asset management and control of street works and repair programmes. A number of tables across the database would have direct or indirect linkages with the GIS for digital map-based query and display. At the same time they would be involved in different kinds of transaction processing activities and the provision of non-GIS related decision support queries. Meeting both GIS and MIS requirements from a hybrid model system on a parallel machine will therefore result in a very mixed load on the parallel database. There is no doubt as to the ability of the parallel database to outperform standard mainframes, in terms of the MIS transaction processing load, as the current performance records are held by the former (DeWitt & Gray, 1992). Similarly, current MIMD machines can offer much better GIS facilities for multiple users than mainframes. The only question that remains, therefore, is whether the mixing of the loads, with numerous different types of queries, some GIS-based, others not, being streamed in by large numbers of users, either will seriously impede the performance of the transaction processing tasks or, alternatively, will have their performance impeded by these tasks. While vendors have claimed that the combination of parallel processing and data warehousing techniques will enable their products to meet both transaction processing and MIS type query demands, independent assessments have indicated that such claims need to be examined carefully, and user requirements matched very closely to system capabilities (Sankar, 1995; Beckett, 1995).

Recent investigations have been undertaken by Tranter *et al.* (1995) to address precisely this point. A test database of approximately 3 GB in size, designed to emulate the structure of major parts of a utility company database, was implemented within parallel server Oracle 6.2 running on a 7-SPARC node, 24 transputer Meiko CS-1. Both Arc/Info and a number of Oracle instances were run on the SPARC nodes. A C program with embedded SQL update statements was used to generate a variety of different transaction loads up to 80 tps, on a continuous basis in the background. A number of database queries were then issued from Arc/Info concurrently on different nodes, using the Arc/Info-Oracle interface. While under very heavy transaction loads there was evidence of interaction with the query load, the tests indicate that performance was robust under a mixed load and suffered far less degradation than would have been observed on a single-processor machine.

The implication of these tests is that there is considerable potential for the use of parallel databases within a hybrid GIS environment. The hardware platform of the MIMD machine also neatly resolves the dichotomy between mainframe-based DBMS and workstation-based GIS. This dichotomy can still be observed in many organisations which had well-established MIS functions prior to the adoption of GIS.

15.5   Conclusions

While a number of areas of future development remain for parallel database management, it is apparent that the technology has now reached a level where it can begin to address the requirements of even the largest organisations involved with GIS. The hybrid model approach may suit those with a mixed processing requirement while the integrated model approach (increasingly with object-oriented extensions) can address the needs of mapping organisations, for whom the digital map datasets themselves are of primary importance.

While the workstation revolution initially favoured decentralisation of both processing and data resources, some of the problems of this approach in terms of coordination and additional system management are now more widely appreciated. With the rapid expansion of broad band high speed WANs, it may now be time to re-assess the potential value of centralised corporate GIS data/processing servers built around the core technology of parallel database management.

15.6   References

Abel, D.J. & Smith, J.L., 1986, A relational GIS database accommodating independent partitionings of the region. *Proceedings of the 2nd International Symposium on Spatial Data Handling*, Seattle, WA, 213-224.

Banerjee, J. & Hsiao, D.K., 1978, Concepts and capabilities of a database computer. *ACM Transactions on Database Systems*, **3** (4), 347-384.

Beckett, H., 1995, Analysis: parallel computing. *DEC Computing*, 18 October, 6-7.

Bhaskar, S.K. & Rosenfeld, A., 1988, Parallel processing of regions represented by linear quadtrees. *Computer Vision, Graphics and Image Processing*, **42**, 371-380.

Bitton, D., Boral, H., DeWitt, D.J. & Wilkinson, W.K., 1983, Parallel algorithms for the execution of relational database operations. *ACM Transactions on Database Systems*, **8** (3), 324-353.

Bundock, M., 1987, An integrated DBMS approach for geographic information systems. In Chrisman, N.R. (Ed.), *Auto Carto 8 Proceedings*, Baltimore, MD, 292-301.

Chance, A., Newell, R.G. & Theriault, D.G., 1990, *An Overview of MAGIK*. Technical Paper 9/1, SMALLWORLD Systems Ltd, Cambridge, U.K.

Chen, P., 1976, The entity-relationship model - towards a unified view of data. *Association of Computing Machinery, Transactions on Database Systems*, **1** (1), 9-36.

Date, C.J., 1986, *An Introduction to Database Systems*. 2nd Edition, Addison-Wesley, Reading, MA.

DeWitt, D. & Gray, J., 1992, Parallel databases: the future of high performance database systems. *Comm. ACM*, **35** (6), 85-98.

DeWitt, D.J. & Hawthorn, P.B., 1981, A performance evaluation of database machine architectures. In Zaniolo, C. & Delobel, C. (Eds), *Proc. 7th Int. Conference on VLDB*, Cannes, France.

Guptill, S.C., 1986, A new design for the U.S. Geological Survey's National Digital Cartographic Database. In Blakemore, M. (Ed.), *Proceedings, Auto Carto*, London, September, Vol. 2, 10-18.

Guptill, S.C., 1987, Desirable characteristics of a spatial database management system. In Chrisman, N.R. (Ed.), *Auto Carto 8 Proceedings*, Baltimore, MD, 278-281.

Hack, J.J., 1989, On the promise of general purpose parallel computing. *Parallel Computing*, **10**, 261-275.

Healey, R.G., 1991, Database management systems. In Maguire, D.J., Goodchild, M.F. & Rhind, D.W. (Eds), *Geographical Information Systems: Principles and Applications*, Longman, Harlow.

Holman, A. & Barton, E., 1991, *The Computing Surface - A Platform for Oracle*. Meiko Technical Paper, Meiko Ltd, Bristol.

Hughes, J.G., 1991, *Object-Oriented Databases*, Prentice Hall, Englewood Cliffs, NJ.

Jakobek, S., 1990, *Scalable Computing in the 1990s*. Paper presented at the UK Oracle User Group Meeting, 18 September.

Lu, H., Ooi, B.-C. & Tan, K.-L., 1994, *Query Processing in Parallel Relational Databases*, IEEE Computer Society Press, Los Alamitos, CA.

Maguire, D.J., Goodchild, M.F & Rhind, D.W. (Eds), 1991, *Geographical Information Systems: Principles and Applications*, Longman, Harlow.

Marx, R.W., 1986, The TIGER System: automating the geographic structure of the United States Census. *Government Publications Review*, **13**, 181-201.

Mason D.C., Corr, D.G., Cross, A., Hogg, D.C., Lawrence, D.H., Petrou, M. & Tailor, A.M., 1988, The use of digital map data in the segmentation and classification of remotely-sensed images. *International Journal of Geographical Information Systems*, **2**, 195-216.

McLaren, R.A. & Healey, R.G., 1992, Corporate harmony: a review of GIS integration tools. *Proceedings, AGI'92 Conference*, Birmingham.

Meiko, 1989, *Computing Surface Overview*. Technical Paper, Meiko Ltd Bristol.

Morehouse, S., 1985, Arc/Info: a geo-relational model for spatial information. In *Auto-Carto VII Proceedings*, Washington, DC., 388-397.

Oracle, 1995, *Oracle Multidimension: Advances in Relational Database Technology for Spatial Data Management*. White Paper No. A30957, Oracle Corporation, Redwood Shores, CA.

Parsys, 1991, *Oracle Parallel Server on the Parsys Supernode 1000 Series*. Technical Paper, Parsys Ltd, London.

Pirahesh, H., Mohan, C., Cheng, J., Liu, T.S. & Selinger, P., 1990, Parallelism in relational database systems: architectural issues and design approaches. In Lu, H., Ooi, B.-C. &Tan, K.-L. (Eds), *Query Processing in Parallel Relational Database Systems*. IEEE Computer Society Press, Los Alamitos, CA.

Rys, M. & Weikum, G., 1994, Heuristic optimisation of speed-up and benefit/cost for parallel database scans on shared-memory multi-processors. *Proc. 8th Int. Parallel Processing Symposium*, Cancun, Mexico, April.

Sankar, H., 1995, *Oracle7 Server Scalable Parallel Architecture for Open Data Warehousing*. White Paper C10271, Oracle Corporation, Redwood Shores, CA.

Smith, N.S, 1987, Testing of relational databases in Ordnance Survey research and development. In *Proceedings of the SORSA Symposium*, Durham, May.

Statler, S., 1993, How parallel should a database be? *Parallelogram*, **56**, 21-23.

Tranter, M.J., Healey, R.G., Dowers, S. & Gittings, B.M., 1995, *Parallel Computing for Corporate GIS*. Paper presented at the 3rd GIS-RUK Conference, Newcastle, April.

Van Roessel, J.W. & Fosnight, E.A., 1984, A relational approach to vector data structure conversion. In Marble, D.F. *et al.* (Eds), *International Symposium on Spatial Data Handling*, Zurich, Vol.1, 78-95.

Watson, M. & Halsall, F., 1990, Concurrent operations on very large cartographic databases. In Freeman, L. & Phillips, C. (Eds), *Applications of Transputers I*, IOS Press.

Waugh, T.C. & Healey, R.G., 1986, The Geoview design: a relational database approach to geographical data handling. In *Proceedings of the 2nd International Symposium on Spatial Data Handling*, Seattle, WA, 193-212.

Worboys, M.F., 1992, A generic model for planar geographical objects. *International Journal of Geographical Information Systems*, **6** (5), 353-372.

16

Algorithms for Parallel Terrain Modelling and Visualisation

P. Magillo and E. Puppo

16.1 Introduction

The wide variety of geometric problems arising in terrain modelling and analysis, as well as the different data formats and the huge sizes of data sets, make this field an excellent arena for the application of parallel computing.

The two major terrain models, i.e. the Regular Square Grid (RSG) and the Triangulated Irregular Network (TIN), have very different structures and must be handled with very different methods. The RSG exhibits a highly regular topology that is directly embeddable in massive parallel computers, such as the mesh and the hypercube, and makes any approach based on data partitioning straightforward. On the contrary, TINs exhibit an irregular topology, which requires the application of more complex parallel computing techniques. Yet, some tasks involve both grids and sparse data, thus yielding interesting hybrid situations that must be faced with a mix of different techniques.

Some problems in terrain modelling and analysis have been investigated in the framework of parallel computation, both in the specific GIS and survey engineering literature and in more general literature on computational geometry and computer graphics. Hence, a variety of solutions have been proposed, from *ad hoc* algorithms based on empirical approaches and heuristics to very general algorithms with high theoretical efficiency.

In this chapter, we review examples of parallel algorithms for different problems, designed under different programming paradigms, for different architectures, and from different points of view, from a very theoretical to a very practical approach.

16.1.1 Programming Paradigms and Architectures

Programming paradigms and computer architectures have been discussed in Chapters 2 and 3 but are briefly reviewed here for the sake of clarity.

Data parallelism denotes a programming strategy for massive SIMD machines, in which one processor per datum is available, and all processors perform the same task on different data in parallel.

Data partitioning refers to any technique that is based on a subdivision of the data set among a pool of processors. This can be applied in the context of both SIMD and MIMD machines.

The *task-farm* paradigm is based on a subdivision of computation rather than a subdivision of data: the total computation is broken into a number of tasks, which are stored in a virtual shared memory. The *farmer* is a processor that generates such tasks and collects the results; *workers*, i.e. the remaining processors, iteratively get tasks from memory and complete them, until an end-condition is met.

A parallel architecture is formed of a number of *processing elements* (PE) that can communicate with each other either according to some fixed topology, or through messages exchanged in arbitrary patterns, or through a shared memory. The following common architectures will be considered:

- Mesh: a SIMD machine formed from a (large) number of processors arranged into a square mesh. Each processor has its own memory and can communicate with its 4-neighbours in the mesh.

- Hypercube: a machine where processors are placed at the vertices of a hypercube and can communicate only along the edges of the hypercube. Advanced hypercube machines, such as Connection Machine's CM-2 (SIMD), the CM-5 (MIMD) and the nCUBE-2 (MIMD), implement fast routing, thus providing communication between arbitrary processors as a primitive operation at reduced cost.

- *Transputer array*: a MIMD machine with a small number of processors (Transputers) with configurable communication topology.

- *Workstation network*: a pool of workstations with powerful processors and wide local memory, communicating with each other through a local area network.

- *CREW/PRAM* is more a theoretical framework than a true architecture: an arbitrarily large number of processors is available, which work independently. Processors interact through a shared memory: each memory location can be read in parallel by any number of processors, while it can be written only by one processor at a time.

Some other special architectures will be briefly described in the text.

16.2 Terrain Models

A *topographic* surface (or *terrain*) σ is the image of a real bivariate function f defined over a domain D in the Euclidean plane:

$$\sigma = \{(x, y, f(x, y)) \mid (x, y) \in D\}.$$

A *Digital Terrain Model (DTM)* is a model of one such surface defined on the basis of a finite set of digital data. Terrain data are usually in the form of measures of altitude or elevation at a set of points $S = \{p_0 ..., p_N\} \subset D$, plus possibly a set of non-crossing straight line segments $L = \{l_0 .., l_M\}$ having their endpoints in S (such lines correspond to lineal topographic features); data points in S can either be scattered or form a regular grid. A DTM built on S (and L) represents a surface that interpolates the measured elevations at all points of S (and along all lines of L). Two classes of DTMs are usually considered in the context of GIS (Petrie, 1990):

- A *polyhedral terrain* is the image of a piecewise-linear function f. A polyhedral terrain model (PTM) can be described on a partition of the domain D into polygonal regions $\{R_1, ..., R_k\}$, having their vertices in S (and such that the lines of L appear as borders of regions). The image of f over each region R_i ($i = 1, ..., k$) is a planar patch. The most commonly used polyhedral terrain models are *Triangulated Irregular Networks (TINs)*, in which all regions are triangles.

- A *gridded elevation model* is defined by a domain partition into regular polygons (squares, rectangles, hexagons, or triangles) induced by a regular grid over D. Functions used on such partitions are usually dependent on the shape of polygons. The most commonly used gridded models are *Regular Square Grids (RSGs)*, in which all regions are squares, and f is either bilinear or constant over each region. In the bilinear case, f is continuous, while in the constant case there are discontinuities along the borders between adjacent regions: this latter model is also called a *stepped model*. In the stepped model, each region is usually considered as an atomic entity: no distinction can be made between different points lying in the same region; regions are called *pixels*, for analogy with digital images, and the properties of terrain (elevation, slope, etc.) are given by enumeration at all pixels.

Let σ be a topographic surface defined as above, and let q be a real value. The set $C_\sigma(q) = \{(x, y) \in D \mid f(x, y) = q\}$ is the set of *contours* of σ at elevation q. If q is not an extreme value of f, then $C_\sigma(q)$ is a set of simple non-intersecting lines, which are either closed or open with endpoints on the boundary of domain D. Given a sequence $Q = \{q_0 ..., q_h\}$ of real values, the collection $C_\sigma Q = \{C_\sigma(q_i) \mid i = 0, ..., h\}$ is called a *contour map* of σ. Contours in digital contour maps are often represented as sequences of points; a line interpolating points of a given contour can be obtained in various ways: from the simplest case of a polygonal chain to spline curves of various order.

If σ is represented by a TIN or by a bilinear RSG, then a contour line at elevation q is a chain of straight-line segments, where each segment is obtained by joining the intersections of the edges of a surface patch with plane $z = q$. If σ is a stepped RSG, then each contour line is a sequence of 8-adjacent pixels, where each pixel lies at elevation q or immediately below it.

RSGs, TINs, and contour maps are the three basic representations for terrain surfaces used in GIS.

16.3 Problems Relevant to Terrain Modelling

A number of basic problems relating to DTMs have been outlined in the literature. Such problems can be broadly classified into three groups:

1. *Terrain model construction and conversion.* Problems involving the creation of a DTM of the types defined above, starting either from raw data or from a different terrain model. This is because the sources of data and application needs can be manifold, with almost all possible conversions occurring.

2. *Topographic characterisation.* Problems involving terrain analysis for detecting local surface properties such as slope and curvature, or surface specific features such as ridges, valleys, ravines, peaks, pits, or drainage networks and basins.

3. *Terrain visibility.* Problems involving information about the mutual visibility of points on or above a terrain surface. Basic visibility tasks involve computation of the visible portion of terrain from a given viewpoint (*viewshed*), or the mutual visibility between different regions. Application problems may require more complex analysis of visibility information to satisfy specific requirements. The viewshed problem can also be considered as a version of the *Hidden Surface Removal* problem for the special case of terrains: hence, this problem has impact also on terrain visualisation.

Only one part of such problems has been investigated in the framework of parallel computation. In the following, we briefly review the three classes above, and we outline related geometric problems. Those problems for which relevant examples of parallel solutions have been presented in the literature are described in detail in later sections.

16.3.1   Terrain Model Construction and Conversion

As mentioned earlier, raw data can come in the form of a point dataset S and, possibly, of a line dataset L. In the case where the points of S are distributed on a regular square grid, an RSG is implicitly provided by the raw data. Since sometimes DTMs are derived through digitising existing contour maps, contours can also play the role of raw data. Hence we have the following possible problems.

1. **RSG from sparse points**. Where S is a set of scattered points and an RSG must be constructed, there are three possible approaches (Petrie, 1990):

   a) *Pointwise methods*: the elevation at each point p of the grid is estimated on the basis of a subset of data that are neighbours of p. The concept of *neighbour* can be defined upon different heuristics, such as: selecting the closest k points, for some fixed k; selecting all points inside a given circle centred at p and of some given radius; subdividing the plane into sectors around p, and selecting the closest point (or points) inside each sector. A possible rigorous geometric approach consists in computing the *Voronoi diagram* of S (Preparata & Shamos, 1985), which can be used to find the nearest neighbours of each grid node through local computation. Given a node p, its nearest neighbours are those data points whose Voronoi regions are modified by the insertion of p as a new point in the diagram: after finding the Voronoi region containing p through an efficient point location technique (see e.g. Preparata & Shamos, 1985), the remaining neighbours are found by navigating the diagram through adjacencies. This is equivalent to updating the diagram with the insertion of p (Green & Sibson, 1977).

   Parallel implementation of such methods is straightforward through data parallelism or data partitioning techniques: in principle, estimation of elevation at all grid nodes can be carried out in parallel independently. However, such an approach can involve processing all points of S for each grid node in order to find neighbours. Approaches based on the Voronoi diagram can improve efficiency: once the diagram has been computed, fast estimation of nearest neighbours can be performed by applying sequential search techniques for each node in parallel (Preparata & Shamos, 1985). Some parallel algorithms for computing the Voronoi diagram will be reviewed in section 16.4.

   b) *Global methods*: the terrain surface is initially modelled through a unique high order polynomial, which interpolates elevation at all points of S. The RSG is obtained by sampling such a polynomial at grid nodes. Such methods have the disadvantage of requiring the solution of very large linear systems to obtain the coefficients of the interpolating polynomials. Parallel methods in linear algebra have been proposed in the literature, which can be applied to this purpose. However, the treatment of algebraic methods is beyond the scope of this chapter.

c) *Patchwise methods*: the domain is subdivided into a number of patches, which can be either disjoint or partially overlapping, and of either regular or irregular shape. The surface of terrain is approximated within each patch through an independent function, and the elevations of grid nodes inside the patch are estimated by sampling such a function.

In some cases, a regular subdivision is adopted, which is composed of squares or rectangles where their vertices do not necessarily correspond to the data, but are usually related to the position of a defined grid. In this case, the problem reduces essentially to allocating data into such a subdivision, in order to obtain for each region all data that contribute to the definition of the corresponding function. Because of the regularity of subdivision, all data can be located in parallel in constant time. Then, all regions can be processed in parallel in order to compute their corresponding functions: this step requires a time proportional to the maximum number of data points inside every region. Finally, elevations at all nodes of the grid can be obtained in parallel in constant time from their corresponding regions.

A different possibility is to start from an irregular tessellation, in particular a triangulation having vertices at data points: in this case, the problem reduces to the case of *RSG from a TIN*, which is described next.

2. **RSG from a TIN.** Some systems first compute a TIN from sparse data points (and, possibly, lines), then convert such a representation to an RSG for convenience (Webb, 1990). Once the TIN is available, the geometric problem underlying such conversion is essentially the location of grid nodes in the triangulation. Again, since each grid node must be located independently, a straightforward parallelisation is possible through either data parallelism or partition of grid nodes. As in the case of *RSG from sparse points* through Voronoi diagrams, it is also possible to preprocess the TIN in order to achieve logarithmic time complexity in point location: parallel techniques have been proposed in the literature to perform such preprocessing efficiently (Dadoun & Kirkpatrick, 1987).

Alternatively, grid nodes can be located while the TIN is constructed: in this case, the only operation performed afterwards is the local evaluation of the elevation at each grid node, which can be carried out independently for each node in constant parallel time. In section 16.5 we will see how a data parallel algorithm originally proposed for *TIN from RSG* conversion can be easily modified to meet this aim.

3. **RSG from contours.** Early (sequential) methods obtain the elevation at each node of the grid by considering the intersection of lines emanating radially from the node with contour lines, and interpolating the terrain profile along each line.

Although, in principle, it is possible to parallelise such an approach through data parallelism on grid nodes, the geometric problems involved (location of nodes in the subdivision induced by contours, and line-contour intersection) tend to make this either inefficient or quite involved. A two-step procedure is adopted more frequently: contours are first converted into a TIN (see *TIN from contours*), then such a TIN is sampled to obtain the RSG (see *RSG from TIN*) (Petrie, 1990).

4. **TIN from points**. A TIN is obtained from sparse points by computing a plane triangulation having vertices at data points. The *Delaunay triangulation* is the most commonly used in GIS applications. Other kinds of triangulation have been proposed, called *data dependent triangulations*, which take into account elevations of data points in the triangulation process (Rippa, 1992).

   Several examples of parallel algorithms for computing the Delaunay triangulation are reviewed in section 16.4. Also, a data parallel algorithm for *TIN from RSG* conversion reviewed in section 16.5 can be easily modified to compute either a Delaunay or a data dependent triangulation from scattered data.

   Where raw data also include line segments, the triangulation should contain such segments as edges. A modification of the Delaunay triangulation, called *constrained Delaunay triangulation (CDT)*, has been studied in the computational geometry literature, which serves this purpose (Aurenhammer, 1991). Several sequential algorithms have been proposed for computing a CDT, but no parallel algorithm has been developed so far, at least to our knowledge.

5. **TIN from RSG**. This conversion is usually aimed at data compression: the adaptivity of the TIN to surface characteristics is exploited to produce a model of terrain that can be described on the basis of a reduced subset of elevation data from an input RSG. Since a TIN interpolating elevations exactly at all pixels might need to include all such pixels as vertices, RSG to TIN conversion always involves approximation. The resulting TIN must be based on a reduced number of vertices, and gives a good approximation of the original RSG. It is possible to impose a priori either the number of vertices of the TIN, or a threshold error for the final approximation. In the latter case, a widely used error measure is the maximum difference between the actual elevation at a datum that is not a vertex of the TIN, and the elevation at the same point estimated on the TIN.

   Several incremental techniques have been proposed in the literature to perform such conversion, either based on a refinement approach - i.e. by iteratively adding vertices to a TIN based on a very small initial set of vertices - or based on a simplification approach - i.e. by iteratively discarding points that give a small contribution to the characterisation of the surface. See Lee (1991) for a survey of sequential methods.

In section 16.5 a data parallel refinement method is reviewed in detail. Some refinement techniques start from a set of characteristic points and lines extracted from the RSG (see, for example, Fower and Little (1979)): hence, algorithms for topographic characterisation discussed in section 16.6 also have an impact on this task.

6. **TIN from contours.** A TIN conforming to a given contour map should be based on a triangulation that contains all lines and points of the contour map as edges. This is another problem of constrained triangulation, for which a large number of sequential algorithms (Aurenhammer, 1991), but no parallel algorithms, have been proposed. Some systems obtain the TIN by first computing a Delaunay triangulation, which usually conforms to most edges of the contour map, and then by correcting mistakes manually through a graphical editor (Petrie, 1990).

7. **Contours from points.** As for *RSG from contours*, this conversion is usually done in two steps, by first building either an RSG or a TIN model from raw data, and then extracting contours from such a model.

8. **Contours from RSG.** If the RSG is a bilinear model, all square regions can be processed in parallel independently: possible intersections of the four edges of each region with contour lines are found in constant time. The contour segments inside each region are obtained by connecting intersection points at corresponding elevations. In a second step, corresponding contour segments from adjacent regions must be sewn together to form contour lines. In the sequential case, this is done through contour following, which does not seem to be an easily parallelisable operation.

   If the RSG is a stepped model, all pixels can be classified in parallel independently. Each pixel is compared with its neighbours: if the elevation of the pixel is smaller than or equal to a given contour value, and at least one of its neighbours has an elevation larger than such a value, then the pixel is classified as belonging to the corresponding contour line. The extraction of contours corresponds to extracting chains of pixels that are pairwise 8-adjacent, and classified at the same value (two pixels are called 8-adjacent if they share either an edge or a vertex). Also in this case, sequential chain-following techniques do not seem easily parallelisable.

9. **Contours from TIN.** This conversion is completely analogous to the bilinear case of the previous example. Contour segments for triangular regions can be

found independently in parallel constant time, while contour extraction needs contour following.

16.3.2 Topographic Characterisation

Surface characteristics such as peaks, pits, ridges, valleys and saddles correspond to local differential properties of the terrain surface: a point of the surface belongs to some characteristic class depending on the structure of the surface in its neighbourhood. For instance, a peak is a point of relative maximum, while a point lies on a ridge if its neighbourhood can be subdivided by a line passing through it, and such that the surface in each half-neighbourhood is monotonically decreasing when moving away from the line.

Hence, the extraction of surface characteristics can be performed by analysing the neighbourhood of points in the domain. In the literature, most methods have been developed for stepped RSG models, because the neighbourhood of each pixel is directly implied from the grid structure. Since computation at each pixel depends only on a constant number of pixels forming its neighbourhood, and all computations can be carried out independently, it is straightforward to obtain a parallelisation through data partitioning or data parallelism (Peucker & Douglas, 1975; Skidmore, 1990). In section 16.6, a task-farm parallelisation strategy for some examples of algorithms of this class is reviewed.

In the case of TINs, candidate points and lines to represent surface characteristics are vertices and edges of the triangulation, respectively. The only example of an algorithm for surface characterisation from a TIN is sequential, and it is based on the local analysis of the dihedral angle formed by adjacent triangular patches (Falcidieno & Spagnuolo, 1991); a parallel version of such a method is currently under development (Spagnuolo, 1995).

More elaborate processing of terrain characteristics can be performed in order to obtain structured information on drainage networks and basins, which have relevance in hydrological analysis and simulation. In section 16.6 we will review a parallel algorithm for labelling drainage basins from an RSG model.

16.3.3 Terrain Visibility

Given two points V and W on a topographic surface σ, we say that V and W are *mutually visible* if the *line-of-sight* (i.e. the straight-line segment) connecting V and W lies above the terrain surface. For any point V in the domain D of the terrain surface, we will call the *viewshed* of V the set

$$v(V) = \{ W \in D \mid V \text{ and } W \text{ are mutually visible} \}.$$

In the literature, different sequential techniques and data structures have been proposed to compute, store, and query a *viewshed* efficiently (see, for example, Cole & Sharir, 1989; De Floriani & Magillo, 1994).

On a stepped RSG, visibility information is usually represented in a discretised way, by marking each pixel as visible or invisible from V. The resulting array of Boolean values is called a *discretised visibility map* of the terrain with respect to viewpoint V. Visibility information can also be encoded in a continuous way, by calculating the visible portions of each face of the terrain. This form is called a *visibility map*, and it is mainly used for polyhedral terrain models, such as TINs.

The computation of the *intervisibility map* between two regions has also been considered, mainly on stepped RSGs. Given a source region and a destination region (i.e. two subsets of pixels of the model), the problem consists of finding, for each point of the destination region, the points of the source region from which it is visible. Such a structure is an extension of the discretised visibility map.

The problem of computing the visibility map of a terrain is connected with the more general *Hidden Surface Removal* (HSR) problem for a generic three-dimensional scene. For simplicity, we can consider a viewpoint which lies at infinity in the positive y-direction and uses the xz-plane as an image plane. Given a viewpoint and an image plane, the *visible image* of a scene (possibly a polyhedral terrain) is a subdivision of the image plane formed by collecting the projections (images) of the portions visible from the viewpoint of each face of the scene. For a viewpoint in finite position inside the domain of the terrain, it is possible to obtain an analogous framework by transforming terrain and viewpoint in a spherical coordinate system, hence obtaining a spherical screen rather than an image plane.

The viewshed problem coincides with the HSR for polyhedral terrains, therefore it has direct impact on terrain visualisation. Conversely, any algorithm for HSR can be used to compute the viewshed.

The visibility map (and, similarly, the visible image), of a polyhedral terrain with n vertices has a worst-case space complexity equal to $O(n^2)$ for a PTM with n vertices. A sequential algorithm exists that can compute it in worst-case optimal $\theta(n^2)$ time (McKenna, 1987). The discretised visibility map from a viewpoint has an $O(n)$ worst-case space complexity for a $\sqrt{n} \times \sqrt{n}$ regular grid, and can be computed sequentially in $O(n\sqrt{n})$ time.

Parallel algorithms for computing both the visibility map (or the visible image) of a PTM (TIN) and the intervisibility map on an RSG will be reviewed in section 16.7.

16.4 Parallel Algorithms for Delaunay Triangulation and the Voronoi Diagram

Intuitively, the *Delaunay triangulation* of a set V of points in $\Re^2$ is a subdivision of the convex hull of V into triangles having their vertices at points of V, and such that triangles are as much equiangular as possible. More formally, a triangulation τ of V is a Delaunay triangulation if and only if, for any triangle t of τ, the circumcircle of t does not contain any point of V in its interior. This property is called the *empty circle property* of the Delaunay triangulation.

An alternative characterisation of the Delaunay triangulation is given based on the *max-min angle property*. Let τ be a triangulation of V. An edge e of τ is said to be *locally optimal* if and only if, given the quadrilateral Q formed by the two triangles of τ adjacent to e, either Q is not convex, or replacing e with the opposite diagonal of Q (*edge flip*) does not increase the minimum of the six internal angles of the resulting triangulation of Q. τ is a Delaunay triangulation if and only if every edge of τ is locally optimal. The repeated application of edge flips to non-optimal edges of an arbitrary triangulation finally leads to a Delaunay triangulation. This procedure is known as local optimisation (Lawson, 1977).

The geometric dual of the Delaunay triangulation is the *Voronoi diagram*, which describes the proximity relationship among the points of the given set V. The Voronoi diagram of a set V of points is a subdivision of the plane into convex polygonal regions, where each region is associated with a point P_i of V. The region associated with P_i is called the Voronoi region of P_i, and consists of the locus of points of the plane which lie closer to P_i than to any other point in V. Two points P_i and P_j of V are said to be Voronoi neighbours when the corresponding Voronoi regions are adjacent. Figure 16.1 shows the Delaunay triangulation and the Voronoi diagram for a point set.

Delaunay triangulations and Voronoi diagrams are among the most studied structures in computational geometry (Aurenhammer, 1991). Several parallel methods for computing them have been proposed (see Table 16.1). Many algorithms have only a theoretical interest, since they succeed in showing improved complexity bounds, while the use of sophisticated techniques and data structures makes them unsuitable for practical implementations (Chow, 1980; Wang & Tsin, 1987; Aggarwal *et al.*, 1988; Cole *et al.*, 1990; Saxena *et al.*, 1990). A few practical algorithms exist, which have been implemented on existing parallel architectures (Davy & Dew, 1989; Cignoni *et al.*, 1993; Clematis & Puppo, 1993; Puppo *et al.*, 1994; Ding & Densham, 1994). While computational complexities of theoretical algorithms are mostly analysed by referring to the CREW/PRAM model of computation, the complexities of practical algorithms are more often evaluated through direct experimentation.

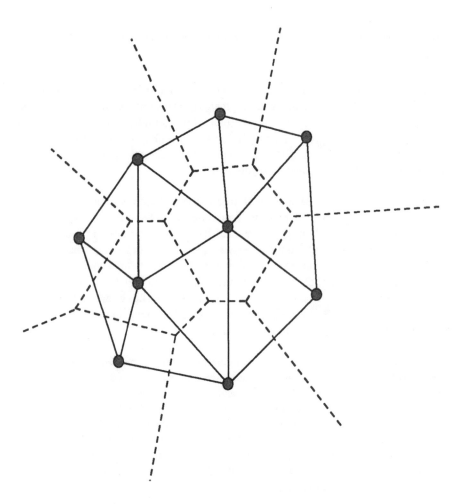

Figure 16.1: The Delaunay triangulation (solid lines) and the Voronoi diagram (dashed lines) of a set of points

All techniques used by sequential algorithms for computing Delaunay triangulations and Voronoi diagrams have been extended in parallel. Chow (1980) first showed that the convex hull in 3D can be computed in parallel in $O(\log^3 n)$ time using $O(n)$ processors, but this approach only produced a theoretical method.

Incremental construction algorithms can be successfully parallelised on fine-grained architectures through data partition. This approach exploits the locality of the empty circle criterion (i.e. the fact that points which are far apart have little or no effect on each other).

Table 16.1: Algorithms for generating the Voronoi diagram (VD) and for Delaunay trangulation (DT)

Reference	Algorithm	Architecture	Platform	Complexity
Chow (1980)	VD through 3D convex hull	CREW/PRAM	—	$O(\log^3 n)$ time with $O(n)$ PEs
Aggarwal *et al.* (1988)	divide-and-conquer VD, parallel merge	CREW/PRAM	—	$O(\log^2 n)$ time with $O(n)$ PEs
Davy & Dew (1989)	divide-and-conquer DT, sequential merge	coarse-grained MIMD	array of Transputers	$O(n)$ time with $O(\log n)$ PEs
Cole *et al.* (1990)	divide and conquer VD, parallel merge	CREW/PRAM	—	$O(\log \log n)$ time with $O(n \log n)$ PEs
Saxena *et al.* (1990)	iterative DT, minimum distance criterion	Orthogonal Tree Network	—	$O(\log^2 n)$ time with $O(n)$ PEs
Clematis & Puppo (1993)	divide-and-conquer DT, sequential merge	coarse-grained MIMD	nCUBE-2	$O(n)$ time with $O(\log n)$ PEs
Cignoni *et al.* (1993)	incremental DT, gridding	fine-grained MIMD	nCUBE-2	—
Cignoni *et al.* (1993)	divide-and conquer DT, anticipated sequential merge	coarse-grained MIMD	nCUBE-2	—
Puppo *et al.* (1994)	incremental DT, local optimisation	fine-grained SIMD	CM-2	—
Ding & Densham (1994)	divide-and-conquer DT, overlapping subsets, sequential glueing	coarse-grained MIMD	array of Transputers	—
Wang & Tsin (1987)	arbitrary triangulation, divide-and-conquer	CREW/PRAM	—	$O(\log n)$ time with $O(n)$ PEs

Cignoni *et al.* (1993) propose an incremental parallel algorithm, which is based on a constructive rule similar to that proposed by McLain (1976). Starting from a Delaunay triangle, at each step a new Delaunay triangle is added, which is adjacent to a pre-existing Delaunay triangle. The third vertex of the new triangle is found using the empty-circle criterion.

The parallel algorithm is based on data partition: the bounding box of the data set is partitioned into m rectangular regions, each of which is assigned to a different processor. A processor constructs all the triangles which are overlapping with the corresponding rectangular region. Triangulations corresponding to adjacent regions are subsequently glued together, while replicated triangles in the output are eliminated. The algorithm was implemented on a MIMD hypercube architecture, the nCUBE-2, and showed good scalability in practical experiments (Cignoni *et al.*, 1993).

A theoretical method for incrementally constructing the Delaunay triangulation in parallel was proposed by Saxena *et al.* (1990). The complexity of the algorithm is evaluated on a special architecture, called an Orthogonal Tree Network. An orthogonal tree network is a $\sqrt{n} \times \sqrt{n}$ array of processors in which each row and each column form leaves of a binary tree. The root of the tree and each internal node are also processors. Most of the processing is performed in the leaves, while internal nodes are used mainly for communication between leaves. Various primitive operations are defined on orthogonal tree networks, sending data along a path on a tree, or computing the minimum or the sum of values stored at the leaves of a tree, which all take $O(\log n)$ time.

The algorithm first finds, for every data point P_i in parallel, the data point P_j, lying nearest to P_i, and draws Delaunay edge $P_i P_j$. Then, for every existing Delaunay edge AB, two points C and D, lying on opposite sides with respect to AB, and having the minimum distance from AB are found, and triangles ABC and ABD are built. The above step is repeated for every newly created edge, until no more edges are created. The algorithm achieves $O(\log^2 n)$ parallel time complexity by using a tree with $O(n)$ processors.

In section 16.5 we will also review a data parallel algorithm that uses an incremental method to build a Delaunay triangulation, which has been implemented on a Connection Machine CM-2 (Puppo *et al.*, 1994). Such an algorithm is presented separately since it was originally designed to perform a *TIN from RSG* conversion, but it can be also used to compute the Delaunay triangulation of an arbitrary set of points.

The *divide-and-conquer* technique is perhaps the best-suited for parallelisation: several algorithms, both theoretical and practical, have been developed based on it (Aggarwal *et al.*, 1988; Davy & Dew, 1989; Cole *et al.*, 1990; Cignoni *et al.*, 1993; Clematis & Puppo, 1993; Ding & Densham, 1994).

A straightforward and practical way to parallelise a divide-and-conquer algorithm consists of assigning each recursive call to a different process, and performing each merge operation on a single processor. Parallel algorithms working with this approach are well suited for coarse-grained MIMD architectures, but they do not scale efficiently on machines with a high number of PEs, since the merging step involves a computational cost which is much higher in the first call than in the rest of the computation.

A first implementation of such an approach on a network of a small number of Transputers was proposed by Davy & Dew (1989). An implementation on a hypercube nCUBE-2 is also reported by Clematis & Puppo (1993), in which the recursive subdivision scheme is improved by alternating horizontal and vertical dividing lines in order to avoid the creation of long and thin strips, which cause numerical problems. Experimental results presented by Clematis & Puppo (1993) show that, for a set of 2^{13} random sites, the efficiency is good up to a size of eight PEs, then it decreases; a theoretical analysis is also given, showing that the algorithm performs well on a coarse-grained architecture, with an $O(\log n)$ number of PEs, where n is the number of input points.

A different divide-and-conquer algorithm proposed by Ding & Densham (1994) adopts a recursive subdivision scheme with overlapping regions. The merging step is avoided because the edges of the recursively constructed partial triangulations are guaranteed to belong to the final triangulation. Thus, the merging stage reduces to gluing partial results together. The algorithm was implemented on a small distributed-memory Transputer array residing in a microcomputer host. Hence the scope for scalability assessment was necessarily limited. (See next chapter also.)

Another parallel algorithm proposed by Cignoni *et al.* (1993) is based on an original interpretation of the divide-and-conquer paradigm, in which merging is anticipated to the subdivision phase. This feature permits the processes generated at each step to run completely asynchronously.

During the splitting phase, both the given data set and the problem solution are subdivided. The data set is split into two equally-sized subsets by using a dividing line *l*. The solution (i.e. the set of triangles composing the Delaunay triangulation of the whole data set) is subdivided into three subsets (see Figure 16.2): the triangles completely contained into the positive halfplane of *l*; the triangles completely contained into the negative halfplane of *l*; the triangles intersected by *l*. The triangles of the first two subsets are computed during the recursive calls of the algorithm. The triangles of the third subset (called the *Delaunay wall*) are constructed directly during the splitting phase: these triangles represent a valid merging set for the two partial triangulations returned by the two recursive calls.

The algorithm was implemented on a MIMD hypercube nCUBE-2. The results of the implementation showed a limited scalability of the algorithm. In addition to the problem common to all parallel divide-and-conquer algorithms with sequential merge, an extra overhead on this special algorithm is due to the fact that some parts of the computation (namely, those related to the computation of the Delaunay wall) are replicated on the PEs. The best results have been obtained with two or four PEs.

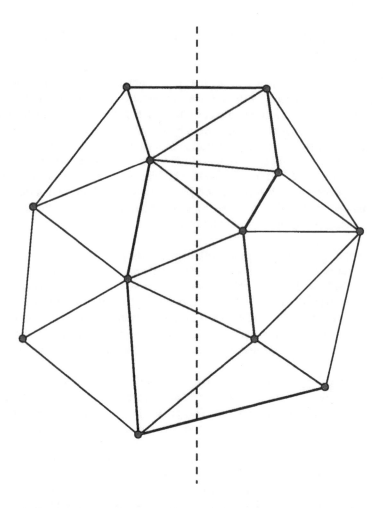

Figure 16.2: The Delaunay wall (dotted) and the two sets of triangles recursively computed (white)

Theoretical divide-and-conquer algorithms for computing the Voronoi diagram have been proposed, which can perform the merge operation in parallel. The method proposed by Aggarwal *et al.* (1988) achieves an $O(\log^2 n)$ time complexity by using $O(n)$ processors in the CREW/PRAM model; in a subsequent work, such complexity was improved to $O(\log \log n)$ time with $O(n \log n)$ processors by means of highly sophisticated structures (Cole *et al.*, 1990).

In spite of the extensive theoretical investigation, an optimal algorithm for Delaunay triangulation or the generation of a Voronoi diagram, which can achieve $O(n \log n)$ parallel time with $O(n)$ processors, has not yet been found. Wang &

Tsin (1987) propose a (theoretical) parallel algorithm based on a divide-and-conquer approach, which matches such time complexity for computing an *arbitrary triangulation* of a set of *n* points in the plane.

16.5 Parallel conversion from RSG to TIN

Puppo *et al.* (1994) proposed a data parallel algorithm, which permits the conversion of RSG data into an approximate TIN model based on a reduced number of vertices. The algorithm is suitable for obtaining a TIN either with a given number of vertices or with a given threshold error.

The algorithm can be considered as a parallel variation of an early sequential refinement technique proposed by Fowler & Little (1979): a first approximation of the terrain consists of an initial Delaunay TIN based on a small number of vertices (in the parallel algorithm, only the four corners of the domain are used). Following this, new vertices are iteratively added, and the Delaunay triangulation is updated, until an end-condition is met (i.e. either the maximum number of vertices have been inserted, or the error gets below the threshold). Figure 16.3 shows a worked example.

Both the sequential and the parallel algorithm use an iterative technique. In the sequential algorithm, one new vertex is selected at each iteration, namely the datum that maximises the error in the current TIN. In the parallel algorithm, a number of new vertices are inserted at each iteration, namely: for each triangle in the current TIN that covers data whose corresponding errors exceed the threshold, the datum maximising the error is inserted.

The parallel algorithm uses two sets of processing elements and it is designed to exploit advanced communication features of data parallel computers, which allow PEs to communicate in arbitrary patterns (Blelloch, 1990). The *triangle set* contains one PE for each triangle in the current triangulation: new PEs are activated dynamically each time new triangles are built. Each such PE stores in its local memory the coordinates of its vertices, and the addresses of the three PEs corresponding to its adjacent triangles. The *point set* contains one PE for each datum in the RSG. Each such PE stores the coordinates of the point and the address of the PE corresponding to the triangle covering it in the current triangulation.

On each iteration, each point that is not a vertex of the TIN gets from the triangle containing it the coordinates of its vertices, and it computes its corresponding error. Then, each point PE communicates its error to the corresponding triangle PE. Each

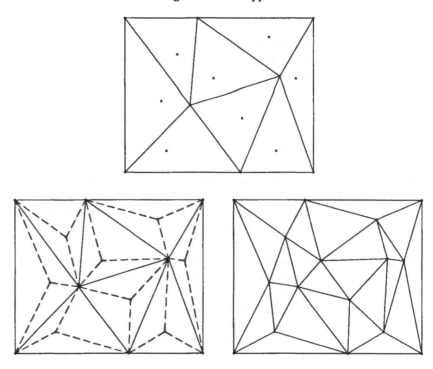

Figure 16.3: Iterative construction of a TIN starting from gridded data

triangle retains the point corresponding to the maximum error received, and considers it as a candidate new vertex.

The most complex part of the algorithm is the triangulation update once new vertices have been selected. In the simplest case, all vertices lie within current triangles. Where candidate vertices lie on edges of the TIN, some of them must be discarded to avoid conflicts between adjacent triangles, which would generate inconsistencies. The triangulation is updated by first splitting each triangle that contains a new vertex in its interior into three, and each triangle that has a new vertex on one of its edges into two. Then, a local optimisation procedure is applied to edges of the newly generated triangles: a sequence of parallel edge flips is performed, until the triangulation satisfies the Delaunay criterion. An edge flip considers a pair of adjacent triangles forming a convex quadrilateral, and swaps their common edge with the opposite diagonal of the quadrilateral. Many flips can be performed in parallel independently, provided that all pairs of triangles involved are disjoint. Conflicts arising from overlapping pairs are eliminated through a technique similar to the one used to resolve conflicts in the splitting phase.

After each change in the triangulation (either split or flip), all data that fall inside the portion of the triangulation which has been modified are relocated, i.e. each point PE computes locally which triangle, in the updated triangulation, covers it. In fact, the relocation phase also provides point PEs with data for computing their approximation error, and the cycle is thus completed.

The algorithm was implemented on a CM-2 and tested on real world data. Results presented show a good speed-up in time, with an especially good quality TIN being produced. Indeed, the parallel selection and insertion of many data at a time allow the algorithm to achieve better precision for a given number of vertices, and a smaller number of vertices for a given threshold error, compared with the results obtained using the sequential method.

The structure of the algorithm makes it modifiable to perform different tasks. It is straightforward to modify the algorithm to compute approximate data-dependent TINs, described by Rippa (1992). Indeed, data dependent triangulations are obtained through the same local optimisation procedure based on a different optimality criterion. Hence, it is sufficient to change the code of the optimality test that controls edge flip.

Another possibility is to use the same algorithm to build a TIN from sparse data. Note that the algorithm does not need input data arranged in a regular grid: the same algorithm can be used for sparse points. Hence, it is sufficient to run refinement until all data points have been included as vertices. In this case, selection of points at each cycle is irrelevant: however, it is convenient to select at each iteration and for each triangle the point lying closest to the centre of gravity of the triangle, in order to minimise conflicts arising from candidate vertices on the edges.

For some applications a TIN is built from sparse points, then it is resampled at the nodes of an RSG (see *RSG from TIN* in 16.3.1). In this case, it is possible to modify the algorithm that computes a TIN from sparse data as follows. Point PEs include both data points and grid nodes. At each iteration, the algorithm selects vertices only from data points, but it relocates grid nodes in addition. In this way, once the TIN is built, all grid nodes already know the triangles covering them, and they can compute their elevation locally in constant parallel time.

The sequential algorithm proposed by Fowler & Little (1979) builds the initial triangulation on a set of vertices selected on topographic features through a parallel algorithm described by Peucker & Douglas (1975), and discussed in the following section. The implementation of such an algorithm in a data parallel framework is straightforward. Hence, a combination of the two algorithms could be implemented easily to improve the selection of vertices included in the TIN.

16.6 Parallel Algorithms for Topographic Characterisation

As mentioned earlier, solutions to problems in topographic characterisation are based on local analysis of the surface. Hence, they are especially well suited to parallel solutions on RSG models. We review two different approaches to topographic characterisation.

The technique proposed by Brunetti *et al.* (1994) is aimed at the evaluation of topographic features such as peaks, ridges, and valleys, or at curvature analysis. This has been designed under the task-farm paradigm for a heterogeneous network of workstations. A common technique is proposed to parallelise two known algorithms for feature evaluation (Peucker & Douglas, 1975; Skidmore, 1990), and an original algorithm for curvature estimation.

The algorithm proposed by Peucker & Douglas (1975) is naturally parallel, since each pixel is classified independently on the basis of the sums of positive and negative differences between the elevations of the pixel and of its 8-neighbours: the pixel can be classified as flat, pass, ridge, ravine, slope, peak, or pit. The algorithm proposed by Skidmore (1990) extends such an approach to a more refined analysis which is aimed at preventing false classifications. From the point of view of parallel programming, this implies only that the portion of data involved in the classification of a pixel can include not only its 8-neighbours, but a window of larger, though fixed, size (e.g. 9×9) around the datum analysed.

The algorithm for curvature estimation permits the classification of each pixel as strong convex, strong concave, strong flat, global convex, global concave or global flat. Classification is based on estimation of Gaussian and mean curvature at each pixel. The algorithm uses a virtual triangulation obtained by adding one vertex at the centre of each square formed by four adjacent pixels, and subdividing such a square into four triangles. The classification of each pixel depends on the eight triangles incident to it: also in this case, only the 8-neighbours of each pixel need to be analysed.

Because all such algorithms exhibit strong locality of computation in regular patterns, a parallelisation under a data parallel approach would be straightforward, and easily implementable on massive mesh or hypercube machines. The alternative approach used by Brunetti *et al.* (1994) is to design a portable scheme that can fit onto a heterogeneous network of workstations. According to the task-farm paradigm, a number of tasks are generated first. Each task simply consists of the classification of a portion of data, corresponding to a segment of consecutive rows of the input grid. The partition into tasks directly corresponds to segmentation of the data; one segment for each task. Each segment is enlarged to a 'chunk' by adding rows above and below it, in order to include in the chunk all data needed to classify pixels of the corresponding segment. Consecutive chunks overlap by one row for Peucker & Douglas algorithms, and for curvature estimation, while they overlap by four rows for Skidmore algorithms with a 9×9 window.

Each worker iteratively gets a chunk from the shared memory, classifies pixels of the corresponding segment, and returns the result to the shared memory. Brunetti *et al.* (1994) discuss a technique which allows the farmer to decide the size of segments automatically, in order to optimise performance and load-balancing. The optimal size depends on the problem size, the number of available workers, their relative performance and the ratio between the size of the window used and of the segment itself.

Experiments on real world data show a good speed-up with respect to sequential algorithms and good scalability of the method. This approach to parallelisation can be considered quite generally applicable to all tasks that involve only local analysis of RSG data. A relevant advantage is that the sequential code can be used directly to implement single tasks, hence making the effort to implement the parallel method limited to communication and task syncronization.

An algorithm proposed by Mower (1992) for labelling drainage basins in RSG data has been designed using a data parallel approach, and it is implemented on a Connection Machine CM-2. While some parts of the algorithm require local processing, as with the algorithm described above, other parts involve more global operations. An example is *wave propagation*, which is an operation especially suitable for mesh architectures. Wave propagation is a general technique that consists of spreading information over a mesh, starting at some set of seeds, and propagating them as waves moving away from these seeds. Information propagation occurs iteratively: at each cycle the PEs lying on the wave spread information to their neighbours that lie outside the region already covered by the wave. Hence, the whole grid can be covered in, at most, $\sqrt{n}$ cycles, where n is the size of the grid. The algorithm embeds RSG data in the PEs, and executes the following steps.

- Elevations are smoothed through local averaging, i.e. the elevation of each pixel is substituted with an average of its current value and of elevations of its 8-neighbours. Smoothing allows the algorithm to reduce the number of false pits due to excessive roughness of data. Such an operation is local and it is performed in parallel constant time.

- For each pixel, slope and exposure are evaluated through local finite differentiation involving only its 8-neighbours. Each pixel whose elevation is smaller than or equal to the elevation of its 8-neighbours is initially marked as a pit, and it is assigned a unique label. This step also involves only local operations.

- Starting at each pit, labels are propagated through waves as follows. Initially, only PEs corresponding to pits are active. Each pit sends its label to each of its uphill neighbours. All PEs corresponding to such neighbours are activated, while pits are deactivated. Then, each active PE sends its label to any of its uphill neighbours that have not been labelled yet, and so on, until all pixels have been labelled. Figure 16.4 shows an example of label propagation. This step provides an initial partition of the RSG into drainage basins.

- For each basin, a 'pour point' is detected, which corresponds to the pixel on the border of the basin having the smallest elevation. Border pixels are detected locally by comparing the label of each pixel with the labels of its neighbours. Then, pour points are found by activating the border pixels of one basin at a time, and performing a global-minimum operation, which is an operation typical of data parallel computing. The complexity of this step could be reduced through a segmented sort operation, which permits all pour points to be found in a single step in time $O(\log n)$ (Blelloch, 1990).

- For each basin, all pixels whose elevation is lower than the elevation as the pour point are raised to the same elevation as the pour point, and their slope is set to zero. Also this operation can be performed in $O(\log n)$ with a single segmented scan operation (Blelloch, 1990).

- Finally, a modified wave propagation is repeated starting at the pour point of the lowest basin. Such a propagation is repeated until the whole grid has been relabelled. In this way, some of the basins detected earlier are now included as sub-basins of larger basins.

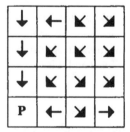 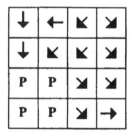 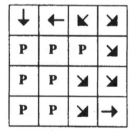

Figure 16.4: The initial situation is shown in (a). Arrows represent drainage directions. P is a pit. (b) and (c), respectively, show the situation after one or two propagation steps. P is unable to propagate its labels into pixels that face away from it

This algorithm is a good, yet reasonably simple, example which shows the application of basic techniques in data parallel programming, including neighbour communication as well as global operations such as global-minimum, segmented scan, and sort. Implementation details and experimental results were not discussed in Mower (1992).

16.7 Parallel Visibility Computation on Terrains

Since visibility problems on a polyhedral terrain model (PTM) are very different from visibility problems on a stepped RSG, we review algorithms for these different problems in two separate sections. Table 16.2 summarises the reviewed algorithms.

Table 16.2: Parallel algorithms for terrain visibility

Reference	Approach	Architecture	Platform	Complexity
Reif & Sen (1988)	visibility map on TIN	modified CREW/PRAM	—	$O(\log^4 n)$ time with $O((n+k)/\log(n+k))$ PEs
Teng *et al.* (1995)	visibility map on PTM	CREW/PRAM	—	$O(\log^2 n)$ time with $O(((k+n\alpha(n))/(\log n))$ PEs
De Floriani *et al.* (1994)	visibility map on TIN	coarse-grained MIMD	nCUBE-2	—
Teng *et al.* (1993)	intervisibility on RSG	fine-grained SIMD	CM-2	—
Mills *et al.* (1992)	intervisibility on RSG	fine-grained SIMD	CM-2	—

16.7.1 Computation of the Visibility Map

Several parallel algorithms for computing the visible image of three-dimensional scenes have been proposed in the literature because this problem has an important interest in computer graphics (Whitman & Parent, 1989; Theoharis, 1989; Franklin & Kankanhalli, 1990; Scopigno *et al.*, 1993). Although the viewshed problem can be considered as a special case of such problems, we review only parallel algorithms that were developed specifically for the viewshed.

The algorithm by Reif & Sen (1988) is mainly of theoretical interest, since the sophisticated techniques and data structures used make it unsuitable for practical implementation. Another theoretically efficient algorithm has been proposed by Teng *et al.* (1995). A practical algorithm has been designed and implemented by De Floriani *et al.* (1994).

Before examining the algorithms, we introduce some definitions and properties related to visibility on polyhedral terrains, which are widely used by visibility algorithms (both sequential and parallel).

The visible image of a terrain covers a simple polygon on the image plane, unbounded below and bounded from above by a chain of segments. This chain is formed by the images of those visible portions of edges of the terrain which, in any direction, mark the furthest limit of the sight (i.e. the 'horizon') from the given viewpoint, and it is called the *profile* of the terrain. Figure 16.5 shows the visible image of a terrain and the corresponding profile.

In particular, the profile is formed by visible portions of blocking edges of the terrain. A *blocking edge*, with respect to a viewpoint V, is any edge e such that its incident face closer to V is seen from above by V (*face up*), while the other incident face is seen from below (*face down*).

Given a viewpoint V, a face (or an edge) A of a scene is said to be *in front* of another face (or edge) B if a ray emanating from V intersects A before intersecting B. If A is in front of B then B is *behind* A. On a terrain, the in front/behind relation between faces or edges in 3D reduces to an analogous relation among the corresponding regions or edges of the underlying subdivision of the 2D domain, with respect to the *xy*-projection of the viewpoint.

A scene is called *sortable* if the in front/behind relation defines a partial order of the faces (edges), which can be completed to a total *front-to-back order* through topological sorting. Every PTM is sortable with respect to a viewpoint whose vertical projection lies outside the domain; however, for an internal viewpoint this is not always true. Delaunay-based TINs have been shown to be sortable with respect to any viewpoint (De Floriani *et al.*, 1991). A non-sortable terrain model can always be made sortable by splitting some of its faces (Cole & Sharir, 1989). An RSG is sortable with respect to any viewpoint.

Reif & Sen (1988) were the first to aim at designing an *output-sensitive* parallel algorithm for visibility computation on terrains, i.e. an algorithm in which the work is proportional to the output size. Their parallel algorithm computes the visible image of a sortable TIN. The *xy*-projections of the terrain edges are recursively partitioned into equally-sized subsets, lying one in front of the other, and the profile of each group, within each level, is computed in parallel. Finally, for each edge in parallel, the intersections of this edge with the profiles in front of it are computed.

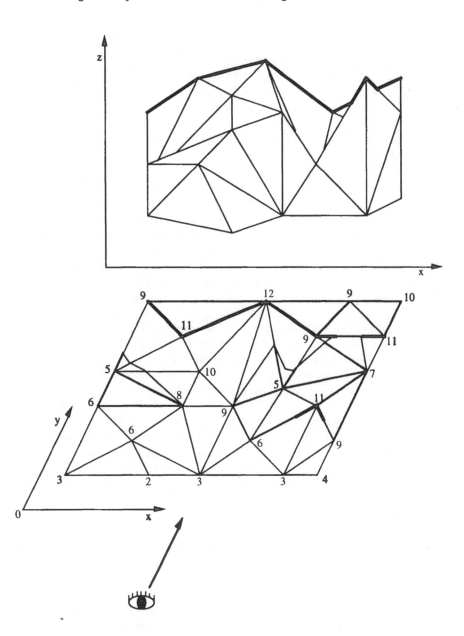

Figure 16.5: The visible image of a polyhedral terrain model and the corresponding profile. Invisible areas are shown dotted on the domain. The portions of edges forming the profile are drawn in thick lines

Reif & Sen show that this approach, combined with the use of special data structures, leads to a complexity of $O(\log^4 n)$ time using $O((n+k) / (\log n))$ processors. However, the model of computation they use is slightly different from the conventional PRAM model, since they assume a pool of free processors from which processors can be requested at run-time.

The algorithm proposed by Teng *et al.* (1995) is based on the sequential algorithm proposed by Katz *et al.* (1991).

The algorithm of Katz *et al.* is designed for computing visible images within a wider class of 3D scenes, and is then specialised for terrains. For a (sortable) PTM with *n* edges, the time complexity is equal to $O((k + n\ \alpha(n))\log n)$, where *k* is the output size. The algorithm follows a divide-and-conquer approach, in which the 'divide' step is made according to the depth order of the objects. Input objects are topologically sorted in front-to-back order with respect to the viewpoint and associated with the leaves of a balanced binary tree *T*, in such a way that a left-to-right traversal of the leaves of *T* corresponds to the front-to-back order of the objects. In the 'conquer' step, tree *T* is traversed twice: during the first (bottom-up) traversal, for each node v, the union U_v of the images of all objects associated with the leaves of the subtree rooted at v is computed. During the second (top-down) traversal, the visible portion V_v of U_v is computed for each node v. The visible portions of each of the objects are thus the sets V_v associated with the leaves of *T*. Figure 16.6 shows sets U_v and V_v computed for a terrain.

In Teng *et al.* (1995), the computation of sets U_v and V_v is parallelised by considering one level of the tree at a time, and performing all the operations related to the nodes at that level in parallel. Using special techniques for the manipulation of monotone chains of segments and efficient process reallocation, the authors show that their algorithm works in $O(\log^2 n)$ time, using $O((k+n\ \alpha(n)) / (\log n))$ processors in a CREW/PRAM model.

De Floriani *et al.* (1994) implemented a parallel algorithm for the computation of the visibility map on sortable TINs, which is based on a data partitioning strategy: a sequential algorithm is applied in parallel to different portions of terrain, arranged radially around the viewpoint.

The sequential algorithm performs a front-to-back traversal of the triangular faces of the TIN, and computes the visibility of each face by testing it with respect to the edges forming the upper profile of triangles already processed. The front-to-back traversal of the triangles is performed by incrementally building a star-shaped polygon around the viewpoint, starting from the triangle containing the viewpoint and adding one triangle at a time, until all triangles have been included in the polygon. Figure 16.7 shows a generic stage of the process, with the current star-shaped polygon and horizon.

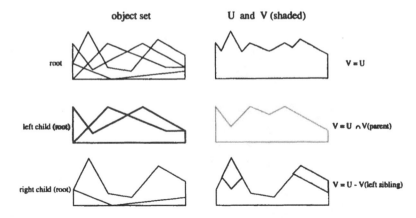

Figure 16.6: The first level of the tree considered by Katz et al.'s (1991) algorithm for a terrain. All U and V regions are profiles, and thus their computation reduces to set operations on profiles

A circular chain of edges is maintained, which form a subset of the boundary of the current star-shaped polygon, containing those edges from which new triangles can potentially be included. At each step, an edge l is picked (i.e. read and deleted) from the current chain, and the triangle t externally adjacent to l is considered. If t can be included in the polygon while maintaining it as star-shaped, then t is included: the visibility of t is immediately computed.

A circular list is maintained, containing all those portions of blocking edges belonging to triangles, which have been examined already, but which can cast a shadow on a triangle not yet examined. Such segments are called *active* blocking edge segments. The visible portions of a face-up triangle t are determined by projecting the relevant part of the list of active edge segments onto the supporting plane of t. Then, the list of active segments is updated by considering the distal edges of t, i.e. the edges which lie on the boundary of the new star-shaped polygon, after t has been included. Face-down triangles are completely invisible. The time complexity of the sequential algorithm is equal to $O(n^2 \, \alpha(n))$.

The parallelisation of the algorithm is based on a data partitioning strategy. The domain is subdivided into radial sectors, which have the viewpoint as their common vertex. Each sector is assigned to a processor, which examines triangles fully or partially contained in the sector. For partially contained triangles, only the visibility of the portion of the triangle lying within the sector is computed.

The algorithm has been implemented on a Hypercube machine nCUBE-2, a coarse-grained MIMD architecture. Several data partitioning criteria are discussed by De Floriani *et al.* (1994). The partitioning criterion used in the implementation is

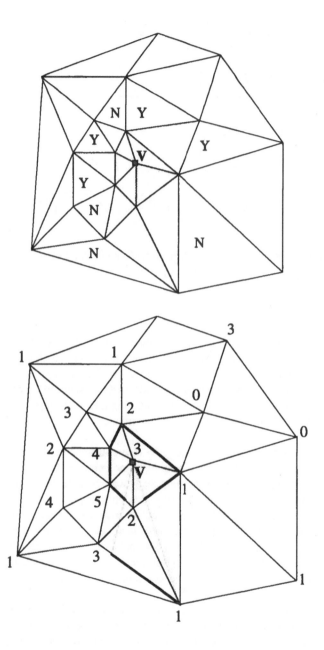

Figure 16.7: The current star-shaped polygon around the viewpoint V (a) and the current profile (b). Each triangle externally adjacent to the polygon is a candidate for inclusion into the polygon; in (a), each such triangle is marked with 'Y' if it can be included while maintaining the star-shape of the polygon, and with 'N' otherwise

based on equal-area sectors. Both a static version and a dynamic version (in which the number of sectors is twice the number of processors and a new sector is assigned to each processor as soon as it terminates its work) have been experimented with. The best results in terms of speed-up and efficiency were obtained with four PEs.

16.7.2 Computing Intervisibility on an RSG

Algorithms for intervisibility computation on stepped RSGs have been developed for practical implementation in GIS applications. Proposed algorithms exploit the fact that the regular spatial structure of an RSG can be directly embedded into a parallel SIMD architecture, such as the mesh or the hypercube. The result of the intervisibility computation for each vertex is indicated by a flag in the corresponding processor.

While describing the algorithms, the pixels of the source region will be called *viewpoints*, while the pixels of the destination region will be called *targets*.

The algorithm proposed by Mills *et al.* (1992) is based on a "walk along the line-of-sight". The line-of-sight joining a viewpoint V to a target T intersects a certain set of pixels. The visibility of T from V can be determined by moving along the line-of-sight from V to T. Each step involves moving to an adjacent pixel in the direction of the line-of-sight, obtaining the elevation from the processor corresponding to that pixel, comparing it with the previous elevation and updating the visibility accordingly. The walk terminates when either the visibility is obstructed before reaching the target pixel or the target pixel is reached. The algorithm walks along each line-of-sight in parallel.

Instead of running this basic algorithm for every pair viewpoint-target pair, Mills *et al.* (1992) propose a more efficient solution, where global communications (between a PE representing a line-of-sight and a PE representing a grid pixel) are replaced by much faster local communications (between processors directly representing adjacent lines-of-sight). Elevation data are shifted through the grid along the four cardinal directions North, East, West and South (NEWS) while performing line-of-sight calculations.

The algorithm has been implemented on a Connection Machine CM-2, and used for experimental evaluation of the effect of errors in viewshed calculation.

Teng *et al.* (1993) propose a data parallel intervisibility algorithm, which combines line-of-sight calculations with a sweep traversal of the source region. This approach exploits the coherence at adjacent viewpoints in order to reduce global communications.

Both the source and the destination region are assumed to be upright rectangles (otherwise, the bounding box of each region is considered). As in the previous algorithm, the underlying terrain is represented by a 2D grid of processors, one for each grid pixel.

Given a single viewpoint, a minimal set of lines-of-sight, covering all the pixels in the destination region, is obtained by collecting all the lines-of-sight from the viewpoint to pixels lying on a *far side* of the destination region. A far side is a boundary edge *l* of the destination region such that both the region and the viewpoint lie on the same side with respect to *l* (see Figure 16.8).

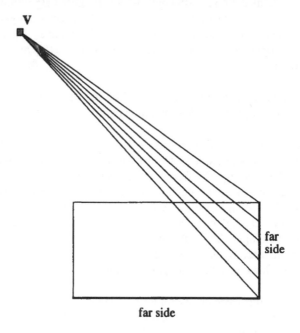

Figure 16.8: The far sides of the destination region and the line-of-sight to one of them

The basic task is computing the visibility for the target pixels along a single line-of-sight. This is done by allocating to the line-of-sight a list of PEs, one for each intersected grid pixel. Every PE gets the elevation of the corresponding grid pixel *C*, then it computes the elevation angle from the viewpoint to the grid pixel *C*, and it determines the visibility of *C* by comparing this angle with the maximum angle of pixels lying on the same line-of-sight, and closer to the viewpoint. The line-of-sight computation is done for all lines-of-sight corresponding to a far side of the destination region in parallel.

For a source region composed of more than one viewpoint, a sweeping algorithm is proposed. The algorithm performs four sweeps. Each sweep considers the set of viewpoints contained in one of the four sectors of the source region obtained by drawing two lines with 45°-slope passing through the centre of gravity of the region. Thus, a northern, a southern, an eastern and a western set of viewpoints are considered (see Figure 16.9).

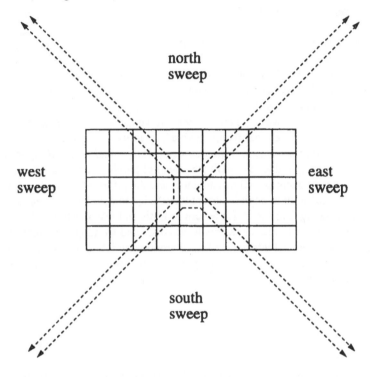

Figure 16.9: The four sets of viewpoints considered in the four sweeps

The eastern subset of the source region is swept from west to east. At each step, a column of viewpoints is examined. Because of coherence, passing from one strip to the next one, the elevation of each pixel can be obtained by local communication. Global communications are necessary only at the initialisation of the first column. The sweep of the other three subsets is performed in a similar way.

If w is the maximum number of pixels along a line-of-sight, and l_s and l_d are, respectively, the number of pixels on the side of the source and of the destination region, the time complexity of the algorithm is estimated in $O(l, \log w)$, when using $O(l, l_d w)$ PEs. Experimental results, obtained on a Connection Machine CM-2, match this estimate.

16.8 Concluding Remarks

Problems in terrain analysis and visualisation have been reviewed from the perspective of parallel computation. Parallel solutions exist in the literature only for some of these problems, while others are still open, or at least not explicitly studied. Overall, it seems that a search for efficient solutions to such problems can be pursued through general techniques rooted in computational geometry and image processing. However, there is often a large gap between theoretical approaches – which seem hardly applicable in practice, largely due to the performance overhead arising from their complexity – and practical algorithms, which are not always able to exploit the power of parallel computation at its best. Moreover, the different paradigms adopted, and the different parallel architectures used, make it difficult to exploit current knowledge to develop an integrated parallel computing framework for terrain analysis and visualisation. These authors recommend that further research should be conducted by adopting a reference architecture of practical interest, and by attempting an integration of techniques on this architecture, as well as by studying as yet unresolved problems under the same unified framework.

16.9 References

Aggarwal, A., Chazelle, B., Guibas, L., O'Dunlaing, C. & Yap, C., 1988, Parallel computational geometry. *Algorithmica*, **3**, 293-327.

Aurenhammer, F., 1991, Voronoi diagrams - A survey of fundamental geometric data structures. *ACM Computing Surveys*, **23** (3), 366-405.

Blelloch, G.E., 1990, *Vector Models for Data-Parallel Computing*, MIT Press Cambridge, MA.

Brunetti, G., Clematis, A., Falcidieno, B., Sanguineti, A. & Spagnuolo, M., 1994, Parallel processing of spatial data for terrain characterization. *Proceedings 2nd ACM Workshop on Advances in Geographic Information Systems*, Gaithersburg, MD, December, 166-173.

Chow, A.L., 1980, *Parallel Algorithms for Geometric Problems*, unpublished PhD Dissertation, Dept of Computer Science, University of Illinois, Urbana, IL.

Cignoni, P., Montani, C., Perego, R. & Scopigno, R., 1993, Parallel 3D Delaunay triangulation. *Computer Graphics Forum*, **12**, 3 (Conference Issue), Hubbold, R.J. & Juan, R. (guest editors), Blackwell Publishers, 129-142.

Clematis, A. & Puppo, E., 1993, Effective parallel processing of irregular geometric structures - an experience with the Delaunay triangulation. *Proceedings AICA 1993 - International Section: Parallel and Distributed Architectures and Algorithms*, 235-251.

Cole, R., Goodrich, M.T. & O'Dunlaing, C., 1990, Merging free trees in parallel for efficient Voronoi diagram construction. *Lecture Notes in Computer Science*, **443**, Springer-Verlag, Berlin, 432-445.

Cole, R. & Sharir, M., 1989, Visibility problems for polyhedral terrains. *Journal of Symbolic Computation*, **17**, 11-30.

Dadoun, N. & Kirkpatrick, D.G., 1987, Parallel processing for efficient subdivision search. *Proceedings 3rd ACM Symposium on Computational Geometry*, 205-214.

Davy, J.R. & Dew, P.M., 1989, A note on improving the performance of Delaunay triangulation. In Patrikalakis, N.M. (Ed), *Scientific Visualization of Physical Phenomena*, Ed: , Springer-Verlag, Hong Kong, 209-226.

De Floriani, L., Falcidieno, B., Nagy, G. & Pienovi, C., 1991, On sorting triangles in a Delaunay tessellation. *Algorithmica*, **6**, 522-532.

De Floriani, L. & Magillo, P., 1994, Visibility algorithms on triangulated digital terrain models. *International Journal of Geographical Information Systems*, **8** (4), 13-41.

De Floriani, L., Montani, C. & Scopigno, R., 1994, Parallelizing visibility computations on triangulated terrains. *International Journal of Geographical Information Systems*, **8** (4).

Ding, Y. & Densham, P.J., 1994, A dynamic and recursive parallel algorithm for constructing Delaunay triangulations. *Proceedings 6th International Symposium on Spatial Data Handling*, Edinburgh, September, 682-696.

Falcidieno, B. & Spagnuolo, M., 1991, A new method for the characterization of topographic surfaces. *International Journal of Geographical Information Systems*, 5 (4), 397-412.

Fowler, R.J. & Little, J.J., 1979, Automatic extraction of irregular network digital terrain models. *Computer Graphics*, 13 (3), 199-207.

Franklin, W.R. & Kankanhalli, M.S., 1990, Parallel object-space hidden surface removal. *Computer Graphics*, 24 (4) (Conference Issue), 87-94.

Green, P.J. & Sibson, R., 1977, Computing Dirichlet tessellations in the plane. *The Computer Journal*, 21, 168-173.

Katz, M.J., Overmars, M.H. & Sharir, M., 1991, Efficient hidden surface removal for objects with small union size. *Proceedings 7th ACM Symposium on Computational Geometry*, 31-40.

Lawson, C.L., 1977, Software for C^1 surface interpolation. In Rice, J.R. (Ed.), *Mathematical Software*, Vol. 3, Academic Press, 161-164.

Lee, J., 1991, Comparison of existing methods for building triangular irregular network models of terrain from grid digital elevation models. *International Journal of Geographical Information Systems*, 5, 3, 267-285.

McKenna, M., 1987, Worst case optimal hidden surface removal. *ACM Transactions on Graphics*, 6, 19-28.

McLain, D.H., 1976, Two dimensional interpolation from random data. *The Computer Journal*, 19 (2), 178-181.

Mills, K., Fox, G. & Heimbach, R., 1992, Implementing an intervisibility analysis model on a parallel computing system. *Computers & Geosciences*, 18 (8), 1047-1054.

Mower, J.E., 1992, Building a GIS for parallel computing environments. *Proceedings 5th International Symposium on Spatial Data Handling*, Charleston, SC, 3-7 August, 219-229.

Petrie, G., 1990, Modelling, interpolation and contouring procedures. In Petrie, G. & Petrie, T.J.M. (Eds), *Terrain Modelling in Survey and Civil Engineering*, Whittles Publishing-Thomas Telford, London, 112-127.

Peucker, T.K. & Douglas, D.H., 1975, Detection of surface-specific points by local parallel processing of discrete terrain elevation data. *Computer Graphics and Image Processing*, **4**, 375-387.

Preparata, F.P. & Shamos, M.I., 1985, *Computational Geometry: an Introduction*, Springer-Verlag. Berlin.

Puppo, E., Davis, L.S., DeMenthon, D. & Teng, A.,1994, Parallel terrain triangulation. *International Journal of Geographical Information Systems*, **8** (2), 105-128.

Reif, J. & Sen, S., 1988, An efficient output-sensitive hidden surface removal algorithm and its parallelization. *Proceedings 4th ACM Symposium on Computational Geometry*, 193-200.

Rippa, S., 1992, Adaptive approximations by piecewise linear polynomials on triangulations of subsets of scattered data. *SIAM Journal on Scientific and Statistic Computing*, **13** (1), 1123-1141.

Saxena, S., Bhatt, P.C. & Prasad, V.C., 1990, Efficient VLSI parallel algorithm for Delaunay triangulation on orthogonal tree networks in two and three dimensions. *IEEE Transactions on Computers*, **39** (3), 400-404.

Scopigno, R., Paoluzzi, A., Guerrini, S. & Rumolo, G., 1993, Parallel Depth-Merge: a paradigm for hidden surface removal. *Computer Graphics*, **17** (5), 583-592.

Skidmore, A.K., 1990, Terrain position as mapped from a gridded digital elevation model. *International Journal of Geographical Information Systems*, **4** (1), 33-49.

Spagnuolo, M., 1995, personal communication.

Teng, Y.A., DeMenthon, D. & Davis, L.S., 1993, Region-to-region visibility analysis using data parallel machines. *Concurrency: Practice & Experience*, **5**, 379-406.

Teng, Y.A., Mount, D., Puppo, E. & Davis, L.S., 1995, Parallelizing an algorithm for visibility on polyhedral terrain. *International Journal of Computational Geometry and Applications*, World Scientific Publishing Company, in press.

Theoharis, T., 1989, Algorithms for parallel polygon rendering. *Lecture Notes in Computer Science*, **373**, Springer-Verlag, Berlin.

Wang, C.A. & Tsin, Y.H., 1987, An $O(\log n)$ time parallel algorithm for triangulating a set of points in the plane. *Information Processing Letters*, **25**, 55-60.

Webb, H., 1990, Creation of digital terrain models using analytical photogrammetry and their use in civil engineering. In Petrie, G. & Petrie, T.J.M. (Eds), *Terrain Modelling in Survey and Civil Engineering*, Whittles Publishing-Thomas Telford, London, 73-84.

Whitman, S. & Parent, R., 1989, A survey of parallel hidden surface removal algorithms. In *Parallel Processing and Advanced Architectures in Computer Graphics*, ACM Press, 157-173.

17

Spatial Analysis

P.J. Densham and M.P. Armstrong

17.1 Introduction

Methods of spatial analysis often require substantial amounts of computation to generate results. The computational intensity found in the methods that we present in this chapter has several root causes. First, there is an increasing tendency for analyses to use highly disaggregated datasets that contain large numbers of observations. As the size of a dataset increases, so does the amount of computation required to analyse it. A second, and related, cause is that many types of spatial analysis require the calculation of large numbers of distances; these distances serve as a kind of data 'raw material' for different types of analyses. A further cause of computational complexity is that a large number of alternatives must be generated, sorted and evaluated when analysing geographical problems using the methods described in this chapter.

This computational complexity can frustrate even the most determined efforts to support interactive, exploratory analyses on serial computers. It is indeed fortunate that many methods of spatial analysis are well-suited to implementation in parallel environments either because data can be partitioned geographically or because functional computational units can be processed independently. In fact, we are not aware of any method of spatial analysis that cannot be adapted to a parallel processing environment. Note, however, that different problems have their own unique sets of processing requirements that may best be met by different parallel architectures. Thus, it is important to invoke appropriate strategies for problem decomposition and to match the processing characteristics of the problem to those of the target computer system.

In this chapter, rather than provide a comprehensive overview of parallel processing and spatial analysis, we use three areas of spatial analysis to illustrate these concepts and demonstrate how different approaches to parallelism yield different levels of performance. In section 17.2, we discuss parallel implementations of Getis's statistic, $G(d)$. This statistic measures spatial association in point sampled data and requires the calculation of large numbers of inter-point distances. In section 17.3 we turn our attention to network analysis and describe two parallel implementations of Dijkstra's (1959) shortest path algorithm.

Although this algorithm also calculates large numbers of inter-point distances, it does so in discrete space on a network rather than in the continuous space representation used by G(d). Two methods used in terrain analysis are discussed in section 17.4: spatial interpolation and hill-shading. Various parallel architectures and machines are used to implement these algorithms and results for a range of problem sizes are presented. Finally, we draw some conclusions from our experience in decomposing and implementing these algorithms.

17.2 Spatial Statistics

Several recent papers have demonstrated that significant improvements in processing time can be achieved when parallel processing is applied to computationally intensive measures of spatial association (Armstrong *et al.*, 1994a, 1994b; Armstrong & Marciano, 1995; Rokos & Armstrong, 1996). The measure used in this section of the chapter to illustrate these improvements [G(d)] measures the strength of spatial association among real-valued data collected at points (Getis & Ord, 1992). G(d) does not provide a global assessment of spatial structure, such as that provided by well known spatial autocorrelation coefficients including I, but rather provides statistical information about local spatial structure, measured at different distance lags. In equation form, G(d) is given by:

$$G(d) = \frac{\sum\limits_{i=1}^{n}\sum\limits_{j=1}^{n} w_{ij}(d)x_i x_j}{\sum\limits_{i=1}^{n}\sum\limits_{j=1}^{n} x_i x_j}, j \text{ not equal to } i, i = 1, ..., n, \ j = 1, ..., n, \tag{1}$$

where the w_{ij} (d) are binary weights that indicate whether site j is within radius d of site i (w_{ij} (d) = 1) or not (w_{ij} (d) = 0).

Execution times for G(d) can become substantial because a large number of distances must be calculated to determine the weights in equation 1; in its original form, the problem is of order n^2. Parallel software to calculate this statistical measure was ported to three types of parallel computer systems to evaluate their relative effectiveness. These systems vary in their internal architectural arrangements as well as in the raw speed of their components and interconnections (see Chapter 2 for a discussion of parallel architectures).

Encore Multimax: The Multimax is a shared-memory multiprocessor system that uses a 100 megabits per second bus to provide the primary communication path among system modules. Each 32-bit processor can access a cache of 32 kilobytes of fast static RAM. The computations were performed using an Encore configured with 14 NS32332 processors and 32 megabytes of shared memory. Systems such as this are often referred to as symmetric multiprocessors (SMP). However, the scalability of these architectures can become problematic when more than 10 processors are used because there is contention for the shared memory resource.

KSR-1: The KSR-1 most closely resembles a scaleable parallel system. The one used at the Cornell (US) Theory Centre was built from 64 superscalar (64-bit) processors that have integer and floating point units with pipelined adder/multipliers. Physical storage is composed of a collection of 32 MB local caches, one for each processor (cell). Each cell also has an additional, faster 0.5 MB hardware subcache. However, in addition to the local cache, the system implements a virtual shared-memory architecture which allows the distributed local caches to behave logically as a single, shared address space. Note, however, that the local caches are implemented in a hierarchical structure. If a cell requests access to a memory location that is not in its subcache, then an attempt is made to fetch data from the local cache. If the required data are not found in the local cache, a request is placed into an open slot on the local communications ring and searching continues amongst the hierarchically-organised caches. Given that memory access times increase with non-local references, efficient programs must maximise access to the local caches. A data value can be accessed in the local cache of a processor in only 18 clock cycles; on the other hand, if a data value must be accessed from a memory location on a different local processor, then 175 clock cycles are needed. With this latency penalty, performance is linked to an effective exploitation of the cache hierarchy.

MasPar: MasPar machines (MP-1 and MP-2) are massively parallel SIMD computers that are configured with up to 16,384 (16 K) processors (Blank, 1990; Nickolls, 1990). Each system (MasPar, 1993) is constructed using a two-dimensional processing element array (PE array) and an array control unit (ACU). A 16 K processor system, therefore, would contain 16,384 PEs that are arranged in a 128 by 128 grid. Each PE has its own memory and shares data with other PEs through a global router. During computation each PE simultaneously executes the same instruction stream that is sent from the ACU, unless it has been instructed to remain idle. The MP-2 is a second generation system that is constructed from components that have higher performance characteristics than the MP-1. The two systems that were used to compute results reported here had the following PE configurations:

1.   MP-1 with 16,384 (16K) PEs,

2.   MP-2 with 4,096 (4K) PEs.

17.2.1 Encore Spatial Statistics Results

Table 17.1 summarises the execution times (in seconds) for selected runs of the G(d) statistical program (written in Encore Parallel FORTRAN) when it was used to analyse datasets containing 256 and 1,600 random points. The amount of memory available precluded the analysis of larger data sets. The number of processors used to implement the test is shown by the columns. The reported run times do not include the time required to perform input-output. The table shows a clear trend for the 1,600 point problem: as additional processors are added, total run times decrease from 11 minutes to just less than one minute. This decrease can be examined in light of the speed-ups that are achieved. A 'perfect' speed-up (efficiency = 1.0) is realised when the speed-up value is equal to the number of processors used. In this case speed-ups remain high across the range of available processors.

Though these reduced run times demonstrate the feasibility of applying parallelism to achieve improved performance for applications in spatial statistics, it is also interesting to examine the results to see whether the size of the problem plays a role in performance. Although one might expect solution times to decrease linearly as additional processors are added, this trend does not extend to small problems. When the 256 point dataset was analysed, the speed-up values levelled off as more processors were added (Table 17.1). For such small problems, the computational overhead required to establish the different parallel processes caused a decline in relative performance. Also note that, in all the parallel cases shown in Table 17.1, speed-ups for the 256 point problem are lower than those of the 1,600 point problem.

These results indicate two problems that are often encountered in parallel processing applications. First, from Table 17.1 it is clear that efficiencies decrease with the number of processors. This problem with scalability is the Achilles heel of the architecture: as the number of processors used grows, bus contention occurs because an increased number of processors require access to a common shared memory. Thus, while run times decrease, they do so at a lower rate as additional processors are added. The second problem is also illustrated well in Table 17.1: speed-ups are far lower for small problems because the overhead penalty required to establish parallel processes is large relative to the total amount of computation required to compute a solution to a problem.

Despite these general limitations, it is evident that the execution of spatial statistical algorithms on parallel machines can result in a substantial decrease in run times. However, the advantage of parallel processing is only realised when problems are large enough to warrant splitting them into separate tasks such that the time required to initiate the parallel tasks (overhead) is recovered through the efficiency of parallel computation.

Table17.1: Run times in seconds, speed-ups and efficiencies for 256 and 1,600 point data sets using from 1 to 14 Encore processors (Source: Armstrong et al., 1994a)

Dataset		Number of Processors							
		1	2	4	6	8	10	12	14
256	Run Time	16.7	9.1	5.3	4.0	3.3	3.1	3.0	2.8
Points	Speed-up	1	1.8	3.1	4.2	5.0	5.5	5.6	5.9
	Efficiency (%)	100	90	78	70	63	55	47	42
1,600	Run Time	691.5	354.5	183.7	125.0	95.3	78.8	67.6	58.6
Points	Speed-up	1	1.9	3.8	5.5	7.3	8.8	10.2	11.8
	Efficiency (%)	100	95	95	92	91	88	85	84

17.2.2 KSR Spatial Statistics Results

The KSR-1 is a more modern computer than the Encore. It has a larger addressable memory and is constructed from components that are far faster. Consequently, larger problems can be handled on this machine and problems of 1,000 (1 K) and 10,000 (10 K) points are discussed in this section. Though the increased size and speed of the KSR-1 is impressive, similar problems are encountered as on the Encore.

The effect of the overhead penalty of process establishment can be seen in the results shown in Table 17.2. In the row for the 1 K runs, for example, although runtimes tend to decrease as additional processors are used (7, 16 and 29 processors), efficiency declines steeply. When 59 processors are used (see the shaded cells in Table 17.2), the time required to establish the individual processes on this compliment of processors becomes large relative to the total amount of time required to compute the results and, consequently, runtimes actually increase. Note also that the efficiency plummets to 18 per cent. When the 10 K problem size is analysed, on the other hand, the results indicate that the additional processors, despite a small decline, are being used efficiently across the range of processor configurations.

Table 17.2: Run times in seconds, speed-ups and efficiencies for 1,000 and 10,000 point dataset sizes and different numbers of cells for the KSR-1 (Source: Armstrong et al., 1994b)

Dataset		Number of Processors				
		1	7	16	29	59
1,000	Run Time	32.7	6.1	3.9	2.7	3.0
Points	Speed-up	1	5.3	8.3	11.9	10.8
	Efficiency (%)	100	77	52	41	18
10,000	Run Time	5919.7	730.4	411.5	239.5	124.7
Points	Speed-up	1	8.1	14.4	24.7	47.5
	Efficiency (%)	100	116	90	85	80

In Table 17.2 there is also evidence of a superlinear speed-up (10 K problem with 7 processors). This is due simply to the cache structure of the architecture of the KSR-1. As the problem was decomposed from 1 to 7 processors, a large proportion of the processing could be done independently in a way that exploited the cache coherency of the architecture. Note, however, that cache coherency is affected by the underlying spatial distribution of data points and their allocation to different processors.

17.2.3 MasPar Spatial Statistics Results

The implementations of the G(d) code on the two SIMD computer systems illustrate the effect of different numbers of processors and the size of each PE's memory on execution times when datasets ranging in size from 256 to 1,024 points were analysed (Armstrong & Marciano, 1995). The analyses of the 1,024 point dataset (Table 17.3) executed in less than a second on both processor configurations. In addition, the execution times do not increase rapidly with the problem size. When the systems are compared, it is evident that the larger (16 K) MP-1 configuration yields the best results. This result demonstrates the importance of matching the spatial characteristics of the problem, in this case represented as a matrix, to the architecture of the system. Though the newer MP-2 is built from faster components, its performance is notably worse for the 1,024 observation data set. Because the MP-1 has a larger number of processors, arrays do not need to be disassembled into as many submatrices that are processed in sequential order. Thus, when an array is larger than the number of processors, it is mapped onto more virtual layers on the MP-2 than the MP-1 and the effect of this mapping on execution time is seen most clearly for the largest (1,024 point) problem.

Table 17.3: Execution times (in seconds) for two MasPar computer systems and different problem sizes

	Number of Points			
	256	**512**	**768**	**1,024**
4K PE MP-2	0.095	0.196	0.337	0.779
16K PE MP-1	0.071	0.136	0.190	0.298

17.3 Building Shortest Paths in Parallel

Many methods of spatial analysis use measures of proximity as fundamental variables. In network analysis, for example, proximity can be defined in terms of the presence or absence of a direct linkage between a pair of locations, or by using some metric of interaction across a network (such as distance, travel time, or cost). In the latter case, a shortest path algorithm identifies the route between two points on a network that minimises the value of the metric. In network analysis these algorithms are used to find shortest paths:

- between any given origin and destination - a *route*;

- from any given origin to all destinations - a *skim tree*; and

- from all origins to all destinations - an *origin-destination matrix*.

The amount of computation required by a shortest path algorithm increases with the size of the network, although the exact relationship depends on both its spatial structure and the product being generated (route, skim tree or origin-destination matrix). In the worst case, where the production of an origin-destination matrix requires shortest paths between all possible pairs of origins and destinations, computation increases with the square of the number of nodes on the network.

Dijkstra's (1959) shortest path algorithm produces routes, skim trees and origin-destination matrices. The iterative algorithm consists of two major steps. The first step selects an origin node and labels all other nodes with the current length of the shortest path from the origin (initially set to infinity) and the identifier of the preceding node in the path (initially a null value). The second step, a loop, uses a 'shortest-first' strategy that searches outwards from the origin along links in the network, updates some or all of the node labels, and identifies a node to 'close'. Closing a node signifies that the shortest path to it from the origin has been identified in that iteration. The algorithm continues to iterate through the loop until all nodes are closed (there are as many iterations as there are destination nodes), and the shortest paths from the origin to all destinations have been built. The result is a skim tree from which individual routes can be extracted. A skim tree also contains the information in a single row or column of an origin-destination matrix - to build a complete matrix, the algorithm is applied to each origin in turn.

Dijkstra's algorithm is often used to build the origin-destination matrices that underlie spatial interaction models (Batty, 1978), various linear programming models (Killen, 1983), and location-allocation models (Goodchild & Noronha, 1983; Densham & Armstrong, 1993). This widespread use has forced software designers to grapple with the computational loads that the algorithm places on serial computers with limited RAM and processing throughput. Two strategies have been applied: the first exploits the fact that the maximum distance over which interaction occurs may be much less than the extent of the network by constraining the length of paths from the origin to some maximum value; the second strategy partitions the network, processes each partition in turn, and then stitches together the results for the entire network. Goodchild & Noronha's (1983) PLACE Suite, for example, implements both strategies. The second strategy exploits a feature of the algorithm that makes it attractive for use in a parallel processing environment: the skim tree for each origin on the network can be generated independently of any other.

17.3.1 Decomposing the Algorithm

The number of parallel algorithms for shortest path problems described in the computer science literature (including Frederickson, 1991; Gallo and Pallottino, 1988; Hong *et al.*, 1991; Mateti and Deo, 1981; Price, 1982; Smith *et al.*, 1989) is indicative of the importance of this method of analysis. However, the difficulty of decomposing this problem has meant that very few of these algorithms are designed for distributed-memory MIMD computers.

The problem is difficult to decompose because its spatial domain can take on different forms (Ding *et al.*, 1992; Ding, 1993). Whilst a lattice network has a regular structure (geometry and topology), most real networks are irregular. When all the nodes in the network are origins and destinations, the network is homogeneous; however, most real networks have heterogeneous nodes because some are origins, others are destinations, and the rest indicate the presence and location of intersections and other features. Finally, if paths are built that traverse the entire network, processing is global in scope; however, if the lengths of the path are constrained, then processing may be of regional, neighbourhood or even local scope. Two methods can be used to decompose a spatial domain with these characteristics:

1. *Non-overlapping partitions.* The domain is divided into non-overlapping partitions that are processed to generate incomplete paths. These incomplete paths must be integrated with those from other partitions to yield full paths.

2. *Overlapping partitions.* The domain is divided into partitions that overlap by enough to permit a full set of shortest paths to be generated for each partition. Where processing is of limited scope, the amount of overlap can be quite small.

Two parallel algorithms are discussed in this chapter. The first parallel version of Dijkstra's algorithm uses non-overlapping partitions (Gilbert & Zmijewski (1987) describe a series of algorithms that use similar decompositions). First, the network is recursively divided into n parts to account for the distribution and heterogeneity of the nodes and to balance workloads. Each partition is assigned to a processor that builds shortest paths from every origin in the partition; if the path length is unconstrained, all of these paths will extend to the partition's boundaries. When all the partitions have been processed, the incomplete paths in each partition are assembled into full paths. The amount of work required to assemble full paths depends on the number of connecting links, m, between any pair of partitions: the smaller the value of m, the more quickly the full paths can be assembled. Although this approach yields good speed-ups when many processors are applied to a large network, it requires large amounts of memory to store the incomplete paths before they are assembled into full paths.

The second algorithm uses overlapping partitions of the domain. The degree of overlap is determined by the maximum allowable path length to ensure that full paths can be generated for every origin node in a partition. Because each processor creates a full set of shortest paths, they can be transferred directly to hard disk rather than stored in RAM.

This algorithm works as follows:

1. The network is partitioned using dynamic recursive alternating bisection (DRAB). DRAB partitions domains with unevenly-distributed elements to equalise processor workloads. The domain elements are divided into two equal parts using a straight line. The elements in each partition are then divided recursively by a straight line that is perpendicular to the one previously used. This process continues until the required number of well-balanced partitions are obtained (Belkhale & Banerjee, 1990; Hinz, 1990; Ding, 1993) that have equal numbers of origin nodes (see Figure 17.1).

2. Shortest paths are generated for every origin node in a partition.

3. Paths are sent to disk from each processor on a first-come first-served basis.

4. Repeat steps 2 and 3 until all processors have finished.

To ensure that a processor's memory capacity is not exceeded, only a subset of the nodes in each partition may be processed before their paths are written to disk. Consequently, the number of nodes processed in step 2 defines the grain size of processing. This algorithm is most suited to shortest path problems with a restricted path length: large path lengths render the partitioning inefficient because each subpart may include the entire network. However, increasing the path length increases the grain size of processing, which improves the efficiency of the algorithm provided that memory capacity is not exceeded.

17.3.2 Implementation and Results

Both parallel algorithms have been implemented in parallel Pascal (3L, 1989) using three of the processors in a Quadputer - a four-node Transputer array on a PC add-

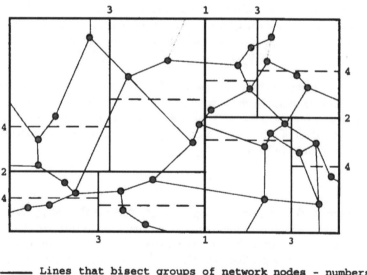

——— Lines that bisect groups of network nodes - numbers ii
      order in which the lines are applied.
——— Links in the network.

Figure 17.1: Dynamic recursive alternating bisection

in card. The master Transputer communicates with the PC host to obtain data from its keyboard and hard disk, partitions the domain, and assigns partitions to other processors. The other two Transputers generate shortest paths for their assigned origin nodes and send results back to the master processor, which writes full paths to the host PC's disk.

Table 17.4: Solution times (seconds) with different search ranges for shortest paths

Version:	Search Range (units)						
	20	25	30	35	40	45	50
Serial	66.2	96.3	134.3	178.8	227.1	284.0	344.4
Parallel							
1. Spatial Partitioning	48.3	68.4	91.7	117.4	146.1	177.1	210.6
Speed-up	1.4	1.4	1.5	1.5	1.6	1.6	1.6
2. DRAB-based decomposition	48.3	68.3	91.5	117.6	146.0	176.9	210.2
Speed-up	1.4	1.4	1.5	1.5	1.6	1.6	1.6

Several sets of results from these algorithms are presented. To examine the effects of varying the grain size by changing the maximum path length and to isolate other factors, including imbalanced workloads, a regular network is used. This network, a square lattice, contains 900 quad-valent nodes (each node is connected to four neighbours) with link lengths of five units. The results in Table 17.4 show that increasing the path length from 20 to 50 units, in increments of five units, yields improved speed-ups. An irregular network of a real area is used to examine the speed-up and efficiency of the parallel algorithms and the effects of grain size on solution times. The longer the path length, the greater the speed-up and the efficiency of the algorithm (Table 17.4). The super-linear speed-up in Table 17.5 arises because partitioning the network in parallel is much more efficient than the process used in the serial version of the algorithm.

In addition to generating paths that are analysed by other models, shortest path algorithms are used to generate displays of solutions to models. The solutions to location-allocation models, for example, consist of one or more facility locations and an associated set of demand locations that is allocated to each facility. Such solutions can be depicted by highlighting the route through the network that links each demand node to its allocated facility (Armstrong & Densham, 1995). Displays of this kind require considerable amounts of processing to generate because a shortest path algorithm is used to identify network links that connect demand and facility locations. Although these links are identified when shortest paths are built, a location-allocation model only requires the distance between each demand node and feasible facility location rather than the actual path. The paths normally are discarded because they require large amounts of storage and, consequently, they are not available for generating displays, requiring that the shortest path algorithm be run again. Parallel shortest path algorithms reduce processing time because: first, the links between each facility and its demand set can be generated independently of those for other facilities; and, second, the sets of origin (facility) and destination (demand node) linkages for which shortest paths must be built is known from the solution to the location-allocation model (Densham & Armstrong, 1993). Thus, parallel shortest path algorithms can be used to fill several different roles in support of visual interactive locational analysis (Densham, 1996).

Table 17.5: Solution times (seconds) on an irregular network

	Search Ranges					
	1 km	2 km	3 km	4 km	5 km	6 km
Single Transputer	38	76	146	248	379	527
Parallel Algorithm	32	53	89	135	190	251
Speed-up	1.2	1.4	1.6	1.8	2.0	2.1
Efficiency (%)	59.5	71.5	82.0	92.0	99.2	105.0

17.4 Terrain Analysis

17.4.1 Parallel Interpolation

Interpolation can be a computationally intensive activity when large problems are encountered. Several approaches to performing interpolation have been developed (see, for example, Lam, 1983). Because the computational complexity of inverse-distance weighted interpolation has long been recognised (MacDougall, 1984), several approaches to computing results have been developed. In this section of the chapter, a worst-case approach is used first to demonstrate how computation times can be reduced through the use of parallel processing. An additional approach, one that reduces the required number of distance calculations, demonstrates the role that algorithm choice can play on performance. A serial algorithm that performs two-dimensional inverse-distance weighted interpolation using each approach was decomposed into a form suitable for parallel processing (Armstrong & Marciano, 1994, 1996). In these experiments the problem size, as well as the number of processors used in parallel, was systematically increased.

Cell values in inverse-distance weighted interpolation are computed in the following way:

$$Z_j = \frac{\sum_{i=1}^{N} W_{ij} Z_i}{\sum_{i=1}^{N} W_{ij}} \qquad (2)$$

where:

Z_j is the estimated value at grid location j,

Z_i is the known value at control point location i, and

W_{ij} is the weight that controls the effect of control points on the

calculation of Z_j.

The W_{ij} value is often specified as d_{ij}^{-a}, where d_{ij} represents the distance between locations i and j, and a is a parameter that normally assumes a value of 1.0 or 2.0 (Hodgson, 1992). Z_j is often estimated for a local neighbourhood and a limit is often set on the number of nearest neighbour points (k) used to compute its values (Hodgson, 1992). Different approaches to reducing search for the k-neighbours form the basis for interpolation algorithms that are less computationally intensive than the brute force approach. Hodgson (1989) and Clarke (1990: 207) describe approaches that restrict search to a window that is placed over each grid cell; only those control points that are in the window are used to compute interpolated cell values. In a similar vein, Hodgson's (1989) 'learned search' method exploits the spatial structure that is present in the distribution of control points for most problems to reduce the total number of distance calculations required.

In the remainder of this section we report on a comparative evaluation of two sets of interpolation analyses. The first set of analyses used a brute force approach to interpolation. The second set of experiments was performed using the local interpolation algorithm described by Clarke (1990).

17.4.1.1 Brute Force Interpolation Results

A FORTRAN implementation of a two-dimensional, 'brute force' interpolation algorithm (MacDougall, 1984) was translated into versions that ran in the Encore and KSR-1 parallel computing environments (Armstrong & Marciano, 1994). These 'brute force' results represent worst case scenarios against which the local approach to parallel algorithm enhancement (described in the next section) can be compared. In each case a 240 by 800 grid was interpolated from a variable number of control points.

From Table 17.6 it is evident that when one Encore processor is used, more than one day of run time is required to compute the 240 by 800 grid using the 'brute force' approach. However, as additional processors are added, the total run time decreases markedly to 2.5 hours when 14 processors are used. Moreover, the processors are being used efficiently. This algorithm is 'embarrassingly parallel'; calculations for each interpolated cell can be isolated cleanly for independent processing. However, in spite of this high level of efficiency, the 2.5 hours required when 14 processors were used still represents a substantial computational burden. To assess the effect of a faster set of processors, the 'brute force' algorithm was ported to the KSR-1 and the same problem was analysed.

Table 17.6: Computation times (hours) and efficiencies when different numbers of Encore processors are used to compute the 240 by 800 grid from 10,000 control points and k = 3 using a 'brute force' interpolation algorithm (Source: Armstrong & Marciano, 1994)

	Number of Processors							
	1	2	4	6	8	10	12	14
Run Time	33.3	17.0	8.5	5.7	4.2	3.4	2.8	2.5
Efficiency	1.0	0.99	0.99	0.97	0.99	0.98	0.99	0.95

In Table 17.7 it can be seen that using a larger complement of faster processors causes run times to decrease dramatically. In addition, the processors continue to be used efficiently. However, in spite of these improvements, the two and one-half minutes required when 58 processors were applied remains an impediment to interactive analysis of geographical data sets. Consequently, a local approach to interpolation was tested to determine whether results could be obtained in less than five seconds.

Table 17.7: Results from a 'brute force' algorithm used to interpolate a 240 by 800 grid with 10,000 control points and k = 3 on a KSR-1 system (Source: Armstrong & Marciano, 1994)

	Processors			
	1	**7**	**15**	**58**
Seconds	7719	1185	591	151
H:M:S	2:08:39	0:19:45	0:09:51	0:02:31
Speed-up	1.0	6.5	13.1	51.1
Efficiency	100	93	87	88

17.4.1.2 Interpolation using Local Search

Clarke (1990: 205-214) describes a procedure in which, for each interpolated cell, the search for control points is (normally) restricted to a few square contiguous subsets of a study area. This approach was chosen to illustrate the effect that local search has on results because its implementation is well documented.

Table 17.8 illustrates the substantial reduction in run times when Clarke's algorithm is used with one processor, with times decreasing from 7719 seconds to 15.6 seconds. Indeed, the Clarke algorithm with one processor outperformed the brute force algorithm when 58 processors were used. However, with one processor interactive analyses could be tedious for large problems and when large numbers of nearest neighbours are used to interpolate values. In the worst case for the one processor runs (1,000 control points and 8 nearest neighbours), more than 8 minutes was required to perform the required calculations. It should be noted that the problem used in these experiments is not excessively large and it is likely that researchers will encounter much larger problems on a regular basis.

402 P.J. Densham and M.P. Armstrong

Table 17.8: KSR-1 runs of the Clarke algorithm on a 240 by 800 grid using one processor. The number of control points increases from 1,000 to 40,000 and the number of k-nearest neighbours increases from 1 to 8. Times are in seconds (Source: Armstrong & Marciano, 1996).

Control Points	Number of k-Nearest Neighbours				
	1	2	4	6	8
1,000	115.6	168.5	277.1	384.2	490.6
10,000	15.6	21.8	33.8	45.5	57.1
40,000	5.8	7.8	11.3	14.8	18.1

The results for the parallel implementation of the Clarke algorithm, in this case focusing on the 10,000 control point problem, are shown in Table 17.9. Clearly, run times have been reduced substantially, with no case requiring more than 1 minute of execution time. However, a more important point becomes apparent when the bottom row of Table 17.9 is examined: all runs were completed in less than two seconds when 50 processors were used.

Table 17.9: Clarke algorithm using a 240 by 800 grid with 10,000 points. Comparison is made between the number of k-nearest neighbours and the number of processors. Times are in seconds (Source: Armstrong & Marciano, 1996).

Processors	Number of k-Nearest Neighbours				
	1	2	4	6	8
1	15.6	21.8	33.8	45.5	57.1
7	3.6	5.1	7.9	10.7	13.5
15	1.9	2.6	4.1	5.5	7.0
50	0.5	0.7	1.1	1.5	1.9

While the results obtained from the local search algorithm are clearly superior to the brute force approach, it is interesting to note that the more sophisticated

algorithm introduces dependencies into the structure of the interpolation process
and consequently, efficiencies are reduced. The serial and parallel results shown in
Table 17.9 demonstrate that the interpolation process is affected by the number of
neighbours that are used to interpolate the grid. Another factor that can affect
performance is the number of control points used during interpolation. Tables
17.10 and 17.11 contain results obtained when the starting configuration of control
points for the 240 by 800 grid is increased from 1,000 to 40,000. Again, the effects
of parallelism are clearly evident in each column of the three tables. The runs with
1,000 control points demonstrate that the computing load increases as a function of
the number of k-nearest neighbours used. However, run times decrease
substantially as additional processors are used to compute results. Similar trends
occur for the 40,000 control point runs. Overall, the fastest run time of 0.20
seconds was obtained for the 40,000 control point problem when one nearest
neighbour was used with 50 processors.

Table 17.10: KSR-1 analysis of 1,000 control points for a 240 by 800 grid using
different number of k-nearest neighbours and processors (time in seconds)
(Source: Armstrong & Marciano, 1996)

Processors	Number of k-Nearest Neighbours				
	1	2	4	6	8
1	115.6	168.5	277.1	384.2	490.6
7	29.2	43.5	70.7	98.8	128.1
15	15.7	23.3	39.1	53.9	69.0
50	4.7	6.5	10.7	14.4	18.3

17.4.2 Hill Shading

The purpose of analytical hill-shading is to model how light is reflected from
surfaces. To determine where shadows will occur, and the form that they will take,
a hill-shading algorithm must account for a terrain's relief, the location of the light
source, and atmospheric scattering effects. Brightly lit slopes typically face a light
source and reflect light strongly. However, slopes that are in shadow may face
away from the light source and be self-shadowing, or they may face the light source
but be in shadows cast by other terrain features. In both cases, any reflection is due
to atmospheric scattering effects. In this section we focus on a hill-shading
algorithm (Ding, 1992) that expresses the reflectance at a location on a digital
elevation model (DEM), $R(x, y)$, as a function:

$$R(x, y) = f\ (az, el, \gamma, x, y) \tag{3}$$

where

> az is the azimuth of the light source;
> el is the elevation angle of the light source;
> γ is the ratio of sky scattering over direct radiance; and
> x, y are the coordinates of the target pixel.

The hill-shading algorithm integrates three models: a component that detects self-shading using a bi-directional reflectance distribution function (BRDF); a component that detects cast shadows using a coordinate transformation approach; and a component that simulates atmospheric scattering effects using a ball model.

Table 17.11: KSR-1 analysis of 40,000 control points for a 240 by 800 grid using different number of k-nearest neighbours and processors (time in seconds) (Source: Armstrong & Marciano, 1996)

	Number of *k*-Nearest Neighbours				
Processors	**1**	**2**	**4**	**6**	**8**
1	5.8	7.8	11.3	14.8	18.1
7	1.4	1.9	2.8	3.6	4.4
15	0.7	0.9	1.6	1.8	2.3
50	0.2	0.3	0.4	0.5	0.6

The self-shading component of the algorithm uses a BRDF to determine how brightly a surface reflects light in a given direction (see Horn, 1981, 1982). A DEM is treated as a series of surface elements and, given the locations of the viewer and the light source, the reflectance of each element can be determined from the relationship between its surface normal and the angle of incidence of light rays. The surface normal of a pixel is calculated from the slopes between it and each of its eight neighbouring pixels.

Opaque objects that occur between a light source and a target surface cast shadows on that surface. To detect cast shadows, therefore, every pixel in the DEM must be checked to establish whether or not it is visible from the light source. This is done by applying two coordinate rotations to the DEM (Ding, 1992). The first rotates the DEM around its Z axis in a clockwise direction so that the Y axis parallels the azimuth of the light source. The second transformation makes the Y axis parallel to rays from the light source by rotating the DEM around its X axis. The result is that the DEM is oriented so that the Z values can be used to determine where cast shadows will occur.

The atmospheric scattering model is based on a ball radiance model (see Ding, 1992) that determines the nature and extent of scattering effects on different types of terrain. For any pixel, scattering effects are a function of its exposure to the sky; the solid angle of exposure is estimated from the slopes between a pixel and each of its eight neighbours. The output of the atmospheric scattering model is an index for each pixel that ranges between 0 and 1: plains generate a value of 0.5 whilst peaks have greater values that depend on their size and shape.

To decompose the three components of the integrated hill-shading algorithm into parallel processes, their computational characteristics must be determined (Ding & Densham, 1994a). The atmospheric scattering and self-shading components are both implemented as neighbourhood operations on a DEM. With the exception of the pixels or cells along the edges of the DEM, each element is processed in exactly the same way. In this context the DEM is a regular and homogeneous domain that can be divided into equally-sized partitions, each of which can be assigned to a separate processor. Unfortunately, the cast-shading component is an irregular and homogeneous problem. A global operation that encompasses the entire DEM is required to detect cast shadows. Although the two coordinate rotations applied to the DEM require the same number of calculations to be carried out for each cell, and this is easily done in parallel, the detection of cast shadows requires that the rotated DEM is partitioned along the azimuth of the light source - both to minimise communication and to obviate data dependencies. If a simple, regular partitioning of the DEM is used, a feature in one partition may cast a shadow over features in other partitions. Partitioning the DEM along the azimuth yields regions of equal size but irregular shape for most positions of the light source (Figure 17.2).

The three models are independent and can be solved sequentially, in any order, or simultaneously; their results are then combined to produce a shaded image. The atmospheric scattering and self-shading components are decomposed using geometric parallelism: the raster is partitioned into equally-sized subregions that yield balanced processor workloads. To prevent the need for processors to exchange information about elements along the edges of their respective partitions, each partition overlaps its neighbours by half the size of the processing window (one cell). This marginal increase in data overhead obviates the need for communication and improves processing performance.

The atmospheric scattering and self-shading components can use the same, regular partitions. However, if these partitions are also used for the cast-shading component, extra processing and communication are required to accommodate features that cast shadows over other partitions. If an irregular partitioning is used,

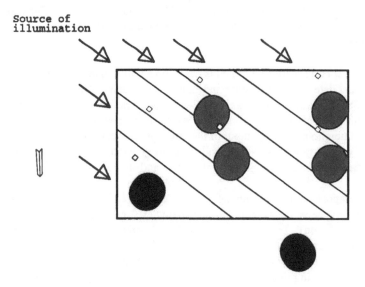

Figure 17.2:Equal area partitioning of a DEM along the azimuth of a source of illumination

on the other hand, the results are more difficult to integrate with those from the other components but less computation is required to detect cast shadows than to detect self shadows and estimate atmospheric scattering (Ding & Densham, 1994a).

An irregular partitioning is used for the cast-shading component because cast-shadows can be detected by a small subset of the processors in a MIMD computer while the others work on the atmospheric scattering and self-shading components in a data parallel fashion. When processing for all three components is complete, the cast-shadows are distributed to be integrated with the results from the other components. The algorithm has been implemented on a four-node Transputer array. One of the Transputers is designated the master processor and controls data partitioning and transfers, coordinates processes, and oversees the integration of results from the three components. This role means that the master processor is idle when the worker Transputers are detecting self-shadows and estimating atmospheric scattering effects. Thus, the master detects cast shadows while the workers are busy with their self-shading and atmospheric scattering tasks; when the workers have completed their tasks, the master distributes the cast shadows to them and receives back partitions of the final, shaded image. This approach implements the algorithm as a set of loosely-synchronous parallel processes (Ding & Densham, 1996).

Results are presented for three DEMs in Table 17.12. The first DEM, of Keating Summit, PA, consists of 128 by 128 pixels, each with a resolution of 30m by 30m. A further pair of DEMs consist of 256 by 256 pixels each, with resolutions of 30m by 30m, and represent part of the Hudson River around West Point, NY, and the western part of Crater Lake, OR. To assess the performance of the parallel algorithm, a serial version was implemented that runs on a single Transputer. This serial version took 14.1 seconds to shade the Keating Summit DEM. When the parallel version of the algorithm was run on the same DEM using three and four processors, it required 7.7 seconds and 6.1 seconds respectively to produce the same shaded image. The efficiency of the algorithm is greater with only three processors, at 62%, than with four processors, at 58%, because the grain size is larger when fewer processors are used. For a given size of DEM, solution times decrease when more processors are used but the efficiency of the algorithm also decreases. The results for the West Point and Crater Lake DEMs show similar effects: the efficiency of the processing is higher, even with four processors, because these DEMs support larger grain sizes.

Table 17.12: Solution times for three DEMs with various numbers of processors

Data Set	No. of Processors	Solution Time (sec)	Efficiency
	1	14.1	Not applicable
Keating Summit	3	7.7	62%
	4	6.1	58%
West Point	1	99.7	Not applicable
	4	31.1	80%
Crater Lake	1	98.7	Not applicable
	4	29.6	83%

17.5 Conclusions

Any parallel implementation of an algorithm is affected by the strategy used to divide both the program and data into a collection of concurrently executing processes (see, for example, Armstrong & Densham, 1992; Lilja, 1994). Among the factors that must be considered are the architecture of the computer system, the number of available processors, the amount of data, and the size of the code block that is allocated to each processor. In addition, the geometrical form of the allocation process can play a role in determining the level of performance. When problems that exhibit asymmetrical or clustered geometries are encountered, the balance of computation across the available set of processors can be disrupted. In such cases, one processor might continue to run while all others would wait, idly, for it to complete its task, thus reducing the efficiency of the parallel program.

If, after careful consideration of these factors, algorithms are designed to exploit the characteristics of the target architecture, substantial reductions in total processing time can be achieved - as the results reported in this chapter show. In some cases we also observe speed-ups that are close to linear when additional processors are used. However, this is not a general trend, as some problems contain elements that exhibit spatial or temporal dependencies which cause performance to fall off as additional processors are used.

In general, the approaches described in this paper can be used to improve the performance of other types of computationally intensive spatial analysis (Openshaw, 1994, 1995) from both human and physical geography. Several researchers have investigated how to decompose such methods into parallel processes. Mower (1994), for example, implemented three DEM-based (a regular and homogeneous domain) drainage basin analysis algorithms in which each DEM cell was assigned to a different virtual processor. Other researchers have found that visibility analysis is a computationally intensive process that responds well to parallel processing (DeFloriani *et al.*, 1994; Mills *et al.*, 1992). Delaunay triangulation is a method that is used in both human and physical geography; researchers have used both vector (Ware & Kidner, 1991; Puppo *et al.*, 1994) and raster data models (Ding & Densham, 1994b) to support their parallel algorithms (see also Chapter 16).

Spatial analysis increasingly is being undertaken to support decision-making by groups. The computational loads associated with providing group members with both individual, private workspaces and a shared collaborative workspace eclipse those of single-user systems. Each group member works to explore a problem, understand its structure, and develop scenarios on their own workstation. The methods of spatial analysis described in this chapter can be run in serial on these workstations, or they can be decomposed and run as parallel processes across a network of workstations (NOW, Anderson *et al.*, 1995). Using NOW to build MIMD computers is well supported with parallel development environments, including Linda (Carriero & Gelernter, 1990) and MPI (Dongarra *et al.*, 1996). Thus, individual group members can access the workstations of others in the group to reduce the processing time of their analyses. Moreover, when group members work together in a shared workspace, their workstations can be used in concert to provide parallel processing support for the analysis and display of problems.

17.6 References

3L Ltd, 1989, *Parallel Pascal User Guide.* 3L Ltd, Livingston, UK.

Anderson, T.E., Culler, D.E. & Patterson, D.A., 1995, A case for NOW (Networks of workstations). *IEEE Micro,* February, 54-64.

Armstrong, M.P. & Densham, P.J., 1992, Domain decomposition for parallel processing of spatial problems. *Computers, Environment, and Urban Systems,* **16** (6), 497-513.

Armstrong, M.P. & Densham, P.J., 1995, Cartographic support for collaborative spatial decision-making. *Proceedings, Twelfth International Symposium on Computer-Assisted Cartography* (Auto Carto 12), 49-58.

Armstrong, M.P. & Marciano, R., 1994, Inverse distance weighted spatial interpolation using parallel supercomputers. *Photogrammetric Engineering and Remote Sensing,* **60** (9), 1097-1103.

Armstrong, M.P. & Marciano, R., 1995, Massively parallel processing of spatial statistics. *International Journal of Geographical Information Systems,* **9** (2), 169-189.

Armstrong, M.P. & Marciano, R., 1996, Local interpolation using a distributed parallel supercomputer. In press, *International Journal of Geographical Information Systems,* **10** (5).

Armstrong, M.P., Pavlik, C.E. & Marciano, R., 1994a, Parallel processing of spatial statistics. *Computers & Geosciences,* **20** (2), 91-104.

Armstrong, M.P., Pavlik, C.E. & Marciano, R., 1994b, Experiments in the measurement of spatial association using a parallel supercomputer. *Geographical Systems,* **1** (4), 267-288.

Batty, M., 1978, *Urban Modelling: Algorithms, Calibrations, Predictions.* Cambridge University Press, Cambridge.

Belkhale, K.P. & Banerjee, P., 1990, Recursive partitions on multiprocessors. *Proceedings of the Distributed Memory Computing Conference,* I.E.E.E. Computer Society Press, New York, 930-938

Blank, T., 1990, The MasPar MP-1 architecture. *Proceedings of the Thirty-Fifth IEEE Computer Society International Conference* - Spring COMPCON 90, San Francisco, CA, 20-24.

Carriero, N. & Gelernter, D., 1990. *How to Write Parallel Programs: A First Course.* The MIT Press, Cambridge, MA.

Clarke, K.C., 1990, *Analytical and Computer Cartography.* Prentice Hall, Englewood Cliffs, NJ.

DeFloriani, L., Montani, C. & Scopigno, R., 1994, Parallelizing visibility computations on triangulated terrains. *International Journal of Geographical Information Systems,* 8 (6), 515-531.

Densham, P.J., 1996, Visual interactive locational analysis, in P. Longley and M. Batty, eds., *Spatial Analysis: Modelling in a GIS Environment.* Cambridge: GeoInformation International, in press.

Densham, P.J. & Armstrong, M.P., 1993, Supporting visual interactive locational analysis using multiple abstracted topological structures. *Proceedings, Eleventh International Symposium on Computer-Assisted Cartography* (Auto Carto 11), 12-22.

Dijkstra, E., 1959, A note on two problems in connection with graphs. *Numerishe Mathematik,* 1, 101-118.

Ding, Y., 1992, An improved method for shadow modeling based on digital elevation models. *Proceedings, GIS/LIS '92,* 1, 178-187.

Ding, Y., 1993, *Strategies for Parallel Spatial Modelling Using MIMD Approaches.* Unpublished PhD Thesis, Department of Geography, State University of New York at Buffalo.

Ding, Y. & Densham, P.J., 1994a, A loosely synchronous, parallel algorithm for hill shading of digital elevation models. *Cartography and Geographic Information Systems,* 21 (1), 5-14.

Ding, Y. & Densham, P.J., 1994b, A dynamic and recursive parallel algorithm for constructing Delaunay triangulations. In Waugh, T.C. & Healey, R.G. (Eds), *Advances in GIS Research,* Proceedings of the Sixth International Symposium on Spatial Data Handling, Volume 2, 682-695.

Ding, Y. & Densham, P.J., 1996, Spatial strategies for parallel spatial modelling. In press, *International Journal of Geographical Information Systems.*

Ding, Y., Densham, P.J. & Armstrong, M.P., 1992, Parallel processing for network analysis: decomposing shortest path algorithms on MIMD computers. In

Proceedings, Fifth International Symposium on Spatial Data Handling, International Geographical Union, Charleston, SC, 682-691.

Dongarra, J.J., Otto, S.W., Snir, M. & Walker, D., 1996, A message passing standard for MPP and workstations. *Communications of the ACM*, **39** (7), 84-90.

Duguay, C.R., 1985, *Topographic Shadows on Landsat Images: a Review and Alternate Approach Using Digital Elevation Models.* Unpublished Master's thesis, Department of Geography, State University of New York at Buffalo.

Frederickson, G.N., 1991, Planar graph decomposition and all pairs shortest paths. *Journal of the Association for Computing Machinery*, **38** (1), 162-204.

Gallo, G. & Pallottino, S., 1988, Shortest path algorithms. *Annals of Operations Research*, **13**, 3-79.

Getis, A. & Ord, J.K., 1992, The analysis of spatial association by use of distance statistics. *Geographical Analysis*, **24**, 189-206.

Gilbert, J.R. & Zmijewski, E., 1987, A parallel graph partitioning algorithm for a message-passing multiprocessor. *International Journal of Parallel Programming*, **16**, 427-449.

Goodchild, M.F. & Noronha, V., 1983, *Location-Allocation for Small Computers.* Monograph No. 8, Department of Geography, The University of Iowa, Iowa City, IA.

Hinz, D.Y., 1990, A real-time load balancing strategy for highly parallel systems, *Proceedings of the Distributed Memory Computing Conference*, I.E.E.E. Computer Society Press, New York, 951-961

Hodgson, M.E., 1989, Searching methods for rapid grid interpolation. *The Professional Geographer*, **41** (1), 51-61.

Hodgson, M.E., 1992, Sensitivity of spatial interpolation models to parameter variation. *Technical Papers of the ACSM Annual Convention*, Vol. 2, American Congress on Surveying and Mapping, Bethesda, MD, 113-122.

Hong, C., Narasimhan, L.S. & Ratliff, H. D., 1991, *Finding the Shortest Path in Huge Planar Networks.* PDRC Report Series 91-04, Georgia Institute of Technology, Atlanta, GA.

Horn, B.K.P., 1981, Hill shading and the reflectance map. *Proceedings of the Institute of Electrical and Electronics Engineers*, **69** (1), 14-47.

Horn, B.K.P., 1982, Hill shading and the reflectance map. *Geo-Processing*, **2**, 65-146.

Killen, J.E., 1983, *Mathematical Programming Methods for Geographers and Planners*. St Martin's Press, NY.

Lam, N.-S., 1983, Spatial interpolation methods: a review. *The American Cartographer*, **2**, 129-149.

Lilja, D.J., 1994, Exploiting the parallelism available in loops. *IEEE Computer*, **27** (2), 13-27.

MacDougall, E.B., 1984, Surface mapping with weighted averages in a microcomputer. *Spatial Algorithms for Processing Land Data with a Microcomputer*. Lincoln Institute Monograph #84-2, Lincoln Institute of Land Policy, Cambridge, MA.

MasPar, 1993, *MasPar Parallel Application Language (MPL) User Guide*, Software Version 3.2, Document Part Number 9302-0101, Revision A5.

Mateti, P. & Deo, N., 1981, *Parallel Algorithms for the Single Source Shortest Path Problem*. Technical Report CS-81-078, Computer Science Department, Washington State University, Pullman, WA.

Mills, K., Fox, G. & Heimbach, R., 1992, Implementing an intervisibility analysis model on a parallel computing system. *Computers & Geosciences*, **18**, 1047-1054.

Mower, J.E., 1994, Data-parallel procedures for drainage basin analysis. *Computers & Geosciences*, **20** (11), 1365-1378.

Nickolls, F.R., 1990, The design of the MasPar MP-1: a cost effective massively parallel computer. *Proceedings of the Thirty-Fifth IEEE Computer Society International Conference* - Spring COMPCON 90, San Francisco, CA, 25-28.

Openshaw, S., 1994, Computational human geography: towards a research agenda. *Environment and Planning A*, **26**, 499-505.

Openshaw, S., 1995, Human systems modelling as a new grand challenge area in science. *Environment and Planning A*, **27**, 159-164

Price, C.C., 1982, A VLSI algorithm for shortest paths through a directed acyclic graph. *Congress Numerantium*, **34**, 363-371.

Puppo, E., Davis, L., DeMenthon, D., & Teng, Y.A., 1994, Parallel terrain triangulation. *International Journal of Geographical Information Systems*, 8 (2), 105-128.

Rokos, D. & Armstrong, M.P., 1996, Using Linda to compute spatial autocorrelation in parallel. *Computers & Geosciences*, 22 (5), 425-432.

Smith, T.R., Peng, G. & Gahinet, P., 1989, Asynchronous, iterative, and parallel procedures for solving the weighted-region least cost path problem. *Geographical Analysis*, 21, 147-166.

Ware, A. & Kidner, D., 1991, Parallel implementation of the Delaunay triangulation within a Transputer environment. *Proceedings, EGIS'91*, 1199-1209.

18

Parallel Processing Approaches in Remotely Sensed Image Analysis

G.G. Wilkinson

18.1 GIS and Remote Sensing

Amongst the most compelling reasons for wishing to exploit parallel computation in the GIS context are the enormous quantity of data present in typical datasets and the complexity of the spatial operations to be applied to them. These characteristics also apply to the remote sensing context, and furthermore the disciplines of remote sensing and GIS move ever closer:

- remote sensing is used to gather information about the Earth which is normally geo-referenced and which frequently comes to be converted into digital cartographic products - hence remote sensing can be a provider of datasets for GIS

- contextual data from GIS are often used as a means of aiding either automatic or manual interpretation of remotely sensed data (the former through knowledge based system approaches)

- the integrated use of digital information from a GIS with remote sensing data offers enhanced capability for modelling, monitoring and understanding many environmental processes.

Many commercial GIS provide facilities for exchange of data with remote sensing applications software and vice versa.

18.2 Parallel Processing in Remote Sensing - General Issues

18.2.1 Dataset Volumes and Complexity of Analysis

The volume of data captured by Earth observation satellites grows year by year as more satellites are deployed in orbit by various space organisations around the world and as the sensor systems which these satellites carry become more sophisticated. In the late 1990s it is expected that the volume of remote sensing data received and archived per day around the globe will be of the order of several terabytes - particularly once the new generation of platforms in NASA's Mission to Planet Earth come on-stream (Schowengerdt, 1993). Whilst it is unlikely that anyone should choose to analyse a full day's harvest of data from every space-borne sensor system, many applications require the analysis of datasets with impressive volumes - typically several gigabytes of data. This comes about through the growing need to use synergistic multi-sensor data to get optimum information, and also the need to use time series of datasets captured over very extensive geographical areas - particularly now that monitoring 'global change' has become an important priority activity. In addition the demand for greater precision in environmental information requires more advanced processing and less tolerance for error. This increases the complexity of many processing chains used in remote sensing. For example there has been a recent growth in the use of so-called 'artificial intelligence' approaches (Kontoes *et al.*, 1993; Peddle, 1995) which are computationally extremely demanding and this trend can be expected to continue.

18.2.2 The General State of Parallel Processing in Remote Sensing

Although it is relatively clear that processing bottlenecks are already limiting the value of what can be extracted from remotely sensed data, few parallel implementations exist for the normal set of algorithms used in remote sensing processing chains, and few investigators make use of parallel hardware systems. This situation has to change given the problems outlined above. Many of the techniques which will be described in the rest of this chapter for processing image datasets in parallel have emerged from research laboratories dedicated to machine vision and not to remote sensing. However, it is likely that the inter-disciplinary transfer of knowledge and experience will grow in the years to come as the problems of analysing remote sensing data become more acute.

18.3 Architecture and Data Mapping Issues

18.3.1 General Parallel Architecture Concepts for Image Analysis

Chapter 2 has considered in some detail the development of hardware for parallel processing and the distinction between SIMD and MIMD architectures, and between shared and distributed memory. Although all types of parallel architecture are applicable in various ways to the analysis of remotely sensed imagery, the fine-grained SIMD approach is most common in which machines with a large number of processors and relatively small independent memory are used (see for example

Hockney & Jesshope, 1981). This is a direct consequence of the fact that remotely sensed images typically consist of millions of pixels which need to be treated in an identical or similar way. Whilst this is the most common approach, it is not exclusive and many examples can be found of experimental systems for remote sensing which use MIMD machines. Also there is growing interest in the use of extremely coarse-grained networks of cheap workstations to improve throughput in handling remotely sensed data simply because most laboratories engaged in remote sensing already possess such networks and do not need to invest in special purpose hardware to begin parallel processing. However, in this latter case, advantage is usually taken of possibilities either to farm out a few relatively independent processing tasks on separate images or spectral bands or to form a regular decomposition of the image data and distribute blocks across the network and work on them separately at each location.

As for any parallel computing problem, the critical issues in achieving effective speed-up compared with a traditional Von Neumann serial computing approach are:

- the way in which the remotely sensed image data are mapped on to the parallel processor array, and the communication overhead associated with this

- the design of the algorithms to make best use of parallelism wherever possible (either data parallelism or functional parallelism)

- the type of processor interconnection network available and the inter-processor communication rates.

These issues are explored in Sections 18.3.2-18.3.4.

18.3.2 The Structure of Remotely Sensed Images

Figure 18.1 illustrates the structure of a typical remotely sensed image consisting of a two-dimensional raster array of pixels covering an unspecified geographical area in the x and y dimensions. Such images are usually formed by push-broom scanners. Barrett & Curtis (1992) and Curran (1985), amongst others, provide good introductions to image capture in remote sensing.

Most images captured by satellite sensors are multi-spectral giving an extra z dimension to the image dataset. The x and y dimensions of images used in environmental analysis can vary considerably. A full Earth view scene from the Landsat Thematic Mapper sensor, for example, consists of 6167 by 5667 pixels covering 185 by 170 km at 30m resolution per pixel. In some cases users will choose only to take a small subset of this area, and in others they will choose to produce a mosaic of several scenes depending on the geographical area of interest.

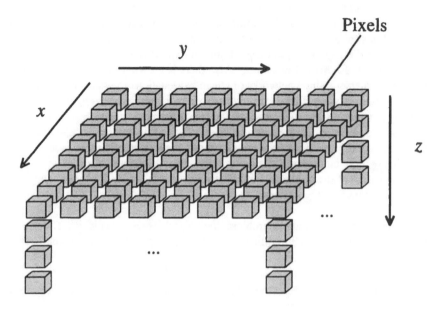

Figure 18.1: Structure of a typical remotely sensed image dataset. The x and y dimensions represent geographical space. The z dimension comes from separate spectral measurements, from different sensors, or from time series at the same location

However, typically, the x and y dimensionality is $\sim 10^2$ - 10^4. The z dimensionality is usually much less, especially if only one scene is considered per geographical area: the Landsat Thematic Mapper, for example, provides 6 separate spectral measurements per pixel at 30m resolution. Traditionally, therefore, remotely sensed images have often been regarded as primarily two-dimensional data objects, or more accurately thin layered wafer-like structures where the inherent data parallelism lies across the regular array of pixels in the x and y dimensions.

However, the current trend is for the z dimension to increase. It is now common for applications to require the analysis of multispectral time series of images for the same geographical location, especially when dynamic phenomena or environmental changes are being investigated. Also, improvements can be made in environmental mapping by using combined images from different satellites and from different parts of the electromagnetic spectrum, such as the Landsat Thematic Mapper and the ERS-1 Synthetic Aperture Radar - once they have been geo-referenced and resampled to the same spatial resolutions (e.g. Wilkinson *et al.*, 1993). Moreover, the number of spectral channels acquired by Earth observation satellite sensors is set to increase markedly - principally through the development of imaging spectrometer systems such as MODIS and MERIS to be carried aboard the EOS and ENVISAT satellite platforms (Barnes, 1993). These sensors will be capable of capturing 36 spectral measurements per pixel. With such images the z dimensionality could easily increase to $\sim 10^2$ if several of them are combined as a time series. This trend is leading to a thickening of the 'wafer model' and is increasing the degree of inherent parallelism in the z dimension.

18.3.3 Fine-grained SIMD Parallel Architectures for Image Processing

The high dimensionality of typical remote sensing image datasets has resulted mainly in the use of fine-grained SIMD architectures for parallel processing. Such architectures may differ in the way in which neighbouring processing elements (PEs) are interconnected. Connectivity architectures are important in image analysis because several kinds of image processing operations require access to the neighbouring pixels in the x and y dimensions - such operations include convolutions, various forms of filters, derivation of image texture measures, image segmentation, region labelling, etc. Ballard & Brown (1982) provide a comprehensive introduction to image processing operations.

18.3.3.1 Mesh Architectures

The simplest form of fine-grained processor interconnection is either the linear or the two-dimensional mesh array (Figure 18.2). A number of SIMD machines have been developed of this type during the last 15 years, such as the Linear Array Processor (LAP), the AMT Distributed Array Processor (DAP, see Gostick, 1979), the CLIP series developed at University College London (Duff & Fountain, 1986) and the Goodyear Massively Parallel Processor (MPP) which was originally developed for NASA for Landsat image processing (Batcher, 1980).

18.3.3.2 Pyramid Architectures

The pyramid form of processor interconnection allows for additional connections to successive layers of processors built on top of a basic mesh-connected array but with fewer processors. The successive layers in the resulting pyramid structure are appropriate to multi-resolution analysis (Figure 18.3). Such systems have been analysed primarily for computer vision applications (Stout, 1988) and have not received much attention so far in remote sensing, although it is clear they would have considerable potential.

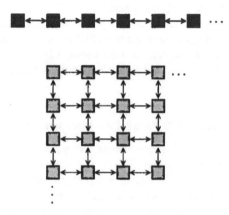

Figure 18.2: Linear array (above) and two-dimensional (below) mesh connected processing elements. Neighbours are shown 4-connected. It would be possible also to link 8-connected neighbours, i.e. along diagonals (as used in the 'CLIP' series)

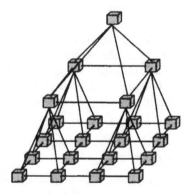

Figure 18.3: Pyramid connected processing elements for a fine-grained SIMD computer (suitable for multi-resolution image processing and machine vision algorithms)

18.3.3.3   Hypercube Architectures

In the hypercube form of connectionism, two geometrical hypercubes of N processing elements have connections between each of their corresponding corners (Figure 18.4). Each of these hypercubes is constructed from two lower dimensional hypercubes. Hypercube processor connectionism has the advantage that some of the connections cover 'long distances' and therefore image processing problems for which long distance context information is needed can be handled more efficiently than by simple mesh connectionism (Cypher & Sanz, 1989). Typical examples of hypercube computers are the Connection Machine (Hillis, 1985) and the Cosmic Cube (Seitz, 1985).

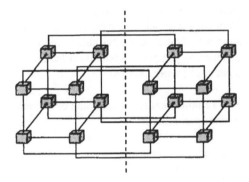

Figure 18.4: Hypercube processing element connectionism (a four-dimensional hypercube is created by connecting corresponding processing elements in two three-dimensional cubes - left and right of the dashed line)

18.3.4   Image Partitioning for Parallel Processing

In general, as indicated in section 18.3.2, remotely sensed images have x and y dimensionalities of the order 10^2-10^4, and a z dimensionality of the order 10^1-10^2. This markedly exceeds the dimensionalities of the majority of fine-grained mesh or pyramid connected SIMD computers, which typically contain ~10^2 processing elements (PEs) in x and/or y dimensions and ~10^0-10^1 PEs in a third z dimension (we will not consider in this analysis the mapping of three-dimensional remotely sensed image 'wafers' to fine-grained hypercube machines, since the concept of mappable x, y, z dimensions is not so applicable). Therefore it is usually impossible to process concurrently all data elements from a complete remotely sensed image and it is necessary to divide it into sub-blocks or 'partitions' which contain few enough data elements for parallel processing. The dividing process may be used to create either regular or irregular geometric partitions of the original dataset and methods have been proposed for both. Chapter 10 has examined the issue of raster data management in parallel systems. Here we shall highlight a few of the issues which are particularly important in remote sensing.

The manner in which remotely sensed image data are partitioned has a significant bearing on the overall efficiency of the computation. For this reason the choice of how to make the partitioning has to take account not only of the image dataset size and architecture of the hardware, but also of the nature of the operations to be carried out on the pixels and the need, when processing one data element, for access to others at various distances in the x, y, and z dimensions and whether or not such other elements can be easily accessed without re-mapping the imagery into memory. Ideally such re-mapping operations, i.e. I/O, need to be kept to a minimum for maximum efficiency. Such considerations are not restricted to SIMD computers but apply also to the use of coarse-grained MIMD architectures. The main difference between these two cases is that for SIMD machines it is advisable to minimise the frequency with which the memory of the processor array has to be re-loaded, and in the MIMD case it is advisable to reduce inter-process communication as much as possible, especially if the processes are running across coarse-grained networks of workstations.

A few operations on remotely sensed images can be carried out on individual pixels alone. Such operations can include:

- calibration and radiometric rescaling

- per-pixel classification

- image contrast enhancement.

However, many more operations require that access should be made to neighbouring pixels in the x, y geographical dimensions. Such operations include:

- segmentation

- geometrical correction and resampling; edge / lineament detection

- edge preserving smoothing and noise suppression; various forms of image filtering and spatial generalisation for input into GIS

- derivation of texture measures and texture-based classification; relaxation and use of spatial context information for feature labelling

- SAR speckle noise removal.

In order to carry out such operations efficiently the main consideration in partitioning the image dataset in the x, y plane is to minimise the number of pixels which lie on partition boundaries - i.e. to minimise the total boundary length. Also, in order to minimise the frequency of memory mapping (fine-grained case) or inter-process communication (coarse-grained case), the image data should be mapped into the machine architecture in such a way that:

1) In the fine-grained SIMD case: the processor array contains an x, y region larger than a single partition so that all required data elements which might be needed from the neighbourhood of the pixels on the periphery of the partition are readily available. The partition size is thus determined primarily by the size of the processing element array and the range of any neighbourhood operations (Figure 18.5).

2) In the coarse-grained MIMD case: a single processor should store pixels from an x, y region larger than a single partition (again so that all required data elements which might be needed from the neighbourhood of the pixels on the periphery of the partition are readily available). Schemes for accomplishing this are discussed in Chapter 10.

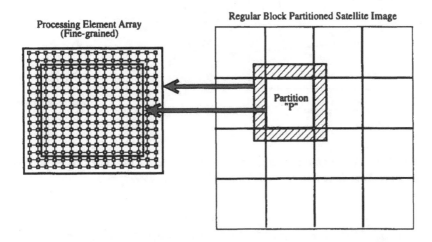

Figure 18.5: Image partitioning for fine-grained SIMD array processor. The shaded area on the right hand denotes neighbouring pixels which are needed in the processing of block partition P. To reduce I/O overheads data from these pixels should also be mapped to the PE array

18.3.5 Irregular Content-based Partitioning of Images

In some cases it may be appropriate to partition remotely sensed images on the
basis of the features they contain in a preprocessing analysis stage. Such an
approach has been implemented by Schoenmakers *et al.* (1992) who have
developed an integrated satellite image segmentation approach based on combined
edge detection and region growing. An initial strong edge detection and linking
process is used as a means of dividing the image into well distinct, but irregular,
land cover polygons inside of which the optimal order independent (but
computationally intensive) 'best-merge' region growing procedure can be carried
out to derive a more detailed segmentation. Because such areas are disjoint they
can be passed to separate MIMD processors for the region growing stage. In
practice an image block corresponding to a bounding rectangle is passed to each
processor and a 'mask' is used to eliminate from the region growing process any
pixels which do not belong to the polygon (Figure 18.6).

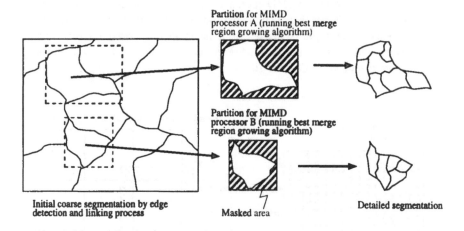

*Figure 18.6: Content based image partitioning approach for image segmentation
on an MIMD machine (based on Schoenmakers et al., 1992)*

18.4 Parallelising Algorithms for Remote Sensing: Experience and Possibilities

18.4.1 Preprocessing

Large operational applications of satellite remote sensing often demand the analysis
of hundreds of images. The preprocessing of these images can often be a lengthy,

though routine, computational procedure. The preprocessing steps are typically (though not necessarily in this order):

- radiometric calibration

- atmospheric correction

- cloud detection and removal (for land applications)

- geo-coding

- removal of geometrical distortions

- filtering/de-striping

- spatial resampling.

Many of these procedures contain stages which can be carried out in parallel. It is therefore possible to use MIMD machines to execute them.

Such an approach was adopted, for example, by Silber *et al.* (1993) who used a Telmat T-Node parallel computer, consisting of a re-configurable network of 64 asynchronous transputers, to streamline the preprocessing of NOAA Advanced Very High Resolution Radiometer (AVHRR) imagery used in the European Community's operational MARS project (Monitoring Agriculture by Remote Sensing). In this work, the various AVHRR image preprocessing steps were organised into 45 slave tasks plus a controller task and a data manager task - the last storing intermediate results and passing them to the slaves in response to requests as and when they were ready.

Another example of the potential for parallel computation in preprocessing was described by Armstrong & Marciano (1994) who used a KSR1 machine with 64 processors to perform spatial interpolation of gridded data using control points. Such interpolation procedures are commonly used when rectifying satellite images to specific cartographic projections and when it is necessary to re-sample imagery from one sensor to a different spatial resolution to match another. Wilkinson and Burge (1982) also reported the use of a parallel machine (Distributed Array Processor) for geometric and radiometric preprocessing operations on meteorological satellite imagery.

Parallel processing has also been used in remote sensing for numerically intensive stereo image matching for creation of Digital Elevation Models (DEMs) - for example from SPOT imagery (Muller, 1988) and from SAR imagery (Lewicki *et al.*, 1993).

18.4.2 Image Transforms

Many kinds of transform can be performed on multispectral remotely sensed imagery to enhance features and aid classification. Typically these involve coordinate transformations in N-dimensional spectral feature space (e.g. Principal Components Transform, Tasseled Cap Transform). Such transformations are usually achieved by deriving a transformation kernel matrix and carrying out a multiplication between the kernel and the N-dimensional vectors made up from the individual spectral samples for each pixel. Matrix and vector operations are typically well-suited to parallel computation and such transforms can be speeded up significantly by appropriately making use of the data parallelism in the imagery and the functional parallelism in the matrix operations. Such an approach was adopted, for example, by Savoji & Wilkinson (1982), who used an SIMD machine to accelerate the computation of the Karhunen-Loeve transform (equivalent to Principal Components).

18.4.3 Classification

Classification is one of the most fundamental operations performed on multispectral remotely sensed images and consists of assigning pixel samples or segments to model classes whose spectral signatures are derived from known ground truth data. Essentially in classification it is necessary to divide spectral feature space into separate hyper-regions within which pixels are assigned to a unique class. The definition of such hyper-regions may be performed by a variety of methods such as: statistical analysis of pixels from ground truth areas and construction of equiprobability surfaces (the Maximum Likelihood Classifier), by defining rectangular boxes in feature space (the Parallelepiped Classifier), by defining regions corresponding to the minimum distance to the statistical mean of prototype classes (Euclidean Distance Classifier) - see for example Thomas *et al.* (1987). Interestingly, little attention has been given to the parallelisation of these procedures even though they can be tedious on serial machines when large numbers of classes are involved.

18.4.3.1 Neural Networks

Artificial neural network algorithms are growing in popularity for the classification of remotely sensed imagery since they are distribution-free, give good accuracy and appear well-suited to multi-source data analysis - see for example Benediktsson *et al.* (1990), Civco (1993), Paola and Schowengerdt (1994), Wilkinson *et al.* (1995). The most common type of neural network employed so far in remote sensing has been the multi-layer perceptron system in which a network of interconnected perceptron processing elements are trained to effect a mathematical mapping from the image data space to a classification space (Figure 18.7).

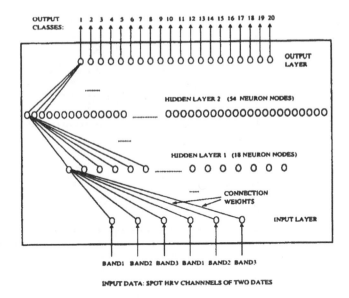

Figure 18.7: Multi-layer perceptron neural network for land cover classification from SPOT HRV imagery. The network, which is essentially a parallel distributed system, effects a transformation of two-date three-channel SPOT data into twenty land cover classes (from Kanellopoulos et al., 1992).

Such networks are inherently parallel distributed algorithms (Rumelhart *et al.*, 1986). However, the most commonly used training algorithm for such systems - the so-called 'back-propagation algorithm' - is essentially a sequential iterative process which can take hundreds of iterations to converge towards a useful result (Figure 18.8).

18.4.3.2 Parallelising Neural Networks

In view of the overall utility of neural networks and their inherent parallel architecture, some efforts have been made to devise methods to overcome the limitations of the sequential training algorithms. Yoon *et al.* (1991) examined different ways of mapping neural network training procedures to parallel hardware. One such approach involves storing multiple copies of the neural network in different processors and allowing each copy to train with a reduced subset of the overall set of training patterns. Locally computed connection weight changes inside each network are averaged at regular intervals into global weight changes which are then broadcast to each processor.

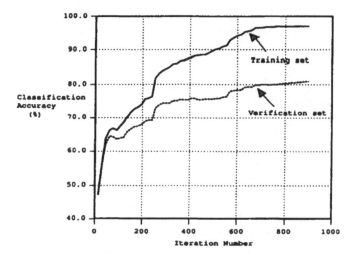

Figure 18.8: Training sequence for the multi-layer perceptron network of Figure 18.7. This process took approximately 25 hours' CPU time on a serial SUN-4/260 workstation at the time of the original experiment (from Kanellopoulos et al., 1992).

An alternative approach has been the development of network architectures in which parts of the classification task are separated, i.e. a set of smaller networks are blocked together into a larger one and they can each be trained in parallel. Such a possibility is offered by the blocked back-propagation algorithm (BBP) developed by Liu & Xiao (1991) for classification of remotely sensed images.

Apart from algorithm adaptations to take advantage of parallel computers, there is currently much interest in the development of special purpose parallel neural hardware systems. These have considerable potential in remote sensing in the long term. A typical machine is the SYNAPSE-1 developed by Siemens (Figure 18.9) which has been subjected to trials on remote sensing data in the context of the European Community's SYNAP project.

18.4.4 Segmentation

Segmentation is concerned with the transformation of an initial array of pixels into a disjoint set of regions (comprising spatially connected pixels), each of which obeys a spectral homogeneity constraint. This procedure is useful since it can improve overall classification accuracy and leads to a spatial simplification of the image which is helpful in raster-vector conversion (i.e. when the image product has to be entered into a GIS). Many image segmentation algorithms have been devised in the past 2-3 decades, most of which are computationally intensive since they rely on an essentially sequential search procedure to split or merge adjacent pixels according to the homogeneity constraint (Cross *et al.*, 1988). Such a process has a strong similarity to the 'image component labelling problem' which has received a lot of attention in the image processing field since it is central to many applications and computationally tedious (Alnuweiri & Prasanna, 1992; Cypher *et al.*, 1990).

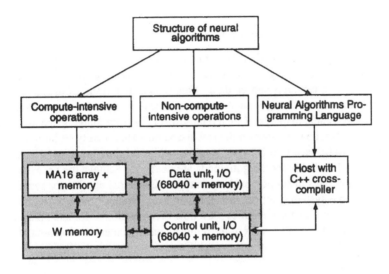

Figure 18.9: SYNAPSE-1 neurocomputer system concept developed by Siemens AG. The machine uses an array of MA16 Neural Signal Processor VLSI chips and has a special Neural Algorithms Programming Language (from Ramacher et al., 1994).

Many of the commonly used segmentation algorithms are essentially order-dependent since the final result can depend on the starting point for the search, which is a significant disadvantage. However, the iterative best merge region-growing technique (Beaulieu & Goldberg, 1989) is one which is essentially non-order-dependent since it involves selection of the 'most similar' pair of pixels to merge in the whole image in each iteration. This approach, whilst optimal in many respects, is extremely tedious as the whole image has to be examined in each iteration. It is therefore an algorithm which needs parallelisation.

Tilton (1990) developed the Iterative Parallel Region Growing (IRPG) algorithm which was eventually implemented at the NASA Goddard Space Flight Center on a

MasPar MP-1 machine - an SIMD computer with an array of 128 by 64 mesh connected PEs (Tilton, 1991). In the IRPG approach, a number of similar pairs of pixels are merged in parallel in different parts of the image (the pairs must be sufficiently well-separated so that order dependence does not become a problem). Subsequently the best pair of region pairs are merged in recursively defined overlapping sub-images until the process converges, e.g. according to an entropy criterion. Such approaches are efficient and appear to give good results (Le Moigne & Tilton, 1995).

18.4.5 Feature Space Visualisation

Besides background numerical operations such as image classification and segmentation, there is a growing need for the use of parallel processing in aiding visualisation of remotely sensed data. This is a partly a result of the increase in the z dimensionality of datasets mentioned earlier and the need for user interaction with complex graphical displays of such data -especially from imaging spectrometer systems. Also, new interaction devices based on virtual reality technology are beginning to be used in remote sensing requiring real time graphical volume rendering. Parallel processing methods are needed to achieve the necessary computation rates to support such operations. Graphical accelerator systems such as the Kubota system (Levine, 1994), which uses parallel 'frame buffer modules' and 'geometric transform engines', are now being exploited for such purposes - e.g. as part of a virtual reality system for remote sensing under development at the European Commission's Joint Research Centre.

18.4.6 Parallel Modelling of Radiation Transfer in Vegetation Canopies

One of the most difficult analytical problems in satellite remote sensing is the inversion of detected radiances to obtain precise geophysical parameters characterising the land or ocean surface. Although they are closely related, this process is different to image classification. Signal inversion implies translating detected radiances into continuous geophysical quantities whereas classification concerns itself only with transforming image radiances into discrete classes. Inversion is frequently a complex process without unique solutions. In order to gain insight into the meaning of observed radiances -in order to improve inversion techniques - it is useful to model the production of those radiances at the photon level. This approach has been adopted in several studies, though it is generally a computationally very demanding procedure and appropriate therefore for parallel computation.

Verstraete & Govaerts (1994) developed a computational model of the scattering of light in vegetation canopies with the aim of improving the understanding of the bi-directional reflectance function for various natural conditions. The reflected radiance arriving at a satellite sensor viewing the Earth in the optical or near infra-red region of the electromagnetic spectrum represents the combination of photons originating at the Sun which are scattered by the canopy and the ground surface. In order to understand the behaviour of different vegetation canopies in scattering solar radiation it is necessary to model the full reflectance function of a complex three-dimensional object. A practical approach for doing this is to use the technique of ray tracing - i.e. to trace either the forward or the backward trajectories of single photons from source to destination (e.g. sun to satellite via the canopy) and to integrate the effect of large numbers of such trajectories into a general reflectance function. Govaerts (1995) reported using ~5 by 10^7 separate rays in modelling typical canopy reflectance functions. Apart from the large number of such rays needed to model the reflectance function, the computational complexity is increased by the fact that each ray may undergo several interactions with surfaces within the canopy as photons are successively scattered by one surface and then impinge on another. Moreover, the canopy needs to be modelled as a full three-dimensional object which is itself complex, though models based on fractal or related techniques such as L-systems (Govaerts & Verstraete, 1995) can simplify this process.

Govaerts & Verstraete (1995) developed the 'Raytran' package to implement their ray tracing model. Raytran was specifically adapted for processing using a single sequential UNIX machine, a cluster of workstations or a distributed memory parallel processor. The Raytran code was split into three layers as follows:

1) tracer layer

    - responsible for all I/O operations, to trace rays and extract statistics about them

2) job sharing and message buffering layer

    - concerned primarily with distribution of jobs between different processing nodes

3) communication layer

    - implements communication between the processors via MPI (Message Passing Interface).

To date, Raytran is one of the few simulation packages used in remote sensing which has been adapted specifically to take advantage of parallel hardware.

18.5 Related Topics in Parallel Image Processing

18.5.1 Visual Programming Tools

An interesting development concerning parallel processing in remote sensing has
been the recent development of a visual programming tool which allows users to
design processing chains for multi-sensor data fusion and execute them on an
SIMD machine (Ionescu *et al.*, 1993). Whilst such systems are in their infancy, the
possibility of automatically generating code for parallel machines from sets of
object oriented software tools clearly has long term potential in streamlining the
way in which software systems are developed for remote sensing and mapped to
parallel hardware.

18.5.2 Benchmarking Parallel Image Processing

Although this chapter has revealed so far the wide range of uses of parallel
processing in remote sensing, the relative speed-up factors attainable compared to
sequential computation have not been examined. One of the reasons for this is that
there is currently no agreed benchmarking procedure for parallel computers in
remote sensing. The speed with which algorithms can be executed rarely reaches
the theoretical maxima, and in many cases depends significantly on the size of
images in use and the relationship of their sizes to the architectures of the parallel
machines in question. Ideally some good benchmarks are needed. Bonnin *et al.*
(1994) have indicated some requirements for such benchmarks in image
processing, for example they should include:

* algorithms at different semantic levels

* data-driven and interpretation-driven (top-down) algorithms

* data conversion algorithms.

Whilst such requirements are not exhaustive, they indicate that realistically parallel
computers must be tested against a range of problems which occur in either remote
sensing or machine vision. This is because their characteristics and computational
requirements can vary so considerably that high speed-up factors for one part of an
image analysis task may be quite irrelevant for another.

18.5.3 The Future Potential of Opto-electronic Image Processing

Although this chapter has concentrated on conventional forms of electronic
parallel computers, it would not be complete without mention of the potential of

optical parallel computing. It has recently been demonstrated that opto-electronic systems with reconfigurable optical interconnects are capable of performing many conventional image processing or understanding operations in parallel - and at extremely high speed (Eshaghian *et al.*, 1994). A number of contributing technologies are already mature, although usable systems have yet to be developed.

18.6 Conclusions

Remote sensing is a discipline in which data volumes are enormous and processing chains relatively complex. As such it is a discipline in which parallel computation has a natural home. As imagery gathered from space becomes more complex and is gathered at ever greater cost, it is becoming more essential that adequate means are utilised to exploit it to the full in elucidating the state of planet Earth - parallel computing is one such means whose utility can be expected to grow considerably in the next few years following the pioneering groundwork which has already been done.

18.7 References

Alnuweiri, H.M. & Prasanna, V.K., 1992, Parallel architectures for the image component labeling problem. *IEEE Trans. on Pattern Analysis and Machine Intelligence*, **14** (10), 1014-1034.

Armstrong, M.P. & Marciano, R., 1994, Inverse-distance-weighted spatial interpolation using parallel supercomputers. *Photogrammetric Engineering and Remote Sensing*, **60** (9), 1097-1103.

Ballard, D.H. & Brown, C.M., 1982, *Computer Vision*, Prentice Hall, Englewood Cliffs, NJ.

Barnes, W.L. (Ed.), 1993, Sensor Systems for the Early Earth Observing System Platforms. *Proceedings Volume 1993*, SPIE, Bellingham, WA,.

Barrett, E.C. & Curtis, L.F., 1992, *Introduction to Environmental Remote Sensing*, 3rd Edition, Chapman & Hall, London.

Batcher, K.E., 1980, Design of a massively parallel processor, *IEEE Trans. on Comput.*, **C-29**, 836-840.

Beaulieu, J.-M. & Goldberg, M., 1989, Hierarchy in picture segmentation: a stepwise optimization approach. *IEEE Trans. on Pattern Analysis and Machine Intelligence*, **11** (2), 150-163.

Benediktsson, J.A., Swain, P.H. & Ersoy, O.K., 1990, Neural network approaches versus statistical methods in classification of multisource remote sensing data. *IEEE Transactions on Geoscience and Remote Sensing*, **28** (4), 540-552.

Bonnin, P., Pissaloux, E.E. & Dillon, T., 1994, Towards a definition of benchmarks for parallel computers dedicated to image processing/understanding. *Microprocessing and Microprogramming*, **40** (10), 783-788.

Civco, D.L., 1993, Artificial neural networks for land-cover classification and mapping. *Int. J. Geographical Information Systems*, **7** (2), 173-186.

Cross, A.M., Mason, D.C. & Dury, S., 1988, Segmentation of remotely-sensed images by a split-and-merge process. *Int. J. Remote Sensing*, **9** (8), 1329-1345.

Curran, P.J., 1985, *Principles of Remote Sensing*, Longman, London.

Cypher, R. & Sanz, J.L.C., 1989, SIMD architectures and algorithms for image processing and computer vision. *IEEE Trans. on Acoustics, Speech and Signal Processing*, **37** (12), 2158-2173.

Cypher, R.E., Sanz, J.L.C. & Snyder, L., 1990, Algorithms for image component labelling on SIMD mesh-connected computers. *IEEE Trans. on Computers*, **39** (2), 276-281.

Downey, I.D., Power, C.H., Kanellopoulos, I. & Wilkinson, G.G., 1992, A performance comparison of Landsat TM land cover classification based on neural network techniques and traditional maximum likelihood and minimum distance algorithms. In: Cracknell, A.P. & Vaughan, R.A. (Eds), *Proceedings of the 18th Annual Conference of the UK Remote Sensing Society: Remote Sensing from Research to Operation*, Dundee, Remote Sensing Society, Nottingham, 518-528.

Dreyer, P., 1993, Classification of land cover using optimized neural nets on SPOT data. *Photogrammetric Engineering and Remote Sensing*, **59** (5), 617-621.

Duff, M.J. & Fountain, T.J., 1986, *Cellular Logic Image Processing*, Academic Press, NY.

Eshaghian, M.M., Lee, S.H. & Shaaban, M.E., 1994, Optical techniques for parallel image computing. *J. Parallel and Distributed Computing*, **23**, 190-201.

Gostick, R.W., 1979, Software and algorithms for the distributed array processor. *ICL Tech. J.*, **1**, 116-135.

Govaerts, Y., 1995, *Modeling the scattering of light in 3D vegetated scenes: development and parallelization of a new Monte Carlo ray tracing code*, private communication.

Govaerts, Y. & Verstraete, M.M., 1995, Applications of the L-systems to canopy reflectance modelling in a Monte Carlo ray tracing technique. In: G.G. Wilkinson, I. Kanellopoulos & J. Megier (Eds), *Fractals in Geoscience and Remote Sensing*, Office for Official Publications of the European Communities, Luxembourg, Publication EUR 16092, 211-236.

Hillis, D., 1985, *The Connection Machine*, MIT Press, Cambridge, MA.

Hockney, R.W. & Jesshope, C.R., 1981, *Parallel Computers*, Adam Hilger, Bristol.

Ionescu, D., Andronic, C. & Goodenough, D., 1993, Object oriented tools for multisensor data fusion on an SIMD computer. *Proc. International Geoscience and Remote Sensing Symposium (IGARSS `93)*, Tokyo, IEEE Press, Piscataway, NJ, Vol. III, 1128-1130.

Kanellopoulos, I., Varfis, A., Wilkinson, G.G. & Mégier, J., 1992, Land cover discrimination in SPOT imagery by artificial neural network - a twenty class experiment. *Int. J. Remote Sensing*, **13** (5), 917-924.

Kontoes, C., Wilkinson, G.G., Burrill, A., Goffredo, S. & Mégier, J., 1993, An experimental system for the integration of GIS data in knowledge-based image analysis for remote sensing of agriculture. *Int. J. Geographical Information Systems*, **7**, 3, 247-262.

Lee, C.-K. & Hamdi, M., 1995, Parallel image processing operations on a network of workstations. *Parallel Computing*, **21**, 137-160.

Le Moigne, J. & Tilton, J.C., 1995, Refining image segmentation by integration of edge and region data. *IEEE Trans. on Geoscience and Remote Sensing*, **33** (3), 605-615.

Levine, R.D., 1994, Volume rendering with the Kubota 3D imaging and graphics accelerator. *Digital Technical Journal*, **6** (2), 34-48.

Lewicki, S., Lee, M, Chodas, P. & DeJong, E., 1993, Stereo processing of Magellan SAR imagery performed on a transputer architecture. *Proceedings of the*

International Geoscience and Remote Sensing Symposium IGARSS '93 Tokyo, IEEE Press, Piscataway, NJ, Vol. IV, 1786-1788.

Liu, Z.K. & Xiao, J.Y., 1991, Classification of remotely-sensed image data using artificial neural networks. *Int. J. Remote Sensing*, **12** (11), 2433-2438.

Muller, J.-P.A.L., 1988, Key issues in image understanding in remote sensing. *Phil. Trans. Roy. Soc. Lond.*, **A**, 324, 381-395.

Peddle, D.R., 1995, Knowledge formulation for supervised evidential classification. *Photogrammetric Engineering and Remote Sensing*, **61** (4), 409-417.

Paola, J.D. & Schowengerdt, R.A., 1994, Comparisons of neural networks to standard techniques for image classification and correlation, In: *Proceedings of the International Geoscience and Remote Sensing Symposium (IGARSS '94)*, Pasadena, CA, IEEE Press, Piscataway, NJ, Vol. 3, 1404-1406.

Ramacher, U., Raab, W., Anlauf, J., Hachmann, U. & Wesseling, M., 1994, *Synapse-1 - A General Purpose Neural Computer*, Siemens AG, Corporate Research and Development Division, Munich.

Rumelhart, D.E., McClelland, J.L. & PDP Research Group, 1986, *Parallel Distributed Processing - Explorations in the Microstructure of Cognition*, MIT Press, Cambridge, MA, 2 volumes.

Savoji, M.H. & Wilkinson, G.G., 1982, Karhunen-Loeve transform on an array processor - applications to multi-spectral and single band images. *Proceedings of the International Conf. on Electronic Image Processing*, York, IEEE Press, London, 225-229.

Schoenmakers, R.P.H.M., Wilkinson, G.G. & Schouten, T.E., 1992, Multi-temporal image segmentation using a distributed-memory parallel computer. *Proceedings of the International Geoscience and Remote Sensing Symposium IGARSS '92*, Houston, TX, IEEE Press, Piscataway, NJ, Vol. 2, 1114-1116.

Schowengerdt, R.A., 1993, Tools for terabytes. *Proceedings of the 12th Pecora Memorial Conference on Land Information from Space Based Systems*, Sioux Falls, SD, American Society for Photogrammetry and Remote Sensing, Bethesda, MD, 243-246.

Seitz, C., 1985, The cosmic cube. *Communications of the ACM*, **28** (1), 22-33.

Silber, C., Heidrich, D., Bierlaire, P. & Kanellopoulos, I., 1993, Processing of AVHRR imagery using the T.Node parallel computer. *Proceedings of the*

International Geoscience and Remote Sensing Symposium IGARSS `93, Tokyo, IEEE Press, Piscataway, NJ, Vol. IV, 1659-1661.

Stout, Q.F., 1988, Mapping machine vision algorithms to parallel architectures. *Proc. IEEE*, **76** (8), 982-995.

Thomas, I.L., Benning, V.M. & Ching, N.P., 1987, *Classification of Remotely Sensed Satellite Images*, Adam Hilger, Bristol.

Tilton, J.C., 1990, Image segmentation by iterative parallel region growing. *NASA Information Systems Newsletter*, February, **19**, 50-52.

Tilton, J.C., 1991, A tool for interactive exploration of a hierarchical segmentation. *Proceedings of the International Geoscience and Remote Sensing Symposium (IGARSS `91)*, Espoo, Finland, IEEE Press, Piscataway, NJ, Vol. II, 1099-1101.

Verstraete, M.M. & Govaerts, Y., 1994, *Modelling the scattering of light in arbitrarily complex media: motivation for a ray tracing approach.* Technical Report EUR 15790 EN, Joint Research Centre, European Commission, Brussels and Luxembourg.

Wilkinson, G.G. & Burge, R.E., 1982, Using the ICL-DAP for satellite climatology. *Computer Physics Communications*, **26**, 469-471.

Wilkinson, G.G., Folving, S., Kanellopoulos, I., McCormick, N., Fullerton, K. & Mégier, J., 1995, Forest mapping from multi-source satellite data using neural network classifiers - an experiment in Portugal. *Remote Sensing Reviews*, **12**, 83-106.

Wilkinson, G.G., Kanellopoulos, I., Mehl, W. & Hill, J., 1993, Land cover mapping using combined Landsat Thematic Mapper imagery and ERS-1 Synthetic Aperture Radar imagery. *Proceedings of the 12th Pecora Memorial Conference on Land Information from Space Based Systems*, Sioux Falls, SD, American Society for Photogrammetry and Remote Sensing, Bethesda, MD, 151-158.

Yoon, H., Nang, J.H. & Maeng, S.R., 1991, Neural networks on parallel computers. In Soucek, B. and the IRIS Group (Eds), *Neural and Intelligent Systems Integration*, Wiley, NY.

Part Five

Conclusions

19

Towards Parallel Libraries for Geographical Algorithms

**M.J. Mineter, R.G. Healey, B.M. Gittings
and S. Dowers**

19.1 Summary of Findings

Previous chapters have presented an overview of parallel computing technology and explored its application within Geographical Information Systems (GIS). To explore the problems and potential benefits of parallel methods, a variety of examples have been presented, ranging from polygon overlay to digital terrain modelling. A number of general points have emerged from these studies, which are of general applicability for the development of parallel methods in GIS. These are now examined in turn.

Firstly, there are issues arising from characteristics of geographical data. Typically, there is considerable variation in complexity across a dataset, causing problems for load-balancing between processors. Further, the volumes of data are such that techniques to overcome I/O bottlenecks are necessary.

Secondly, different data formats give rise to distinct challenges. The multiple inter-related components of vector-topological data, the large sizes of raster datasets and the alternative methods of encoding them, and the overall heterogeneity of spatial datatypes (Chapter 17) are examples of this. Each format has implications not only for data decomposition, but also for subsequent processing and the structure of the required software modules.

Thirdly, the partitioning of data implies additional algorithmic development to handle both the decomposition at the beginning and the collation of the dataset at the end of whichever operation is performed on it. Emphasis in the existing literature tends to be placed on the decomposition stage. However, the final collation stage is of particular importance especially for all operations that produce vector topological data as output. The implications of this have been explored and a solution suggested, designed in a manner which allows reuse of functionality and which can potentially speed algorithmic development (e.g. Partial Topology functions make building topology relatively straightforward). The potential commercial benefits of concentrating on developing such generic reusable functions are significant.

Fourthly, although the sophistication of parallel compilers and languages continues to advance, the GIS programmer is still faced with the need for explicit message passing and careful design to accommodate controlled data decomposition. There remains a gap between the functions that a parallel environment (both language and operating system) will provide automatically and the reasonable expectations of the applications programmer.

The need for this detailed awareness results in approaches to software development which are in some ways reminiscent of the early days of sequential processing. For example, there is a need for a low-level understanding of where and how data are held, communications techniques and user-interface software (such as a layered approach to software development, state-event analyses and multi-threading). These approaches have to be re-discovered and re-applied in, for example, the rather different contexts of generalising raster data and populating a vector GIS. Whilst these techniques, together with complex sort/joins of vector data, allow a nostalgic and fruitful wander down memory lane for the more senior of programming colleagues, the time and expertise required are prohibitive to many. It is thus scarcely surprising that parallel GIS software has yet to make its presence felt outside a narrow range of grid-based applications.

Fifthly, given the significant expertise in both GIS and parallel processing techniques required to develop new algorithms, and the development time involved, a pragmatic approach to the introduction of parallel methods into GIS needs to be adopted. Commercial vendors may wish to reuse existing sequential algorithms within a parallel context, by decomposing an overall task into sequential components, running concurrently on different parts of the dataset, on different processors. This method was utilised in Parts Two and Three of this book in explaining the processing of sub-areas in each of the three example operations (raster-to-vector conversion, vector-to-raster conversion and vector polygon overlay). Adopting an approach based on a standard data interchange format may allow a parallel engine to be used for specific compute-intensive tasks, while other tasks are handled using commercial, off-the-shelf GIS packages in a workstation environment.

Finally, although there are a number of problems to be overcome, it is apparent that parallel technology opens the way to a range of possible new algorithms, which could operate on multiple data layers simultaneously (many serial algorithms for polygon overlay are restricted to two datasets at a time) or which allow concurrent operations on different data types, e.g. raster and vector. The benefits of these and similar developments for complex, data-intensive analyses are examined in more detail in the next section and proposals are made for an approach to parallel GIS software design using portable reusable libraries.

19.2   The Need for Portable, Reusable Parallel Libraries

The use of parallel processing to handle geographical information can be both technology and user driven. Prominent in creating the technological impetus is the growing availability of cheap parallel platforms, in the form of multi-processor workstations and PCs in the potential use of under-utilised networked workstations as a parallel resource and the establishment of software standards, especially MPI. The existence of parallel GIS, with performance an order-of-magnitude greater than current systems, would act as a significant catalyst to promote more sophisticated kinds of analysis. Data mining, exploratory data analysis and other types of application, previously run in batch mode, that could now be run interactively, are obvious potential beneficiaries of enhanced processing power.

There is no doubt that a demand for this improved performance is already evident: the potential of integrated GIS dataservers, the desire to have instantaneous on-screen results to respond to telephone queries received at Call Centres, and an unwillingness to wait for batch-oriented reports all signpost the future. There is also the growing realisation of the need to represent the fuzziness of real-world data more effectively. There is an increasing requirement to move beyond static portrayals of error and uncertainty (and even these are not well represented in current commercial GIS packages). Such editions could include the modelling of chaotic behaviour and dynamic visualisations where the user can change the parameters of a potentially highly complex model, in real-time.

In addition to the requirements of these richer and hence more compute-intensive data models, the demand is evident in areas such as visualisation, real-time environmental monitoring, the serving of data across the Internet (or intranets), the integration of techniques from remote-sensing, GIS, and environmental modelling, and the development of decision support systems based on large datasets. Techniques from Artificial Intelligence are also beginning to be exploited by, for example, the Centre for Computational Geography in the University of Leeds (CCG, 1997), and Openshaw & Openshaw (1997) who have implemented a number of applications using both neural networks and genetic algorithms on parallel computers.

In many of these areas sequential algorithms are still being developed, which might appear to be an argument against the adoption of parallelism at present. However, appropriately designed parallel software libraries have the potential not only for accelerating existing algorithms but also for facilitating the prototyping of new ones, with the advantages of parallelism in their wake. We would propose moving away from the current situation in which sequential and parallel algorithms are developed separately, with the inevitable duplication of effort, to an integrated approach which takes parallelism into account at the outset. In this way, all algorithms should now be designed to be parallel, and will therefore be able to take advantage of parallel hardware, yet be able to operate quite efficiently in sequential mode on single processor machines.

Our example operations have shown the potential for defining reusable, portable modules, encapsulating complexity and thereby opening parallel GIS to the non-specialist. Thus parallel libraries, analogous to those of Chapter 4, but specific to geographical data and applications, are undoubtedly the key to opening the technology to more general use. In the same way that a regular decomposition library can facilitate parallelisation of grid-based algorithms, so similar libraries supporting raster, vector-topological, and digital terrain and other types of data are fundamental requirements if parallel technology is to be utilised more widely in GIS. At present there are also several major international initiatives, spanning the academic and commercial sectors, where a significant potential role for new parallel libraries can be identified. The first of these is the Open GIS Consortium (OGIS, 1996), with its emphasis on open standards and layered software for GIS applications. Their enthusiasm for *middleware*, which isolates users from specific hardware platforms and software systems and helps clients and servers communicate in today's heterogeneous distributed computing environments, fits well with an application-oriented, high performance *results engine*.

Secondly, planned future developments in the SQL3 standard (ISO SC21/WG3-ANSI X3H2, 1994) will facilitate the storage of user defined data types, together with their associated processing *methods*. Clearly the latter could be either serial or parallel in implementation, but this can be largely hidden from the application programmer by means of suitable software encapsulation and the use of well-defined Application Programmer Interfaces (APIs).

Thirdly, the emergence of Network Computer (NC) technology raises the prospect of at least a partial shift away from PCs, with their perceived high cost of ownership, back towards a lower-cost client and more powerful server arrangement. Experiments with fully-distributed computing environments have proved costly, and the trend is now towards a *data warehouse*, with centralised management, much in the same manner as the oft-deprecated traditional mainframe. The fundamental tenets of database management suggest that such a model has considerable merit; the Network Computer will allow this centralised model, while retaining the level of interactivity of a client-server environment, but without the significant disadvantages of current clients. One of the main reasons for the demise of the mainframe was a perceived or actual lack of performance. A parallel server becomes mandatory if a more centralised approach is to be resurrected effectively. Variants of NC 'cartridge' technology will allow different types of application software modules to be 'plugged in' to the server, in a more developed form of the kind of user defined datatypes and methods envisaged for SQL3.

Close attention to these emerging standards and technologies will enable future developers of parallel libraries to target their efforts to the areas of greatest user need and market opportunity.

In summary, parallel hardware is now available in a wide range of configurations, some incurring minimal additional costs (such as exploiting a network of workstations as a parallel computer), and the demand already exists for high-performance geographical applications.

These demands are set to grow within the next few years, as the complexity of research problems and the size of datasets increase, as user expectations continue to rise and as the ability of hardware engineers to squeeze yet more performance out of a single silicon chip approaches physical limits. The software technology required to underpin parallel GIS applications is not yet available, and therefore software development costs will remain significant in the short term. However, a disciplined focus on the design and implementation of parallel GIS software libraries provides the best route forward when both financial resources and the necessary expertise are in perennial short supply.

19.3 References

CCG,1997, The Centre for Computational Geography, http://www.geog.leeds.ac.uk/research/ccg.html.

ISO SC21/WG3-ANSI X3H2, 1994, *Working Draft: SQL Part 1: SQL/Framework*-http://epoch.cs.berkeley.edu:8000/sequoia/schema/STANDARDS/SQL3/sqlpart1.txt.

OGIS, 1996, The Open GIS Consortium, Inc. Vision and Mission Statements, http://www.ogis.org/vision.html.

Openshaw, S. & Openshaw, C., 1997, *Artificial Intelligence in Geography*. Wiley & Sons, Chichester.

Glossary

64 Bit architecture - a processor in which the default size of data items and memory addresses is 64 bits

AEL - Active Edge List

Algorithm - a defined sequence of steps necessary for the completion of a task

American National Standards Institute (ANSI) - the organisation responsible for the setting of standards within the US

Area - a fundamental two dimensional spatial feature. Name of a corresponding topological record type in the NTF Level 4 data model

ATM - a high-speed networking protocol

Attribute Value - the descriptive component of a spatial object

AVHRR - Advanced Very High Resolution Radiometer, on National Oceanic and Atmospheric Administration satellites

Backplane - an interconnection between components of a computer

Backpropagation - a type of neural network architecture

Benchmarking - performance assessment through comparative or absolute testing

Boolean - a value which is either 'true' or 'false'

British Standard Institute (BSi) - the organisation responsible for the setting of standards within the UK

Cache - an intermediate store of data or instructions held in close proximity to the CPU, for the purposes of boosting performance by reducing the number of relatively slow accesses to main memory

Central Processing Unit (CPU) - the central component of a computer responsible for calculation and the processing of data

CEPG - Coincident Event Point Group

Chains - an NTF Level 4 topology record used to define the boundaries of Faces in terms of a list of Edges

Checkpointing - the ability to stop and then restart a program

CMYK - a colour representation method based on the components Cyan, Magenta, Yellow and Black

Coverage - a thematic map layer

CSTools - a messaging library provided by Meiko for its parallel computers

danglers - an error typically found when building topology, in which a node at the end of one edge fails to establish a connection with another edge

Database Administrator - the person responsible for the structuring, user authorisation and performance monitoring of a DBMS

Database Management Systems (DBMS) - software to store, retrieve and manipulate data

Data Model - a high-level representation of the data in a system, independent of the mode of implementation

Datatypes - forms of data representation e.g. integer, real

DCT - Discrete cosine transform

Delaunay Triangulation - a method for tesselating space into triangular units

Digital Cartographic Database - a database composed of spatial data in raster or vector format

Digital Elevation Model (DEM) - a terrain representation composed of an array of point-heights (also known as an Altitude Matrix). Used in the US as the generic alternative to a DTM

Digitisation - the process of capturing information in vector format

Digital Terrain Model (DTM) - accepted within Europe as the generic term for any representation of the surface of the earth held in a computer

Edge - an NTF Level 4 data record defining topology at the boundary of two polygons, and bounded by a Node at either end

EPCC - Edinburgh Parallel Computing Centre

ERS-1 - a European Space Agency satellite mission

Face - an NTF Level 4 data record used in defining topological relationships between Areas and Edges

FLOPS - FLoating point OPerations per Second

Generalisation - the reduction of detail in a spatial representation

Geographical Information System (GIS) - a computer-based system for the capture, management, analysis and display of spatially-referenced data

Geoms - an NTF Level 4 record used to define the vertices associated with Edges and Node

GFLOPS - Giga FLOPS, a measure of performance - see FLOPS

Graphical Interchange Format (GIF) - a raster encoding format, used for images, rather than spatial data

Gridded Data - data in a cell-based raster representation

HHCODE - helical hyperspatial code

HPF - High Performance FORTRAN

HSR - hidden surface removal

Hypercube - a class of parallel computer architecture in which there are 2^n processors, each of which has n connections

IDG - Identifier Generator, a process proposed in the design of Chapter 9, for use in creating vector topology

Intervisibility - a determination of whether two points on a terrain surface can see each-other

JPEG - a lossy raster encoding format, used for images, rather than spatial data

LANDSAT - a programme of satellite missions primarily aimed at remote sensing of earth resources

LANDSAT TM - Thematic Mapper , the fourth in the series of LANDSAT satellites

Localisation - design of algorithms to minimize processing based on proximity relations between data values in space

Locational accuracy - the extent to which a spatial feature is correctly referenced to its position on the earth's surface

Lossless Compression - a compression which allows the original image to be reconstructed perfectly

Lossy Compression - a compression which allows the original image to be reconstructed only approximately

Lempl-Zif-Welch (LZW) Encoding - a raster data compression technique used, for example, in GIF

Meiko - A Bristol (UK) based manufacturer of parallel computers, based on transputer and SPARC arrays

MERIS - the medium resolution imaging spectrometer to be launched on the ENVISAT-1 mission by ESA in 1999

MIMD - Multiple Instruction, Multiple Datastream architecture

MIMD-DM - Distributed memory MIMD

MIMD-SM - Shared memory MIMD (formerly known as "symmetric multiprocessing MIMD")

MISD - Multiple Instruction, Single Datastream architecture

Model - a representation of reality

MODIS - the moderate resolution imaging spectroradiometer planned for launch by NASA in 1998

Monotonic sections - groups of lines, pairs of which can intersect each other only once. A monotonic section of a line has no turning point in x or y.

Message Passing Interface (MPI) - new, standard, low-level interface for message passing in parallel programming

MPI Forum (MPIF) - the group responsible for specification of the MPI standard

Massively Parallel Processor (MPP) - a computer with a large number of processing nodes linked via an internal network

Multi-processor - a computer composed of multiple CPUs, often used to perform multiple single-processor tasks.

Multi-streamed Workload - a series of independent tasks, capable of simultaneous running, which can take advantage of a multi-processor architecture

Multispectral - that which covers a number of spectral bands, a remote sensing term

Neutral Transfer Format (NTF) - previously known as 'National Transfer Format' a BSi standard for the representation of spatial data in the UK

NUMA - Non Uniform Memory Access. an architecture in which processors have access to the entire memory of the computer, but not all parts of it can be accessed at the same speed due to a hierarchical organisation

OCCAM - an early parallel language used on the Meiko transputer array

OEL - Ordered Edge List

Open GIS (OGIS) - an organisation that aims to develop open systems standards for GIS applications

Occam Programming System (OPS) - a programming environment for use with the OCCAM programming language

Parallel Processor - a multi-processor architecture, where each of the processors is capable of sharing a component of the same task

PE - processing element, one of the procesors within a parallel processor

Pipeline - a computational sequence that proceeds in steps, with the output from each step serving as input to the next

Polygon - the fundamental vector representation of a two dimensional spatial feature as a closed sequence of geoms

PRAM - parallel random access machine, an idealised model of a parallel computer that assumes zero synchronisation cost and memory access overhead

Prefetching - retrieval of data before they are actually required for processing, so they are available immediately for the next phase of computation

Pseudocode - a structured series of processing instructions that outlines how an algorithm would be represented in computer code

PTF - Partial Topology Functions. A library proposed in Chapter 9 for building vector-topology in parallel applications where each processor holds a partition of the complete dataset.

Parallel Utilities Library (PUL) - a set of software libraries for commonly required parallel programming tasks, developed at the Edinburgh Parallel Computing Centre, e.g. PUL-RD for regular decomposition problems; PUL-TF for problems requiring a task farm solution

Pixel - the fundamental unit within a raster representation

Parallel Virtual Machine (PVM) - a public domain message passing software library distributed by Oak Ridge National Laboratory

Quadtree - an indexing and compression technique used for raster spatial data

Redundant Array of Inexpensive Disks (RAID) - a means of combining multiple modestly-sized disk drives together for better performance or enhanced data integrity

Raster - a spatial representation which divides space into discrete and equally-sized rectangular areas known as pixels

Rasterise - the process of generating information in raster format

Reduced Instruction Set Computer (RISC) - a microprocessor architecture which makes use of a small number of powerful instructions

Relational DBMS - a DBMS where each of the component classes of data are represented as tables, and the data themselves as rows within these tables

Remote sensing - a method of collecting data about a region using distant sensors, for example, carried on aeroplanes or satellites

Run Length Encoding (RLE) - a compression technique used for raster data, which stores the counts of similar cells, rather than storing each cell individually

SAIF - the Spatial Archive and Interchange Format - the Canadian standard for digital map data storage and transfer

SAR - Synthetic Aperture RADAR, a form of remote-sensing

Scalability - the ability of a problem or architecture to continue to make performance gains with increasing numbers of processors

Scanner - a device used to encode an analogue (eg. paper map) source in a digital raster representation

Scanline - a line of data within a raster

SDTS - US Spatial Data Transfer Standard

SIMD - Single Instruction, Multiple Datastream architecture

SISD - Single Instruction, Single Datastream architecture

SLS - Sweep Line Status, a description of the intersection of the sweep-line with the geometric structure being swept, used in plane sweep algorithms for polygon overlay

SMP - Symmetric Multi-Processor

SPARC - A type of RISC micro-processor produced by Sun Microsystems

Spatio-temporal coordinates - provide a unique location for a feature in time and space

Spin Lock - the continuous checking to determine whether another processor within an array is currently modifying data which the current processor wishes to access

spline - a mathematical device for fitting a curve to a series of points

SPMD - Single Program Multi Data, a subset of MIMD. SPMD requires the same executable image to be run on all processors.

SPOT - a high resolution French Satellite

SRBD - a process disscussed in the designs of this book to divide a raster into rectangular sub-raster blocks for distribution to processors

Superclasses - an object-oriented programming/database term that refers to higher level classes from which attributes or methods can be inherited

Superlinear - adjective applied to processing speedup that increases more than linearly as new processors are added. Usually due to increased cache use

Synchronisation - co-ordination of processing tasks based on message passing between processors

Terabyte - 1,000 gigabytes

Teraflop - 1 million megaflops, where a megaflop is one million floating point operations per second

Tagged Image Format (TIFF) - a raster transfer format, regularly used for images, rather than spatial data.

Transfer format - a data format used for transferring data from one system to another.

Triangulated Irregular Network (TIN) - a data structure used in terrain modelling which permits easy calculation of slope and aspect within a region

Topology - the spatial relationship of one feature to another

Torus - a geometrical structure like a doughnut with a hole in the centre

Transactions-per-Seconds (tps) - a measure of performance concerned with the ability to process real data, and primarily used to assess DBMS

Transputer - a single chip processor produced by Inmos, capable of being combining into a parallel array

Triangulation - see *Delaunay triangulation*

TSO - the topology building, stitching and output module of the parallel software design described in this book

Tuples - rows in database tables

Vector Data Structure - a representation in the form of points, lines and polygons

Vectorise - the process of generating a vector data structure

Vector-to-Raster - a conversion between two of the fundamental data storage formats for spatial data

Viewshed - the area on a terrain which can be observed from a particular point (and the area within which that point can be observed)

Visualisation - the representation of data or information for the purposes of recognising, communicating and interpreting pattern and structure

VLSI - very large systems integration, usually applied to computer chips containing > 100,000 devices

Von Neumann - a term used to describe a serial processing computer architecture

Voronoi Diagram - a geometrical structure used in the construction of digital terrain models

Wide Area Network (WAN) - a network of computers over a distance of some tens or hundreds of kilometres

Index